TECHNICAL DUE DILIGENCE
and
BUILDING SURVEYING FOR COMMERCIAL PROPERTY

Technical Due Diligence and Building Surveying for Commercial Property is the first book to introduce the process of technical due diligence (TDD) and examine the role of the building surveyor within the commercial property sector. The book outlines the processes that the surveyor must go through when performing a TDD inspection and report and, most importantly, covers in detail the typical pathology and defects encountered during TDD.

Performing a TDD survey involves collecting, analysing and reporting on a huge amount of information, often under specific contractual conditions. The book covers everything the surveyor needs to know in order to do a proper job and includes analysis of materials, life cycles and potential defects on an elemental basis, with detail on individual components where necessary.

Coverage includes:

- an introduction to the TDD process and types of commercial buildings encountered
- chapters outlining the life cycle and defects of: structures, roofs, facades, finishes and services
- hundreds of illustrations and photographs of defects, real-world case studies and suggestions for further reading
- a final chapter covering legal issues and technical details.

This book fills a clear gap in the literature and is the first fully illustrated book on TDD dedicated to commercial building stock. It will help students and professionals to understand the process, the science involved and the reasons why defects occur, as well as their evolution and long-term impact.

Adrian Tagg (MRICS) is a lecturer in Building Surveying at the University of Reading, UK. He also runs TechDD Ltd and has over a decade of experience delivering technical due diligence to international investors making acquisitions in Northern Europe.

TECHNICAL DUE DILIGENCE

and

BUILDING SURVEYING FOR COMMERCIAL PROPERTY

Adrian Tagg

Routledge
Taylor & Francis Group

LONDON AND NEW YORK

First published 2019
by Routledge
2 Park Square, Milton Park, Abingdon, Oxon OX14 4RN

and by Routledge
711 Third Avenue, New York, NY 10017

Routledge is an imprint of the Taylor & Francis Group, an informa business

British Library Cataloguing-in-Publication Data
A catalogue record for this book is available from the British Library

Library of Congress Cataloging-in-Publication Data
Names: Tagg, Adrian, author.
Title: Technical due diligence and building surveying for commercial property / Adrian Tagg.
Description: Abingdon, Oxon ; New York, NY : Routledge, 2018.
Identifiers: LCCN 2018004695 | ISBN 9781138745179 (hbk : alk. paper) |
    ISBN 9781138745186 (pbk : alk. paper) | ISBN 9781315180717 (ebk)
Subjects: LCSH: Commercial buildings–Surveys.
Classification: LCC TH4311 .T34 2018 | DDC 692/.5–dc23
LC record available at https://lccn.loc.gov/2018004695

ISBN: 978-1-138-74517-9 (hbk)
ISBN: 978-1-138-74518-6 (pbk)
ISBN: 978-1-315-18071-7 (ebk)

Typeset in Myriad Pro
by Alex Lazarou
Printed and bound in Great Britain by Bell & Bain Ltd, Glasgow

Contents

Acknowledgements

The author would like to thank a number of individuals and companies who have helped to facilitate or contributed to this book.

Individuals

Christian Deprez

José Galeote

Tom Goodhand

Geert Lybeer

Gregory Martin

Christopher Mitchell

Dennis Wilkinson

Paul Wright

Companies

Geert Lybeer bvba

Paragon BC

SECO Belgium SA/NV

Tech DD Ltd

Widnell Europe SA/NV

Further thanks are given for the support and encouragement of family, friends and colleagues (you know who you are).

In memory of Eddie Blake.

Introduction

Defining technical due diligence and commercial surveying

In the context of real estate acquisition, management and disposal, technical due diligence (TDD) is the term used globally to describe what is essentially a commercial building survey. Although it is a widely used term there appears to be a lack of clarity or consistency in what this actually covers; indeed there is often a bespoke nature to TDD instruction, tailored to specific client needs, sector norms or individual investment strategies. 'It's like beginning your geography homework on a Sunday evening knowing it has to be handed in the next day' is the way TDD has been quite poignantly described by a high-profile chartered building surveyor – needless to say, not actively working within this field.

In essence, there is often a huge amount of information to collect, analyse, process and report within very short timescales. This is often done on buildings or portfolios which represent very high investment values. In short, it certainly is not everyone's 'cup of tea' and often the first question upon receiving TDD instruction is 'Where to start?'.

The guidance notes published by the Royal Institution of Chartered Surveyors (RICS) concerned with commercial surveying/TDD provide an essential framework or aide-memoire detailing the principal actions and considerations in this specialist area of operation. There are also numerous published works concerned with building pathology or aspects of commercial surveying. However, there appears to be little available reference information or detailed analysis of the process of TDD in this specific context or the individual construction elements and components which should be inspected. Also, although there are some excellent text or reference books in the field of building surveying, these appear mostly to be in the context of residential properties. Furthermore, there exist significant and important publications, including RICS guidance notes, which deal with specific legal or technical aspects of building surveying from a more 'commercial' perspective. These address individual specialisms, but in isolation, despite them often being 'picked up' as part of the TDD process. In essence, it is important for the TDD process to provide a global overview of a commercial asset yet, where necessary, have the information or detail to go down to component level as and when required. It is a facet of building surveying that embodies a 'Jack of all trades' approach in the provision of advice that encompasses many aspects of commercial surveying practice, but rather than detailing these in isolation, it seeks to deliver a 'joined up' opinion, allowing clients to make an informed decision.

This book will seek to introduce the reader to the process of TDD and, as well as discussing this, it will also analyse the role of the building surveyor within this commercial environment. Importantly, this book will look beyond the process of TDD and into the application of the skills and knowledge required to perform this task. It will also analyse materials, component life cycles and potential defects on an elemental basis as well as at individual component level, detailing the potential impacts on the TDD process as well as offering insights into advising clients.

The nature of TDD and commercial building surveys

TDD and commercial building surveys essentially involve highly bespoke contract instructions due to the often unique nature of the assets or complexity of the transactions. While no two TDD or survey contract instructions are identical, the actual the process is quite 'mechanical'. There is essentially a list of things to check (often driven by the client's requirements). Therefore, the surveying skill set required is the same as that applied to the inspection process of many other facets of building surveying works, such as dilapidations or schedules for planned preventative maintenance. Furthermore, the process of undertaking a TDD or commercial building survey is also highly transferrable between different real estate sectors and different types of building. Indeed, providing the surveyor understands local laws, building culture and language, the process of TDD and commercial building surveys is transferable to different countries.

Principally, there is a requirement to undertake a site visit to record and analyse information before delivering and reporting on the findings. Crucially, this will have to be cross-referenced with historical documents held for the property or asset. The intention is to form an opinion of the past, present and future issues associated with the site/asset under consideration in order to advise the client or investor of the potential risks that they may be exposed to, both immediately and in the longer term.

It is not solely an administrative task in which items on a checklist are noted or confirmed; indeed, the checklist provides the framework for the TDD or survey instruction which requires 'fleshing out' with information obtained during the site inspection. There is a need to apply scientific knowledge or theory to determine the cause, effect and evolution of defects. Ultimately, for investors it comes down to the cost or anticipated expenditure that is required to resolve defects, maintain an asset or 'upgrade' it to comply with legal requirements or commercial standards.

The acquisition process

When a prospective purchaser has an offer accepted by a vendor this is normally subject to a series of conditions, one of which is due diligence. A 'window' of due diligence is further negotiated where the purchaser, occupier or financier will be granted a set period of time to carefully examine the asset. In order to exercise the necessary level of judgement, care and prudence, as well as mitigating potential harm due to the transaction, external specialists are appointed to carry out the analysis. In the context of a commercial acquisition, the TDD or commercial building survey forms part of a wider overall due diligence process. The complete due diligence process will involve a wide range of professionals, each tasked with undertaking a careful and detailed examination of the asset to establish all facts that would be of material interest to the purchaser, occupier or financier. The due diligence process may be typically divided into the following:
• commercial due diligence
• legal due diligence
• technical due diligence
• environmental due diligence.

Commercial due diligence

Essentially, commercial due diligence is concerned with the viability of the asset in terms of acquisition value, yields or rental incomes, capital expenditure (CAPEX) and future disposal.

This work is typically undertaken by chartered surveyors within the general practice faculty (investment/agency) and other sub-specialist advice from accountancy firms who may be undertaking specific analysis in matters such as property tax or company set-up.

Working often in isolation, the commercial due diligence may overlap the TDD in instances where there are technical recommendations which have a commercial impact. This is typically where there are issues of non-compliance with regulations or the cost of defects and investment in the asset.

In most cases the costs identified during the TDD are used to assess the commercial viability of the asset relative to the investment strategy of the investor. Invariably, the cost estimates generated by the TDD are used as leverage in the negotiation process during or post due diligence. It is therefore important that these are fair, reasonable and fully representative of the actual condition of the asset. While there is a team of experts appointed to act for the purchaser in the undertaking of a due diligence there is, in most cases, a similar panel of experts appointed by the vendor to represent their interests in the process. It should be in the best interests of all professionals appointed to act with honesty and integrity.

Certainly, with respect to the commercial and technical due diligence these will mostly be carried out by RICS qualified experts who should abide by their relevant rules regarding professional ethics and conduct. In reality, and in the case of investment or agency surveyors (where a large percentage of the professional fees may be dependent on the successful transaction), there invariably exists a conflict of interest to ensure that the due diligence process is smooth with little risk of either party terminating the deal. This may in some cases lead to aspects of 'commercial negotiation' where technical observations or recommendations are 'dumbed down' or even negotiated out of a deal or effectively cash settled. With respect to the technical experts (chartered building surveyors), acting effectively in the best interests of the building appears the most appropriate way to proceed. In essence, this means recording the condition of the asset and any deficiencies as a matter of fact. Using this evidence-based opinion to recommend the best technical solution (irrespective of vendor or purchaser instruction) adds a valued transparency to the process. This honest and open approach inevitably leads to further instructions and enhanced independent reputation. However, if the surveyor 'sells out' by distorting information or ignoring evidence in favour of facilitating the transaction, this appeasement is likely to do more harm than good.

Legal due diligence

The legal due diligence is a process by which the asset is examined in terms of statutory requirements as well as civil procedures. This is largely a desk-based exercise undertaken by a team of highly specialised lawyers/solicitors who may not even visit the building. They will examine and report on the following typical issues:

- planning permissions
- easements
- boundaries
- guarantees and warranties
- use class
- operating permits
- applicable regulations
  - building codes/regulations
  - fire safety requirements
  - health and safety

- ○   deleterious materials
- ○   energy performance certificates (EPCs)
- lease contracts.

There is an obvious overlap between the legal and technical due diligence on matters that have a legal obligation with a technical application. These are often the most contentious areas of discussion within the due diligence process and while building surveyors (acting for the vendor or purchaser) may act in the best interests of the building, the legal advisors invariably act in the best interests of their client. Legal discussions can be intense and protracted with argument and counter argument. Technical experts will be called upon to effectively give evidence which is effectively used to support or dispel legal argument. In an environment where opinions and findings are transposed or manipulated to suit legal argument, it is paramount that the building surveyor remains objective with evidence-based opinion. This is particularly important where the legal advisors have not visited the site and often have minimal knowledge of building technology or pathology.

Technical due diligence

The TDD is often the only part of the acquisition process which actively spends significant time visiting the asset. This is, essentially, to determine the current condition of the construction elements with a view to analysing this and advising on immediate as well as future costs. The TDD will typically be undertaken by chartered building surveyors and the objective of the visit(s) is to establish facts relating to the past, present and future condition of the asset.

TDD – the past

The past is important as there will be a requirement to look at any historic documents concerned with the asset, such as planning permissions or building permits. These documents will specify the original intended building use and this is crucial as it will have shaped the construction characteristics as well as the material choices. Understanding the past and the historical context of the materials used is highly relevant as there are numerous defects or problems associated with construction techniques or deleterious materials widely used since the end of the Second World War. For example, the use of asbestos as a fire protection material was considered normal for many decades; indeed, its effectiveness in performing this function can hardly be questioned. However, the serious long-term health issues associated with exposure to the material have meant that this unwanted legacy has major implications for building occupation, as well as acquisition or disposal. The historical context of commercial property will be detailed in later chapters but, concerning properties constructed pre-1970s, it is highly probable that these have been the subject of renovation or refurbishment. Post-1970s properties are likely to be the subject of current or future renovations, therefore the historical construction technology is vital when advising during the TDD.

TDD – the present

With respect to the present, it is normal for the TDD to inspect, analyse and report on the present-day condition of the asset. In most cases this is fundamental in establishing current defects and deficiencies which may be used for leverage in negotiations during the acquisition process. It is important to consider that, when advising on the present condition and defects, this includes replacement/repair on a like-for-like basis. This principle is not dissimilar to other surveying specialisms, such as dilapidations or party wall awards. The costs should almost be seen as 'damages' in the context of putting the building back into the relevant state of repair with no betterment. As with dilapidations or party wall awards, where there is a requirement to comply with current legislation there may be instances where such improvements are necessary. Typically, the costs associated with such improvements form part of the due diligence discussion and possible commercial negotiation.

TDD – the future

Probably one of the most difficult and contentious aspects of TDD is assessing an asset and predicting the condition of construction elements in the future. While it is acknowledged that all construction components have a life cycle, there is a science involved in establishing the factors that may reduce this and, in particular, applying this theory in situ to often unique buildings in specific locations. Life cycles and construction elements will be discussed individually by building element later in this book but it should be noted that it is marginally easier to predict future costs when these are bracketed into short-, medium- or long-term considerations.

Immediate works (as defined by the RICS guidance note, 4th edition, 2010) essentially equate to being necessary within the year, short term is one to two years after the acquisition, years three to five are the medium term and years six to ten long term. After ten years it becomes difficult to predict with accuracy the long-term evolution of defects and their subsequent anticipated life cycles with respect to their replacement. For some transactions (typically sale and leaseback of public sector assets) the cost spread can be as long as 27 years and where this is divided into annual categories it becomes almost impossible to make predictions of significant accuracy. The client should be informed of such potential inaccuracies as part of the contract instruction, with phrases such as 'best possible opinion' based on 'information available' at the time of the survey, which could include current knowledge of materials, their performance and other known factors, such as external climate, location or exposure.

Predicting the future evolution of defects and their costs represents the biggest risk to the surveyor in terms of a possible claim for negligence. Present-day defects generally necessitate immediate repair and these costs are often 'taken' by the selling party either doing the works or discounting the acquisition price. The medium- and long-term costs form part of the wider negotiation and it should be noted that 'responsibility' for these (in terms of the TDD process) fluctuates with economic conditions.

In periods of relative economic prosperity, the window of due diligence is often short; as a consequence, the negotiation is also short. Often, with multiple suitors per asset, vendors adopt a 'take it or leave it' approach. Therefore, while immediate costs are considered for negotiation, costs projected to fall after year one are often perceived to be the responsibility of the 'new' owner as part of their future capital expenditure (CAPEX).

The reverse situation occurs when there is an economic downturn, giving rise to increased due diligence time and negotiation. Cost are often analysed in greater detail with short-, medium- and sometimes long-term costs being subject to further negotiation.

Importantly, and irrespective of economic prosperity or downturn, one constant factor is the defects themselves. It is the responsibility of the surveyor also to remain 'constant' and consistent with defect diagnosis, ensuring that their advice or opinion is always evidence based. The surveyor's duty is to provide their client with the necessary information for them to make an informed decision on whether to proceed with the acquisition process or stop.

One method adopted (particularly with vendor due diligence) to reduce the perceived future investment costs is to only present immediate, short- and medium-term costs. Removing the presence of long-term costs (typically those arising in years six to ten) from the TDD report may significantly improve the investment appearance of the building. Many investors typically utilise a five-year investment strategy; therefore, at the time of the purchase, longer term costs may be perceived as less important. The consequence of removing the costs from the long-term projection is that when the 'new' owner decides to dispose of the asset, they often become aware of costs that they were not exposed to with the original TDD. This may also place added pressure on the surveyor where anticipated life-cycle costs and defect projections fall around the five- to six-year point. This may pose a dilemma with the timing of such works, particularly if the anticipated costs are high. Provided there is a sufficient trail of evidence concerning the material, its life cycle and current condition then it may be prudent to adopt a risk-averse approach and show this as a five-year cost if it is on the threshold between the medium and long term.

TDD framework

In principle, the framework for undertaking a TDD or commercial building survey can be quite straightforward, the complexity may lie in what to include, as this is often down to the client's specific requirements. However, the TDD or survey itself comprises a series of actions that form a checklist which can be adopted and adapted to suit different types of assets or assets within different commercial sectors..

When determining the framework and what to include in the TDD or commercial building survey, the new practice standard that the RICS is set to release in 2018 and the following guidance notes/best practice may be considered:

- *Building surveys and technical due diligence of commercial property* (4th edition guidance note, 2010)
- *Technical due diligence of commercial, industrial and residential property in continental Europe* (1st edition, 2011).

These are good sources of information. However, they are guidance notes suggesting the minimum requirements needed to undertake such works. They do not go into significant technical detail as to the problems which may be encountered and, importantly, on how to advise clients. In essence, the actual process of the TDD may be considered to comprise three main components:

- architecture and structure
- services installations (mechanical and electrical – M&E)
- legal/technical (document review).

Architecture and structure

The inspection of the architecture and structure should be carried out in accordance with the guidance notes and should include the following construction elements:

- roofs
- rainwater goods
- walls and cladding
- windows, doors and joinery
- structural frame
- substructure/basements
- floors
- internal walls, partitions and doors
- finishes

- staircases
- sanitary fittings.

    The construction elements listed in the guidance note can be more succinctly divided into the following:

- structure
- roofs
- facades
- internal finishes
- external areas.

However, it is important to expand these construction elements to include specific sub-elements which will form part of the TDD inspection. This can be further illustrated as follows:

- structure
    - substructure
    - superstructure
- roof(s)
    - roof covering(s)
    - parapet walls
    - rainwater evacuation
    - roof lights
    - other
- facades
    - external walls and cladding
    - windows and doors
    - entrances
    - other
- internal finishes
    - offices, retail, classrooms, warehouses, production etc. (dependent on primary function/ tenant use of the building)
    - entrance hall, lobby, reception
    - lift lobbies
    - staircases and emergency escape provision
    - sanitary rooms
    - kitchens, staff welfare rooms
    - technical rooms
    - basement, internal parking areas
    - other
- external areas
    - access routes, entrance
    - external paving, pathways
    - parking
    - landscaping
    - other.

Each one of these elements and sub-elements will be detailed fully in later chapters to include the relevant inspection techniques, criteria and typical defects. The sub-elements listed may be considered to be the general minimum items for inspection and it is acknowledged that, due to the bespoke nature of TDD, there may be more specific or specialist sub-elements which are covered under the terms 'other'.

Services installations (M&E)

The requirement to inspect the services installations as part of the TDD or commercial building survey is crucial since the elements and sub-elements/components associated with this have relatively short lifespans (compared to the 'architecture and structure' category) with high investment costs. A fundamental obligation on the building surveyor (also stipulated in the RICS guidance note) is that they work within the limitations of their professional qualifications/ expertise. Therefore, in most cases the inspection, analysis and subsequent recommendations concerned with the services installations are usually carried out by an M&E consultant. Despite this, the RICS guidance note does stipulate that the building surveyor advising on the TDD is required to have an understanding of the principal components and M&E systems at the building, without necessary being a 'specialist'.

A sound working knowledge of the M&E installations is particularly useful during the inspection process of the roof and basement areas. These locations are typically used to house chillers, compressors, air handling units etc. and it is often at the interface of this equipment with the building fabric where defects or non-compliance issues occur. This again will be further examined and discussed in later chapters.

Legal/technical

As previously discussed, there are many legal requirements concerned with construction, maintenance and refurbishment of investment properties or built assets. Invariably, where there is a legal obligation or requirement, there is a technical application. This is where the legal due diligence interfaces with the TDD and in most cases the legal team undertaking their facet of the due diligence process will have little or no knowledge of the site conditions. It is 'normal' for the legal advisors to undertake essentially only a desk survey or paper audit without actually visiting the building. Therefore, the TDD will have to take into account the principal legal/technical issues with its own review of the relevant documents during the process.

Having analysed the legal/technical documents, the TDD will have to verify (as far as is feasible) whether any specific legal requirements have been implemented or executed. Naturally there are certain specific legal requirements which are difficult to fully verify and this is where it is appropriate to exercise 'limitations' as stipulated in the contract instruction. This will be discussed in full detail in *Chapter 3 – Contract instruction*, together with the 'rules of engagement' associated with undertaking a TDD or commercial building survey.

While TDD instructions may be considered 'bespoke' to individual buildings or building sectors, the following list presents typical legal areas which may require analysis during the TDD process by virtue of the fact that they have a technical application:

- planning and building use/operation
- health and safety considerations
- fire engineering
- accessibility
- environmental considerations
- deleterious and hazardous materials
- sustainability
- cultural heritage.

While investors invariably want guarantees or assurances that built assets have been constructed and are being operated fully in compliance with all relevant legal prescriptions, it is often difficult to fully prove this in the TDD process. Furthermore, the legal advisors often put pressure on the technical experts to verify compliance when this cannot always be achieved. For example, where there is a building permit or planning permission, the legal due diligence and client will seek confirmation that this has been executed in accordance with all relevant conditions and regulations. However, with only a limited time to visit the building (often just one or two days) it is impossible to fully ascertain this fact. Attached to the building permit or planning permission should be a set of authorised plans, which detail all aspects of the property, such as building heights, dimensions, floor areas and a multitude of other legal/technical features. There is simply insufficient time in the TDD process to devote to checking every one of these details as this is a highly labour-intensive task which could take many days or weeks of time on site to verify. It could be argued that more resources could be devoted to this part of the process but it is highly unlikely that clients would be willing to pay for what would be huge amounts of professional time.

Regrettably, TDD may be seen as the last 'tick in the box' of the acquisition process and that, arguably, it commands lower fees than the legal counterpart or those of the agent negotiating the deal. Putting it bluntly, there is quite a high level of professional risk and liability attached to TDD and commercial building surveys for the relatively low levels of fees allocated to the process, often on assets worth tens of millions of pounds or euros. Therefore, the checking of the legal/technical issues is often a compromise offset against the agreed time allocated to undertake the TDD and the fees that the client is willing to pay for this. Such compromises are analysed in more detail in *Chapter 3*, but an apt example is to relate back to the earlier planning permission/building permit mentioned above.

Where a planning permission/building permit and permit plans have been made available for review, a desk study may be sufficient to offer an opinion on compliance. The desk study will explicitly include a comparative analysis between the principal plans certified by the authorities and attached to the planning permission and the 'as-built' plans signed off or certified by the architect/design team. During the desk study any differences may be obvious and these should be noted to ascertain the 'significance'. Such differences may be as obvious as extra constructed sections of the building/site, floor areas, key external envelope dimensions or facade detailing (material choices or numbers of windows/doors etc.). There may be other, more subtle differences, such as the use class of the whole asset or individual sub-sections or floors. Ideally, such a desk study should be completed prior to the site inspection, enabling part of the time spent actually visiting the building to be used to confirm outstanding queries and collect evidence of such breaches. However, it should also be acknowledged that no matter how 'organised' the TDD process is, there is always an element of pressure and an emphasis on undertaking a site inspection at the earliest opportunity. Access to the data room documents often occurs after this initial site appraisal.

In conclusion, regarding the assessment of the legal/technical issues, it is critical that the building surveyor inspects these accordingly (within the agreed scope of works and according to the RICS guidance notes). As with the whole TDD process, it is anticipated that the building surveyor will offer the best possible opinion to the client based upon their understanding of the relevant legal frameworks as seen/executed on site.

Environmental due diligence

There may be a need to have a highly specialist environmental due diligence for specific 'environmental' matters which are beyond the expertise of the commercial, legal or technical due diligences. This should be (and normally is) instructed directly by the client, although it is acknowledged that this may be sub-contracted in certain circumstances by the principal technical consultant, but this should be done in accordance with the RICS guidance note. The environmental due diligence is often concerned with the operation of the asset and any specific risk activities. There are also areas of the principal construction elements which may need to be addressed by the environmental due diligence, such as the presence of asbestos or other hazardous materials. In most cases, this specific audit follows a well-defined checklist of key issues as an evidence-based assessment. Similar to the legal due diligence, there will be an overlap of environmental issues which have a legal requirement or application within a highly regulated area of specialism.

There are also some aspects of technical application concerned with the findings of the environmental assessment which may need to be considered by the building surveyor as part of the TDD or commercial building survey. It is useful to have access to copies of historic environmental assessments or environmental due diligence reports in the data room documents as these often give an insight into key risk or potential 'red flag' items which may come to light as the due diligence process evolves.

Conclusion

TDD and commercial building surveys are an important part of the acquisition process and building surveyors and consultants undertake this work in a highly pressurised environment where an opinion is often required within unrealistic timescales. Therefore it is important to acknowledge (and state in the contract instruction/report) that the findings are based upon the levels of granted access, available technical documents and time allocated to undertake this process.

Importantly, clients or investors do not look at the TDD in isolation but as a part of the wider due diligence process. In most cases it is acknowledged that there is no asset or acquisition which is in a 'perfect' condition without problems or issues. Invariably there is a need for investors to compromise or negotiate the findings of the TDD to close a deal. This is something that is done at the discretion of the buyer or seller and it is not in the remit of the TDD to simply change the survey findings to suit. This is particularly important as the building surveyors' and consultants' duty of care or professional liability is likely to extend a significant number of years beyond the transaction. It is therefore the responsibility of the TDD expert to give the best possible, evidence-based opinion to allow the client to make an informed decision.

In order to provide such an opinion it is it necessary to have an understanding of the TDD process and the client's aim or objectives as well as the ability to agree a contractual framework for undertaking the process. Sound technical knowledge is a key requirement and this should be relative to all of the key construction elements and may also extend to sub-elements of individual components. The TDD process will be discussed in more detail in the next chapter.

Technical knowledge, including defect diagnosis and building pathology, is among the key skills required by the TDD expert. To an extent, the scientific principles of building pathology are the same for some construction materials that are used in residential as well as commercial buildings; this largely applies to 'simple' low-rise commercial property. However, there are lots of other materials that are used specifically in medium- or high-rise properties. These have different characteristics and properties presenting specific building pathology which may be more specific to the commercial sector. The material use, context and associated building pathology of each construction element will also be discussed in later chapters.

References

RICS. 2010. *Building surveys and technical due diligence of commercial property* (4th edition, guidance note), London.

RICS. 2011. *Technical due diligence of commercial, industrial and residential property in continental Europe* (1st edition, best practice and guidance note), London.

Photo courtesy of Widnell Europe.

Process overview 2

Commercial property

In surveying terms there are either 'residential' or 'commercial' property sectors. Residential property is essentially homes or dwellings that are owned by private individuals. In most cases these are owner-occupied properties, which may include properties operated on a 'buy-to-let' basis by individual investors.

In contrast, commercial property is essentially assets or buildings which are operated as a business and, while buy-to-let could be seen as a business operation, these are usually done on a relatively small scale. Therefore, to an extent, the scale as well as the type of business activity also influences whether the property or asset is classified as commercial. It could be suggested that single buy-to-let properties could be classed within the residential sector; however, where there is a portfolio of multiple residential properties or where there is a mixed use element, such as retail, then this may be viewed as a commercial entity.

Commercial properties fall largely into one of six categories:

- offices
- retail
- industrial
- leisure
- public sector/infrastructure (including schools and health care)
- residential.

The commercial sectors and their characteristics will be analysed and discussed in *Chapter 4*.

Clients

Clients investing in commercial property are effectively doing this to generate an income or profit and, while there are many investment vehicles available, property may be seen as a 'safe' option. The returns from commercial property investment may be considered low when compared to the stock market; however, it is less volatile and more 'predicable'.

While there are some wealthy private investors with commercial property portfolios, the majority of clients operating in this sector are investment banks and pension funds. These are usually very large organisations which have different departments dealing with discrete parts of the property life cycle, including acquisition, asset or property management and disposal. The individuals working for investors often have a wide range of academic qualifications, ranging from accounting to engineering, law and other backgrounds. It is a dynamic and diverse business with some individuals seeking to obtain professional qualifications, such as membership of the Royal Institution of Chartered Surveyors (RICS) and others working without such recognition. Professional membership is particularly relevant as the RICS has an extremely diverse range of surveying faculties encompassing all aspects of property, land and real estate. The one common denominator with all professional members of the RICS is that they are governed by strict rules and regulations regarding professional conduct. This is important as the discussions or negotiations between chartered surveyors, irrespective of the clients they work for, should be honest, transparent and objective. There are many other individuals or organisations who are not members of professional bodies who may view real estate investment opportunistically as offering the prospect of financial gain. Accordingly, where the financial incentives are good, this can sometimes 'cloud' the judgement of individuals or organisations, resulting in conflicts of interest or the 'stretching' of moral and ethical boundaries. When dealing with such individuals or organisations, surveyors should not resort to lowering their standards, instead maintaining at all times their obligations to the RICS or their professional body, as well as any contractual obligations they have to their client.

Professionals

Within the scope of TDD or commercial building surveys there are a number of 'professionals' operating in a variety of different roles. The majority of these professionals are likely to be members of organisations registered and regulated by the RICS and may typically include:

- general practice or estate surveyors
- building surveyors
- quantity surveyors.

Other important professionals who may also be included in real estate transactions are M&E engineers, architects, lawyers, accountants, tax advisors and environmental specialists.

It should be assumed that professionals who are members of the RICS will abide by and conduct themselves within the ethical and moral framework of the establishment. However, each individual, operating within their instruction for an acquisition, invariably has different motives or modes of operation.

General practice or estate surveyors are often involved at the start of the property acquisition process as this discipline is typically responsible for bringing together purchasers and vendors, so they can be pivotal in the arrangement or negotiation of the transaction. Their involvement is often at the front end of the process and typically they are involved throughout this stage, up to and beyond completion of an acquisition. Initially, general practice or estate surveyors take an instruction to market a property and this will include an assessment of its rental value or yield (return on investment); this is significant as it will affect the asset value or purchase price. Post completion they are often involved in the marketing and leasing of the property. Logically it may be assumed that they have a vested interest in not only completing the acquisition but also in successfully leasing the building. Both of these actions generate professional fees. However, as these are often percentage based it is advantageous to get the best possible deal. The obvious risk associated with percentage-based fees is the potential for the acquisition price to be exaggerated or the condition of a building to be 'dressed up' to be better than the reality in order to make it more attractive to a buyer. Likewise, investors or agents on the selling side, who seek to diminish or remove costs from the projected capital expenditure budget, also risk giving a misrepresentation of the asset. There are often significant discussions between investors, their representing agents and the technical advisor to clarify the condition of an asset and defects as the parties often view these aspects differently. However, it is the TDD or commercial building surveyor who carries the responsibility and liability for the report findings, and not the agents. Accordingly, the TDD or commercial building surveyor is likely to be more risk averse than the agent.

The TDD or commercial building surveying team are essentially the surveyors charged with actually undertaking the survey and, although predominantly this will be done or led by a chartered building surveyor, it may also involve specialist advice or assistance from quantity surveyors. This is particularly the case when there are complex or large remedial works recommended to rectify defects and quantity surveyors may be most able to offer the best possible cost advice to a client. The use of such specialist knowledge or expertise can be extremely valuable in providing a trail of evidence to establish accurate cost data. In some cases, the consultancy which performs the TDD or commercial building survey may be subsequently appointed to manage or monitor the remedial works, upon successful completion of the acquisition. This is another example of the positive contribution that quantity surveyors can bring to the team, as the subsequent specification and tendering of the works may be a more accurate and smoother process if they have been involved from the outset.

The principal intention of the technical team when performing a TDD or commercial building survey is to be objective and transparent. This differs from the situation of agency or investment surveyors as fees for the TDD or survey are often fixed and paid regardless of whether the transaction completes or not. Invariably, the technical team is tasked with analysing the asset to identify and report on defects or problems. This is, in some ways, a thankless task as every identified defect or problem may need addressing in the subsequent negotiation, often with opinions on each item being subject to counter-opinion from the opposing technical team. One of the facets that the technical team must learn to develop is the ability to break bad news to the client and manage the 'fallout' associated with unearthing defects. TDD or commercial building surveys are an auditing process and it's not unusual that, when undertaken on a commercial property, it may be the first time that this has been done since the property was initially completed. Inevitably, a thorough examination of the legal/technical documents combined with a detailed on-site assessment of the property will discover abnormalities or irregularities as well as defects. Consequently, TDD or commercial building surveys result in costs being generated for either the vendor or purchaser, which may be viewed as bad news.

Constraints and checklists

As alluded to in the *Introduction*, the biggest constraints within the TDD or commercial building survey process are time and access. The 'window' of due diligence is agreed between the vendor and purchaser, and it is up to the technical team to undertake the survey within the permitted time, which can mean mobilising staff at short notice to achieve this.

The time required to fulfil a TDD or commercial building survey instruction should primarily include the site visit as well as a review of the technical documents, writing of the report, attending meetings and revising the report as necessary. Initially, timescales and deadlines may be fixed, but the whole process of due diligence involves specialist legal, commercial and, potentially, environmental analysis. With many different and often overlapping sub-specialisms, the due diligence process is quite fluid and opinions can change as more information is received. With increased amounts of information to process and deadlines fixed there is a tendency to increase the allocation of staff resources to the job but, in the case of TDD, fees are often fixed. Therefore, the contract instruction should be carefully written to include in the scope of works both a fixed fee and a relevant clause to deal with extra documents, visits or meetings which may become necessary.

The contract instruction should also make provision for the survey to be visual, with access provided to all areas. However, in practical terms, it is almost impossible to gain access to all areas during the site inspection and this is due, typically, to rooms being locked, roofs being inaccessible, the presence of private tenant archives and a multitude of other reasons. Where access is not afforded, the surveyor should make note of this and detail it in the report. In many cases the tenants are unaware of the purpose of this site inspection, as landlords often do not disclose to the occupants their intention to sell their property. Naturally, some tenants are suspicious of auditors entering their building or space and, bound by client confidentiality, the surveyor may divulge that they are performing the audit with the owner's permission to establish the current condition and any anticipated maintenance or repair costs. This neither discloses the intentions of the owner nor creates represents a false impression in the mind of the tenant.

Some clients or investors issue long and detailed checklists of their requirements for a TDD or commercial building survey. Indeed, within the RICS Europe guidance note for technical due diligence (RICS, 2011) there is a long checklist appended to the document. In almost all cases it is impossible for a surveying practice to have all of the relevant in-house expertise to pick up every item on a client's checklist. Certain tasks, like verifying lettable floor areas or verifying the individual dimensions of the plot or building heights, are difficult to achieve within the timescales. Therefore prior to agreeing on a client's checklist and appending this to the contract instruction, the surveyor should cross through or remove any items they are unable or unwilling to perform. The surveyor can refer or recommend other consultants who could or should be appointed to undertake these 'specialist' tasks. The worst-case scenario is that the surveyor accepts responsibility for undertaking everything within the client's checklist but is not able to fulfil 100 per cent of the obligations. If a surveyor misrepresents their skill set and ability, this is likely to do more harm than good. When a client tenders for the services of a TDD or commercial building survey they may provide each candidate consultancy with a checklist of requirements. It is unlikely that there will be one individual or organisation capable of doing everything. It is therefore more important that those returning tenders stick with what they can do well and refer or outsource the areas of work in which they have little or no experience. This, in essence, is knowing and sticking to professional limitations and this wholly honest approach is more likely to be appreciated by clients.

Surveyors often use checklists themselves to undertake the site visit and these can prove to be the basis of good surveying practice for those actually undertaking the building survey or TDD. However, while is it quite straightforward to issue a checklist of legal/technical documents to review, sticking strictly to checklists for the site visit can oversimplify and 'mechanise' the survey. The temptation is to follow a checklist and strike off each item as and when it is surveyed; however, the nature of construction elements and sub-elements is such that there is often an overlap between them. It is advisable to be methodical with the visit and the recording of information but, having visited the roof and discovered infiltration to the occupied areas below, it may be necessary to revisit the roof to establish a trail of evidence detailing the cause and effect. Therefore, rigidly sticking to checklists invariably leads to tasks being completed and compartmented with the ticking or filling in of a comments box. The necessity with TDD, commercial building surveys and building pathology is literally to think outside the box.

In essence, checklists can be important in the formation of a skeletal approach to undertaking a site inspection. However, the real skill is to flesh out this framework and connect individual items together in respect of their function, assessing the cause and effect of defects as well as the influence this may have on other building components or functions.

Aims, objectives and outcomes

The aim of individuals or organisations working within the service sector is client satisfaction, which can go a long way to attaining repeat instruction. This is widely evident in the surveying sector where a significant percentage of income is from existing clients. Irrespective of the client and whether the TDD or commercial survey is being undertaken for a vendor or purchaser, the surveyor should always work in the best interest of the building. Some clients appear misguided when appointing an external technical consultant, in that by paying for the consultant's services, they appear to expect a 'good' or 'favourable' report, This often seems to be the case with vendor due diligence or surveys and, while agency or investment surveyors and even legal advisors may seek to manipulate the existing building situation to appear favourable, technical consultants must always remain objective. Subjectivity is something that should be reserved for advertising, art or marketing and, in respect of TDD or commercial building surveys, defects are defects. It therefore is critical that the surveyor does not divorce themselves from this basic principle, irrespective of who they are working for. The aims of the TDD or commercial building survey is to work in the best interests of the building and this includes an objective assessment of conformity to the relevant statutory requirements as well as taking into account the wider responsibility to the general public and society. Professional judgements should not be clouded or influenced by shareholders, yields or the pressure to complete the deal. Accordingly, civil procedure recognises that surveyors have a duty of care to undertake their work diligently. Any breaches of this duty, in favour of completing the acquisition or pleasing a client, are likely to be looked upon gravely in the event of a claim for negligence. Consequently, one important point which can be overlooked when performing a vendor due diligence is that reports placed into a data room from the seller's side are normally expected to be relied upon post-closing. Accordingly, there is often a transfer of liability with the use of a 'reliance letter', which effectively obliges the surveyor to be responsible to the new owner for the findings in their report for many years after the deal has been completed. The consequence of liability and duty of care is that too many surveyors undertake defensive practice; they appear risk averse in their recommendations. Clients appear to like and respect technical advisors who have an opinion, therefore it is vital to ensure that any opinion offered is strong, clear and evidence based.

The objective of a TDD or commercial building survey is to deliver the best possible advice to a client within the timescales permitted and based upon the availability of the legal/technical documents. Essentially, the surveyor should seek to provide sufficient information to allow the client to make an informed decision on how to proceed with a transaction. The findings of the TDD report are used as a tool for leverage or negotiation in the acquisition process. Therefore, there is a requirement to provide evidence-based findings which will survive cross-examination or counter-claim from surveyors appointed by the 'other side'. The reality is that most surveyors operating in the field of TDD or commercial building surveys often act for and against the same investors, depending on which side of the deal they are appointed to represent. Therefore, it pays to be transparent and objective as this gives real credence to opinions, particularly if the surveyor has a reputation for being impartial. The best surveyors adopt a firm but fair approach and this is the real reason why they are likely to get repeat instructions. However, such an approach is not always viewed so favourably by agents, as such thoroughness may be deemed to be a potential risk to the acquisition process if too many defects are noted. Where fees are generated on a percentage basis for successful completion of an acquisition, some agents may be reluctant to recommend surveyors who are likely to deliver honest opinions which have a cost consequence and could potentially influence or risk an acquisition.

Finally, the outcome of the TDD and commercial building surveying process is the delivery of a robust and accurate assessment of a built asset. Inevitably, this will include the identification of defects and issues which often appear to come as a surprise to both the vendor and purchaser, as these are not included in the investment memorandum. Therefore, a certain amount of time and responsibility should be afforded to client care and managing their expectations, as well as any fallout from the process. In truth, it is rare for the surveyor to know the bottom line figures which are acceptable for completion of a deal as this is often settled in private with only the legal or commercial advisors present. The market conditions inevitably affect the process, with a buoyant economy usually encouraging investors to buy at all costs and accordingly the costs of defects appear conveniently to be diminished or just ignored. The reverse approach operates when the economy is struggling and the costs of defects are often examined in great detail. Irrespective of market conditions or whether the surveyor is appointed by the vendor or purchaser, the outcome of a TDD or commercial building survey report should read exactly the same.

TDD – desk survey prior to the site visit

With all TDD or commercial building surveys there is an element of desk survey and this is often the process of analysing the existing documents for the property, from the initial planning permits through to construction, completion and the operation of the asset. To an extent, surveyors are expected to have knowledge concerning the localities in which they are working; however, with increasing TDD instructions involving portfolios of property often located 'out of area', the surveyor often has to research this too.

With most TDD and commercial building surveys, a data room is prepared by the building owner or property manager, which usually comprises a physical collection of files and documents. Increasingly, these are scanned electronic copies of information held in a virtual data room or issued on a CD ROM or portable memory device. Many owners opt for an online resource which requires individual password entry and it is assumed that those 'entering' the data room can be tracked in terms of the time and duration of the 'visit' and documents downloaded for review. Typically, the data room is 'opened' once the window of due diligence commences and usually, due to the large amounts of available documents combined with relatively short timelines for report preparation, it is not possible to review the complete data room before starting the site visit. The surveyor should list and confirm with their client the principal technical documents which will form the basis of the TDD or commercial building survey. This should be verified in the contract instruction (discussed in *Chapter 3*). It's easy to be overwhelmed by the potentially large amounts of technical documents in a data room and the surveyor, in consultation with the client, should confirm an approved list of the critical documents for review.

With large portfolios of multiple properties which require auditing in relatively short periods it is usual for a number of surveyors to work collectively as a team to achieve the objective. Accordingly, as with all teamwork, there should be a team leader to brief the team and coordinate the process, as well as providing the focal point for colleagues and the client. With more than one commercial asset, the quantity of documents can be vast and it can be advantageous to appoint one team member solely to review the documents in a systematic manner. The checklist of documents agreed with the client should form the basis of a checklist for the document review. It is therefore necessary to begin with an audit of the data room itself to establish which documents are present and note those that are missing. Missing documents should be identified and highlighted in the 'red flag' report, as should any significant issues, requirements or omissions in the information contained in the documents.

A summary of the principal information and requirements or conditions in the documents should be placed into the *schedule of condition/defects and costs* along with any recommendations from the surveyor. Ideally, all of this information should be established pre-survey, thus allowing the surveyor undertaking the site inspection to cross-check any specific requirements or observations contained within the documents. However, the reality is that the due diligence process is often too short to dwell upon the documents in advance of the site visit. Therefore, this process happens almost simultaneously when an individual is appointed to fulfil the data room duties. Where there is only one surveyor working on the project, the visit often takes priority. The potentially negative implication of carrying out the site visit before the document check is that it is not always possible to make a second or third visit to verify issues or conditions contained in the documents. This is especially problematic when the asset is located in a different region or even country from the offices of the surveyor. Furthermore, with certain building occupiers or tenants the provision of access can be a difficult issue. This is particularly the case with buildings occupied by organisations in the banking sector, data processing or some areas of specific manufacturing. In most cases, the tenant or building occupier does not know the real purpose of the visit and, bound by confidentiality agreements, the surveyor is not permitted to disclose the reason. Accordingly, some tenants can get irritated or annoyed by multiple visits to their building and the situation is often not helped by agency surveyors who do not appear to understand the role of the building surveyor and the time required to perform a visit. There appears to be a common feeling amongst agency or investment surveyors that visits to large buildings or assets can be completed in a number of hours or a half day, when in reality the ideal amount of time would include a series of visits over several days or even weeks. Therefore, visits for TDD or commercial building surveys are always a compromise to suit all parties involved in the process and this effectively puts everyone under a degree of pressure to perform.

The visit or site inspection

Probably the most important part of the TDD or commercial building survey process is the site visit, so it is surprising how little time is actually allocated to this aspect. The RICS guidance note for building surveys and technical due diligence of commercial property (RICS, 2010) identifies the principal elements of the property which should be inspected. Most surveyors will have developed a systematic approach to cover these. The biggest challenges faced during the visit are areas where access is not granted and actual physical time on site. Regarding access, the guidance note stipulates that areas where access was not granted should be clearly identified in the report, so the surveyor should ensure that they have a copy of the 'current' as-built plans prior to the survey. Marking the plans with a highlighter pen and annotating the plans accordingly to note the areas where access was denied is an excellent way to collate this data. This is particularly important as most commercial properties have a 'maze' of access corridors, technical rooms and archives provision, typically in basement areas. Without accurate as-built plans it will be difficult to confirm the areas inspected and those inaccessible during the survey, which is, in itself, a limitation.

Both *Chapter 1 – Introduction* and *Chapter 3 – Contract instruction* discuss the construction elements which should normally be encompassed within the TDD or commercial building survey visit. The minimum requirement should also be those elements listed in the RICS guidance note. There is, however, no standard or fixed methodology to doing this; surveyors should be responsible for adopting the minimum survey requirements and adapting them specifically to the bespoke nature of the asset they are surveying.

In most cases the time afforded to undertake the site survey or property inspection is usually insufficient. When preparing a fee proposal or contract instruction the surveyor should consider carefully the size and complexity of the building, to apportion the time necessary to undertake the visit. However, invariably most inspections still appear rushed. It's not always the case that larger buildings take more time than smaller buildings, as industrial sheds often have large quantities of surface area but are relatively simple buildings. Multi-storey office buildings or mixed-use developments can be smaller in size but technically more complex, thus requiring more site time.

Individual surveyors will have their own approach to undertaking the site inspection and, often, working in pairs can increase efficiency. However, one 'rule of thumb' is that one surveyor cannot normally visit more than 8,000–10,000 m^2 of commercial office building alone in one day. This principle works on the assumption that the surveyor will spend most time on the roof and in the basement, undertaking a short visit to the internal floors between, focusing on the common areas, shafts and staircases. The inspection of the external facades will usually be done from ground floor level as well as internally for glazing elements. Time should also be devoted to the external areas. It is always advantageous to get a second look at a building and it's often surprising how much information a surveyor can retain from the first visit in terms of knowledge of a property's characteristics. One way to obtain a second inspection is to arrange separate days for the architecture/structure survey and the M&E survey, thus allowing the surveyor to join the M&E surveyor. The purpose of this is to visit the construction elements in the technical rooms as well as revisiting specific areas of the building where there are technical issues, queries or uncertainty.

When access is severely restricted or hampered by insufficient time, this is usually due to the unrealistic timescales of the due diligence process, or pressure exerted by the investment and agency surveyors. This is a limitation which should be noted in the report as well as forming the basic requirement in the fee proposal or contract instruction. The essence of the site visit is for the surveyor to say what they see and if insufficient time or access is granted for this, it is unreasonable for the surveyor to be responsible for matters rising from these constraints. Consequently, this results in surveyors being highly risk averse and practising defensive surveying, which is not always helpful to the due diligence process.

Reporting process

Having accepted an instruction to undertake a TDD or commercial building survey and completed the site inspection, the surveyor should diligently embark upon the reporting process. Sometimes within hours and certainly within days of completing the visit, the client will usually make contact to establish the principal findings or concerns of the surveyor. At this point in time, the surveyor has to be extremely composed and deliver an opinion without jumping to conclusions, as it is often this first opinion that the client will refer to throughout the process. On returning to the office and having spent intense time on site collecting data or evidence relating to the condition of a property, the surveyor should pause and take stock of all the relevant information prior to processing this. A logical approach to the reporting process may include the following:

- red flag items
- executive summary and draft schedule of condition/defects and costs
- full draft report
- final report.

Red flag report

Most clients require a red flag report, presentation or notification and this is usually done within a few days and may involve the first findings of other aspects of the due diligence process, such as those of the legal, commercial and environmental consultants.

The red flag report can be in writing or a more structured presentation made to the client. It should always equate to an opinion on the asset with respect to the initial findings, which may constitute a problem to the long-term closing of the deal. Typically, this will involve observations or evidence which relate to the structural stability of the property or potentially high costs attributable the facades, roofs or services. It may also be simply that insufficient information has been gleaned, or access denied and even missing key legal/technical documents. In most cases, the findings of the red flag presentation are communicated back to the 'other side' and form the initial framework or battle lines for the future negotiation. The red flag report may be also used as a tool to push for extra site visits, additional documents or even an extension of time. Therefore, it is critical that the surveyor succinctly identifies the principal issues and defects in order to pre-warn the client of the likely issues affecting the asset and subsequent negotiation. It is an important hurdle in the process which requires careful thought in a relatively high-pressure environment.

Executive summary and draft schedule of condition/defects and costs

Having delivered an initial appraisal of the building and the associated documents, the surveyor should collate all of the information and place this into an easily readable document which summarises the condition and gives cost recommendations. This is potentially one of the riskiest parts of a TDD or commercial building survey instruction. Clients are invariably focused on the costs and, in particular, those which require immediate or short-term attention. It is therefore not surprising that they often remember the first cost that the surveyor reports. This first cost, or even draft, document should be evidenced based and, to an extent, the surveyor has licence to be more risk averse than usual. All unknown variables concerned with the building condition can be given an estimated budget, until proved otherwise. Being risk averse often means that cost estimates are based on a 'worst-case' scenario and, in most cases, as the TDD process advances towards completion and outstanding issues are resolved, the cost estimate often reduces. The exception to this is when 'new' defects or discoveries are made during the process and costs increase. One important observation is that it is easier to remove or reduce costs during the negotiation and considerably more difficult to introduce new costs mid-process. This is often because both the vendor and purchaser have been informed of defects and the associated cost estimates. They have accordingly drawn their battle lines for negotiation, having subconsciously set a figure at which they are prepared to compromise. Therefore, introducing costs into the process can make either party suspicious, as well as potentially damaging trust in the negotiation. Accordingly, any extra costs must be fully justified and this is one of the principal reasons for any opinion offered being evidence based.

The most appropriate way to illustrate condition, defects and costs is with a schedule and this document can be created as a table in a word-processed document but is much better as a spreadsheet. By inserting, altering or omitting costs in the spreadsheet it will be possible to see the effect on the overall cost planning. The schedule of condition is probably the document most widely used by building surveyors and is employed in a variety of forms for the preparation of maintenance works, dilapidations negotiation and party wall awards as well as acquisition costs. These are relatively simple documents and comprise a series of columns identifying the following:

- reference
- location (of the building element/defect)
- description (short assessment of the building technology)
- condition/defect (cause and symptom)

- repair (remedial treatment/repair or recommendation)
- cost (immediate, short term, medium term, long term)
- photo reference (number relating to the defect/observation on photograph schedule).

There is no set format for this and in some cases the observations and defects can be colour coded to indicate their level of 'risk' or urgency.

Client   Building Address   Surveyor

BUILDING ADDRESS - SCHEDULE OF CONDITION / DEFECTS & COSTS - VISIT DATES 13&14/09/2017 - DRAFT REPORT - REV 1												
Ref	Location	Description	Condition / Problem	Works	Unit	Qty	Rate	Y1 (£)	Y2-5 (£)	Y6-10 (£)	Photo Ref
1.	**Building Address**										
1.1	**Roofs**										
1.1.1	Roof Covering										
	General	Traditional softwood rafters, purlins, joists and battens supporting plain clay tiles with approx. 340mm of mineral wool insulation between the joists and above the level +1 ceiling. Noted was essentially no sarkng felt to the underside of the roof which reduces the resistance to moisture penetration.	The roof is believed to be original and is likely to date between 1920 and 1945, there is evidence of many loose and missing tiles with also evidence of tingles present indicating in situ repairs. The external surface of the tiles was noted to be pitted and worn, the internal inspection identified efflorescence build up to the nibs of the tiles on the North pitch. Generally there were a large amount fine terracotta shards in the roof space attributable to decay and crumbling of the nibs. Mortar joints to the ridge tiles was noted to be cracked and loose. The roof is believed to be 'end of life' .	The following observations and recommendations have been noted:							

Possible layout of a schedule of condition/defects and costs.

The level of detail contained within the schedule of condition/defects and costs can vary considerably and with large commercial assets the TDD or commercial building survey should adopt a holistic approach. While this will make an assessment of the principal construction elements and the key findings regarding defects as well as recommendations, there is no harm in going down to individual component level. This is usually done to provide precise examples of general observations or findings but also where there are specific or unusual individual defects. This 'personalises' the schedule and can illustrate to the client the surveyor's attention to detail, as well as providing evidence of the extent of the survey. Going down to component level can also be particularly 'impressive' when this is referenced with photographic evidence, which can be especially important when facing the scrutiny of a cross-examination by surveyors or experts from the 'other side'.

The majority of clients are investors and it is not unusual for them to be involved in multiple transactions throughout a financial year. In most cases they have little detailed knowledge of construction technology or building pathology, therefore it is important to keep the recommendations straightforward. One aspect that is universally understood by most clients is that of capital expenditure and the potentially detailed schedule of condition/defects and costs may be simplified to an overall cost assessment. This can be achieved by creating a *cost matrix* that is linked to the main spreadsheet, so that on one page of information it is typically possible for the client to see where the main costs are generated per building element or the related time requirements.

Client   Building Address   Surveyor

BUILDING ADDRESS - SCHEDULE OF CONDITION / DEFECTS & COSTS - VISIT DATES 13&14/09/2017 - DRAFT REPORT						
	Roofs	Structure	Facades	Internal Finishes	External Areas	TOTAL
Y1	£10,294.00	£2,685.00	£6,980.00	£9,166.40	£4,119.00	£33,244.40
Y2-5	£28,980.00	£0.00	£90.00	£6,214.00	£883.00	£36,167.00
Y6-10	£1,144.00	£0.00	£15,750.00	£11,094.95	£1,398.00	£29,386.95
TOTAL	£40,418.00	£2,685.00	£22,820.00	£26,475.35	£6,400.00	£98,798.35
Fees 12%	£4,850.16	£322.20	£2,738.40	£3,177.04	£768.00	£11,855.80
VAT (20%)	£9,053.63	£601.44	£5,111.68	£5,930.48	£1,433.60	£22,130.83
TOTAL Per Element	£54,321.79	£3,608.64	£30,670.08	£35,582.87	£8,601.60	£132,784.98
Total Immediate works incl fees & VAT	£13,835.14	£3,608.64	£9,381.12	£12,319.64	£5,535.94	£44,680.47
Total Y2-5 works incl fees & VAT	£38,949.12	£0.00	£120.96	£8,351.62	£1,186.75	£48,608.45
Total Y6-10 works incl fees & VAT	£1,537.54	£0.00	£21,168.00	£14,911.61	£1,878.91	£39,496.06
TOTAL						£132,784.98

Example of a one-page cost matrix.

While the use of a cost matrix and schedule of condition/defects and costs is an excellent method for highlighting the principal issues as a cost risk, there is also a requirement to explain or expand upon the key observations. An executive summary is an excellent way to achieve this, as this should be a succinct summary highlighting the principal observations while going into sufficient detail to elaborate upon the cause and effect of the defects. Typically, an executive summary forms the first section or part of the final report but essentially it should also be structured so that it can be a standalone document. This is why the executive summary can be issued as a covering document for the cost matrix and schedule of condition/defects and costs during the early stages of the process.

As with most reports, there is a requirement to open with an introduction followed by the main body or discussion, with a short conclusion to sum up the findings. Therefore, the executive summary should contextualise the survey by referring to the date and basis of the contract instruction and may touch upon any specific limitations which may have influenced the survey procedure or outcome. It should also include a brief description of the property and its location before summarising the principal observations, element by element. It can be used to detail the positive aspects of the property but should also critically appraise the defects, seeking to offer a simple explanation of the science behind these. It is the first opportunity for the surveyor to express, in writing, the consequences of defects or abnormalities and begin to give the client an understanding of the options or evidence to make an informed decision.

Investors in commercial real estate assets do not always come from a construction or surveying background but are normally financially astute and therefore able to ascertain the consequences of and impact that defects or anomalies can have on a transaction. Therefore, the executive summary should refrain from including too many technical terms or jargon. One 'test' for the document is to ascertain whether this can be read as a standalone text for those without a technical background.

There is no limit on the number of pages or words for an executive summary but typically this should not be too long, with something in the region of six to eight pages being sufficient. Specific defects or major items identified in the executive summary will also be expanded upon in the cost schedule. Although not typically evident in a final report, photos can be inserted into the executive summary to illustrate the key defects.

Full draft and final reports

Most investors and clients should have sufficient information within the executive summary and cost schedule to make an informed decision on how to proceed. In some instances, the TDD or commercial building survey instruction may stipulate or request only these two documents for the reporting process. Where a full report is required, this will not normally deviate from the findings of the executive summary and cost schedule but is an opportunity for the surveyor to go into greater detail. This typically involves more analysis and description of the science behind the defects and the rationale for the recommendations for repair or justification for the timing.

Typically, the full report starts as a 'draft' which evolves, as does the cost schedule, when 'new' information or documents come to light as the process is ongoing. Often up to the end of the due diligence process, or even closure of the deal, the final report can be a draft with various revisions, but at some point this has to become a final version. For many clients, the final report is not always read or analysed in detail; however, it is filed as part of the due diligence process and often only opened up again if a post-closing technical issue comes to light or when the building is the subject of a re-sale. Therefore, the full report should start with the executive summary and then present a logical breakdown of the building in chapters. A complete section towards the front of the report should be dedicated to detailing and discussing the contract instruction as well as the limitations and this is the section of the report where inaccessible areas or missing documents are listed.

The final report should be as comprehensive as possible, and this document may also represent the 'shield' which the surveyor can use as protection from potential litigation or negligence claims. Although the surveyor is tasked with writing a report to satisfy the requirements of the client or due diligence process, in essence the report should also be written to the satisfaction of the surveyor. Therefore, equal measures of diligence should be exercised by the surveyor throughout the process, from the preparation of the contract instruction through to the issue of the final report. This is the essence of technical due diligence.

Further investigations

During the course of a TDD or commercial building survey it is not always possible to establish all of the relevant facts or collect sufficient data to form an evidence-based opinion. These issues or items are typically raised as part of the red flag process and while it may simply be that more time or an extra visit is required to verify certain issues, sometimes there is a requirement for specific further investigation.

The decision to recommend a further investigation should not be taken lightly as in most cases the client has appointed the surveyor as an appropriately qualified expert to give an assessment on the entirety of the asset. However, the recommendation to seek further specialist advice is normally something that can and should be decided relatively early in the due diligence process. Typically, for the TDD or commercial building survey, this concerns hidden areas which require opening up or observations of specific defects where the surveyor is not suitably qualified to investigate and form a qualified opinion. Further investigations typically include some of the following:
- structural movement, cracking, loading capacity
- concrete defects (carbonation, high alumina cement (HAC), alkali silica reaction (ASR))
- M&E tests (heating, ventilation and air conditioning (HVAC), electricity, water supply and evacuation, lifts, fire detection and firefighting)
- environmental issues (hazardous waste, soil pollution, asbestos, acoustics, lux levels)

- fire regulations conformity (stability, compartments, separation, means of raising the alarm, means of escape, firefighting).

In many cases these are legal/technical issues and it may be reasonable to assume that the relevant reports detailing conformity and compliance are located in the data room or are the responsibility of the property manager. Any missing information requires action during the due diligence process to obtain the relevant data. They may also be a post-closing requirement, forming part of the representations and warranties. Often where there is a post-closing remark, the vendor is obliged to take responsibility for any consequences of failing to achieve conformity and this is often covered by a financial bond. Ideally, further investigations and their outcomes should be resolved during the TDD or commercial building survey process and this is certainly something that should be done with structural or life safety issues.

Structural issues usually concern evidence of movement to the building and in most cases this presents as cracking to external or internal walls or cracking and deflection of the structure. *Chapter 6 – Structure* analyses in detail how to undertake an appraisal of the structure as well as using surveying skills, such as crack analysis. However, the reality is that the surveyor has to deliver an evidence-based opinion and in certain circumstances it may not be possible to confirm a diagnosis from a crack alone. Where this is believed to present a structural issue, it makes professional sense to appoint an appropriate specialist.

Many material failures, and particularly defects associated with concrete, initially appear as cracked, spalled or crumbing surfaces and, especially with structural or architectural concrete facade panels, there is the potential for high repair costs. Therefore, it is prudent to appoint a material specialist to advise on the cause, effect and likelihood of this to 'spread' to or become prevalent in other areas of exposed concrete. This again will be discussed in more detail in *Chapter 6* as, with concrete and other composite materials, there are a range of defects which tend to surface when the conditions are right for these to prevail.

Regarding M&E or services installations, the surveyor is obliged only to have a general and peripheral knowledge of this area. This is one part of the TDD or commercial building survey process which is outsourced from the beginning to a suitably qualified professional. Typically, where there are specific issues with the services installations, these can be complex and may require significant amounts of time to investigate. The initial TDD or commercial building survey instruction does not normally foresee in-situ testing. However, where there are problems with the M&E installation it is usual to perform a range of tests on flow rates for air or water as well as temperature and pressure analysis, which can prolong the due diligence process. The surveyor should rely solely upon the expertise of the appointed M&E consultant to determine what tests and processes are required to investigate further problems established with this building element. The client should be advised of the effect of undertaking tests on the due diligence timing and they should make an informed decision to proceed with this, negotiate an extension of time or insist on a post-closing obligation for the installations to be fully functional.

Naturally it is not normally the responsibility of the surveyor to comment upon additional sub-specialisms, such as environmental issues, and these should normally have been covered by the existing building owner or property manager in the course of operational occupation. Having to request information and guarantees during the due diligence process can take significant amounts of time and it is the client who should make an informed decision on how to proceed.

Nearly all TDD and commercial building surveys identify issues relating to fire safety and the surveyor can systematically look at a building and benchmark this against the norms or minimum requirements contained within regulations. The sealing of openings in fire compartments or damaged fire doors may appear obvious defects to remedy, more complex fire engineering issues require closer attention. Surveyors are not normally specifically qualified as fire engineering experts and therefore cannot verify compliance with regulations. In the event of certain specific issues, such as fire separation, structural stability of materials, means of raising the alarm, means of escape and firefighting, it will be necessary to appoint an appropriate expert to confirm compliance. This can create significant delay in the due diligence process and planned closing dates. The client needs to be comfortable with the implications for fire or life safety issues; this is particularly pertinent when there could potentially be a fire or loss of life at a commercial building. Accordingly, there is a responsibility and duty of care for building owners to ensure that their buildings are safe places to work and live, as well as their duty of care to the general public who may pass or access the building. There is a distinct possibility that legislative change may lead to building owners or property managers being charged with corporate manslaughter, as well as the existence of civil procedures to account for damages associated with a failure in the duty of care. Knowledge of potential breaches in fire regulations start primarily with the TDD or commercial building survey and, in order to ensure an audit trail covering the duty of care, it will be necessary for the surveyor to recommend the relevant further investigation, if they are not qualified to deliver this opinion.

TDD and building management

Good and accurate TDD seeks to give the client sufficient data for them to make informed decisions and the schedule of condition/defects and costs is an excellent example of a document that can be transferred to the management of the property. This document can initially form a list of works that can be undertaken post-closing, thus giving an opportunity for the surveyor to continue their work with the client, providing continuity and understanding of the building. It can also be used to form the basis of a schedule of planned, preventative maintenance for the ten years following the acquisition.

The TDD or commercial building survey report is something which will be archived and is typically referred to at the end of the investment period when the client decides to sell the property. The report is also revisited in the event of a defect or issue arising in the building in the future; accordingly, the owner or property manager often refers to this document to see if they have received suitable prior warning. Surveyors are professionally liable for the content of their reports and findings for many years following the closure of an acquisition. Typically, items identified in the course of occupation or property management can take days, weeks or months to investigate. Surveyors appointed during the due diligence window are not afforded the luxury of similar timescales to inspect and report upon the past, present and future of a building. This is what makes TDD and commercial building surveys a high-risk area of professional services and again is why surveyors should be risk averse in their recommendations. The test of the robustness of a contract instruction and technical report for a property may eventually be resolved in court in the event of a claim for negligence. It is anticipated that this will be assessed upon a series of factors surrounding the TDD or commercial building survey instruction. The final report should always note if there was sufficient and reasonable time afforded to visit the property and review all relevant technical documents.

References

RICS. 2010. *Building surveys and technical due diligence of commercial property* (4th edition, guidance note), London.

RICS. 2011. *Technical due diligence of commercial, industrial and residential property in continental Europe* (1st edition, best practice and guidance note), London.

Contract instruction

3

Introduction

The contract instruction is one of the most important aspects of due diligence. While there may be some informal pre-due diligence discussion between the client and due diligence team/surveyor, it is critical to define and agree what will be covered by the due diligence instruction. This indeed is the starting point which sets down the rules of engagement and is also a reference point in the event of dispute, discrepancy of renegotiation of the contract terms.

Most TDD or commercial buildings surveys are instructed on a fixed fee basis, therefore it is important for the consultant to be aware of the client's needs in order to ascertain the work involved and the time that this is likely to take.

The current RICS practice standard and former guidance note *Building surveys and technical due diligence of commercial property* (4th edition, 2010) provides a detailed insight into the key items required within a contract instruction. It is important to acknowledge the guidance note when preparing a fee proposal or contract instruction.

Pragmatically, it is essential that, in order to instruct the TDD, the following key aspects should be agreed:

- outline the contract instruction
- scope of works
- specific services
- limitations
- programme
- fee.

Outline the contract instruction

By means of an introduction to the contract instruction it is essential to identify the contracting parties. Clearly, it should be straightforward for the consultant to identify themselves but often it can be more complex to identify who the client is. This is not only important in terms of who is owed contractual liability but also who will pay the fee invoice.

The outline of the contract instruction should seek to confirm the following:

- name and registered address of 'the consultant'
- name and registered address of 'the client'
- nature of the instruction (building survey, technical due diligence, schedule of condition) – 'the works'
- location (building name/address) of 'the property'.

It is important to summarise what is actually going to be surveyed and covered by the contract instruction. The building size, age and complexity should be considered by the consultant when determining the allocation of staff resources to ultimately calculate the representative fee.

Ideally, a pre-survey site meeting with the client would be a good opportunity for the consultant to gauge the amount of work that may be involved in the auditing process; however, such 'luxury' is rarely afforded. Therefore, it is important to seek verification from the client as to the nature of the property and summarise this in the outline of the contract instruction.

It is prudent to state further that any misrepresentation in the description supplied by the client, which results in extra time on site, may be the subject of extra fees. This is relevant as most clients seem to prefer a fixed fee for commercial building surveys or TDD, which is only appropriate if there is a fixed amount of required work. Details of the provision for extra fees should be included in the contract instruction.

With respect to summarising the key characteristics of the building, this information can be bulleted or tabulated in the contract instruction as follows:

- Number of buildings
- Building size
- Type/number of parking places
- Building age
- Renovated
- Date of renovation

Ascertaining the number of buildings is very important as two buildings, each of 5,000 m², take significantly more time to survey than one building of 10,000 m². Logically, two buildings require inspection and analysis of two different sets of construction elements. Sometimes these may be similar buildings of the same age but often they are contrasting ages and styles with completely different building technology.

Size matters, certainly in the context of building type: 10,000 m² of industrial sheds take time to physically walk around and with roofs often the size of football pitches they can be daunting and imposing. But, broken down into the construction elements, these are often quite 'simple' buildings compared to 5,000 m² of mixed use offices/residential, which are more intricate and complex. It is often not logical to assume that twice the square meterage takes twice the time. Therefore, it is important to consider the complexity as well as the size and use of each building on a case-by-case basis when preparing the fee proposal or contract instruction.

The significance of ascertaining the type and number of parking places cannot be underestimated. If a relatively small city centre office building has allocated basement parking, it is not unusual to find this distributed over several underground levels. Such a layout requires sufficient time allocation for a detailed inspection during the physical survey. The basement is one of the most important areas, where it is often possible to see the 'naked' structure in detail, therefore the contract instruction and fee should take into account the requirement to inspect this.

Building age is often overlooked when preparing fee proposals or contract instructions for commercial building surveys and TDD. Generally, old, non-renovated commercial buildings tend to be afflicted with more defects and consequent costs in comparison to those which are 'new' or renovated. It is therefore essential that the consultant ensures that this is taken into consideration when allocating time and resources to undertaking the site inspection.

Having stated that old buildings are likely to have more defects, it should also be noted that new/renovated buildings are often littered with outstanding snagging items. Frequently, these come to light during early post-acceptance occupation. While undertaking a TDD or commercial building survey on a new build or renovated property it is important to foresee the potentially large amounts of time that may be required to investigate or verify the status of snagging issues.

Renovation is an important factor to consider when allocating time and resource to the commercial survey or due diligence process. While (as expected) the condition of a renovated property should be good, there is likely to be a large amount of as-built documentation and acceptance reports which will need auditing as part of the process. It is always surprising, the 'disappearance' or dissolution of as-built data in the years following renovation. This should not be acceptable in today's climate, where there should be hard copy of the as-built file as well as an electronic copy which should be kept in safe storage for transmission to any future owners.

Scope of works

Having established the details of the contracting parties and a description of the property, a scope of works should be agreed to define precisely what will be included in the commercial building survey or TDD instruction.

The purpose of the TDD or commercial building survey will be to undertake a visual inspection of the property to establish the historic construction detail with a view to drawing conclusions on the present condition and anticipated/potential future defects.

Importantly, the scope of works should state that the due diligence process will be undertaken within the timescale agreed with the client. Such timescale should be calculated to include the necessary visit to the building(s) and review of an agreed selection of technical documents.

Concerning the site or property inspection, this should be divided into an assessment of the following:
- architecture/structure
- M&E.

The intention of the visit is to record the current condition and present *visible defects*, as well as potential future defects based upon the evidence available during the visit. As discussed in *Chapter 1 – Introduction*, the site inspection should seek to analyse the construction elements detailed below.

Architecture and structure

- Structure
  - substructure
  - superstructure.
- Roof(s)
  - roof covering(s)
  - parapet walls
  - rainwater evacuation
  - roof lights
  - other.
- Facades
  - external walls and cladding
  - windows and doors
  - entrances
  - other.
- Internal finishes
  - offices, retail, classrooms, warehouse, production etc. (dependent upon primary function/ tenant use of the building)

- entrance hall, lobby, reception
- lift lobbies
- staircases and emergency escape routes
- sanitary rooms
- kitchens, staff welfare facilities
- technical rooms
- basement, internal parking areas
- other.
- External areas
  - access routes, entrance
  - external paving, pathways
  - parking
  - landscaping
  - other.

Mechanical and electrical

- Services installation
  - HVAC supply and distribution
  - electrical supply and distribution
  - sanitary/water supply, distribution and evacuation (normally excluding sewers)
  - fire detection, alarm and firefighting installations
  - lift installations.

The survey of the M&E installations should be considered as a specific sub-specialism; therefore, this should be undertaken by a suitably qualified consultant. Such specialist knowledge may come from within the TDD organisation but is more commonly provided by an external source. Where an expert sub-specialist is used, it is important that instructions are issued directly by the client or consultant in accordance with the RICS guidance note.

A general condition, which should be included in the scope of works, is that covering the requirement for further investigations. In most cases it is almost impossible to inspect, examine or test all aspects of the building. Therefore, the scope of works should explicitly refer to the consultant recommending, where, necessary valid further investigation. This should not be seen as the consultant extricating themselves from the survey work for which they are qualified, but specifically to address any additional areas or defects which are of a 'specialist' nature.

Furthermore, it is common practice in the scope of works to detail what is covered or excluded regarding tenant works or fit out. These should 'normally' be the subject of general comment in the TDD or commercial building survey report; they are not routinely the subject of detailed inspection for cost analysis, unless specifically requested by the client. Tenants works or fit out are usually covered by existing lease contracts and licences to alter. These will usually be scrutinised by the legal due diligence and there is often discussion between the legal and technical teams regarding covenants for repair and alteration. In the event that the client requires specific detailed analysis of tenant works or operations which are ongoing at the property, this should be specifically noted in the scope of works. This is likely to increase both the time on site and the staff resources required to review the relevant documents associated with this aspect Therefore, an allowance for the costs associated with this work should be included in the fee proposal.

The scope of works should set the conditions of engagement and specific reference should be made to the requirement for access to *all areas of the property*, including those areas outside normal tenant occupation. This typically includes roofs, basements, shafts, technical rooms etc.

With timescales often tight and with a need to visit occupied or locked areas of an unfamiliar building, it is prudent to state in the scope of works that access is to be accompanied. This should be by the building owner/manager (or someone similarly qualified). The reason for accompanied access is twofold. First, with someone who knows the building and its history, it is an ideal opportunity to ask them to direct the visit to areas of known specific defects or issues. As part of the data-collection exercise and considering that time on site is often restricted, the scope of works should indicate that the building owner or property manager will be asked (interrogated) to verify any specific tenant complaints or areas of known defects. By doing this, the consultant is pressing the owner to disclose any defects or issues known to relate to any of the building elements. This should include verification of the party responsible (landlord or tenant) for remediation. Therefore, site time can also be focused or pinpointed to specific areas of defects.

Second, the building owner or property manager should be familiar to the tenants and this may help to allay their concerns about an unfamiliar organisation entering their premises, examining their space and taking photographs. Often the tenant is unaware of the ongoing acquisition or possible future change of ownership and consultants are frequently bound to confidentiality on this issue. Therefore, if pressed by the tenant as to the reason for the visit, a standard reply could be that the purpose is for a condition survey on the property to assess the material life cycle of the construction elements. If the opportunity presents itself during any discussions with the tenant or occupier of the building, it is also advisable to ask if they have any general concerns or issues with their demise. Granted that there is often a fractious relationship between the parties, any 'complaints' should take this into account in determining the reasonableness of these. But, in essence, such discussions will quickly establish if there are significant defects within the building fabric or with the internal climate and comfort levels.

Within the scope of works it is also important to detail what to do about inaccessible areas. Essentially, these will be noted and communicated to the client as part of the TDD or commercial building survey process. There will need to be some provision for the actions to be taken to obtain access and this may state the use of specific access equipment, such as 'scissor lifts' or 'cherry pickers'. In most normal circumstances this constitutes an 'extra' to the contract instruction. The potential cost and time required to implement these measures should be covered by a section of text in the scope of works.

Specific services

The contract instruction should explicitly state the specific services that will be included and, in principle, due diligence service is effectively a technical audit of the property to include the following:
- building survey
  - architecture/structure
  - M&E
- document review.

In order to time manage this process, and also to manage the expectations of the client, it is relatively straightforward to define those things covered by the building survey (as listed in the scope of works). However, it is more complex when ascertaining what documents will be reviewed, as significant time and resources should be set aside for this aspect. The document review is an important part of the TDD or commercial building surveying process.

Documents are received either as hard copies or electronically, and sometimes in both formats. They may be held within independent data rooms, via online links, issued on a CD or

USB stick. In most cases the surveyor will be obliged to sign a confidentiality agreement prior to gaining access to this data.

A great deal of time can be expended on analysing the documents and many clients assume that the surveyor has reviewed the entire data room. In reality, the surveyor should propose to the client (explicitly stipulated in the contract instruction) to review all those documents relevant to the technical due diligence and within the timescales afforded. In order to allocate sufficient resource to undertaking the review of documents it may be prudent for the surveyor to issue the client with a list of the 'critical' technical documents, such as:

- current as-built floor plans, facades and sections
- as-built M&E schemes/plans
- planning permissions/buildings permits (including copies of certified plans)
- planning permissions/building permit plans
- building regulations (approval and acceptance certificates)
- operating permits/licences
- asbestos inventory
- fire safety audits/reports
- acceptance/completion reports
- annual (electrical, gas, water, drainage) test reports
- lift test reports and any other test reports concerning fixed installations (landlord) at the building
- energy performance certificate (EPC)
- access audits
- environmental assessments
- rights of way and easements
- boundary information
- guarantees and warranties
- tenant leases
- leasehold and repairing liabilities.

By requesting a list of minimum or critical documents (from those listed above) the surveyor can set the parameters on which this part of the due diligence will be based. It may be fair to state that without these documents some limitations may have to be placed upon the due diligence report.

During the course of the TDD or commercial building survey it is likely that additional documents may be provided or requested for review by the client or the legal/commercial due diligence teams. Having stated the list of minimum documents in the contract instruction, there should also be some provision within the instruction to revisit the fee proposal to include any extra works associated with additional document review. Furthermore, it should be noted that extra work may also require a possible extension of time in the delivery of the final TDD or survey report. In some ways this is not dissimilar to the approach covered by a building contract in the agreement of a contract variation (for additional works) and the effect this may have on completion of the project.

Concerning the actual building/property survey, the contract instruction should clearly state that this is an in-situ visual inspection of the building elements (stipulated in the scope of works) and that the inspection will be:

- visual only
- no tests will be performed.

The reason why no tests will be performed is essentially based on the premise that to test, say, a radiator valve in one isolated location, would not be representative of the entire allocation of radiator valves, irrespective of the outcome. The RICS guidance note does suggest for repetitive items that sample testing may be done, which may be appropriate for window openings and other manual operations but less suited to services, such as components related to water or air flow. Therefore, this adds further weight to the importance of the building owner/property manager disclosing any known problems or tenant complaints with the provision of HVAC. Furthermore, undertaking on-site tests will take time where time may be 'restricted' in the due diligence process. Therefore, it would be better to make recommendations for further specific on-site testing, should the surveyor have reasons to suspect this may be necessary.

The contract instruction should state that the inspection and report will comment on disabled access to the building, but not include or constitute a disability access audit. Clearly, in the course of a TDD or commercial building survey it is anticipated that there may be an access audit among the data room documents. Irrespective of this, it is completely acceptable for the surveyor to make a general assessment (including some checking of dimensions) of the following disabled access criteria:

- parking provision
- access to the building/property
- main entrance
- horizontal circulation (including means of escape)
- vertical circulation (including means of escape)
- services/facilities provision (sanitary rooms, reception desks, waiting areas etc).

Concerning asbestos specifically, the contract instruction should state fundamentally that the building/property inspection will not amount to or include an asbestos survey or inventory. Of course, the surveyor should comment on the suspected presence of asbestos where obviously visible within the building. However, unless the consultant is an expert in this matter, it will be necessary to state that, in the terms of the due diligence:

> The asbestos inventory will be analysed and reviewed, including the findings and recommendations of the specialist asbestos consultant who has undertaken the inventory.

A similar principle should be adopted regarding fire precautions. The contract instruction should state that the inspection and report will comment on fire safety with a review of the fire safety reports/logs made available in the data room. It may be considered 'normal' practice for the surveyor to comment on any obvious defects or deficiencies (where visible) concerning the following fire safety issues:

- fire detection and means of escape
- fire resistance, compartmentation, separation and smoke ventilation
- firefighting provision, including extinguishers, hose reels and automatic sprinklers.

Fire engineering is a highly specialist field of operation and where there appears to be insufficient reports or certificates in the data room documents it will be appropriate that the surveyor recommends further specialist investigation and report.

The contract instruction will have to address other important legal/technical issues, such as planning permissions or specific operating permits. In these cases, clients often assume that the surveyor can/will verify that the building/property is in compliance with such permits. Therefore, it is important that the contract instruction defines how these legal/technical issues will be analysed.

Invariably, the starting point must be to visit the data room documents to see what information is available. Where there are missing documents or insufficient information to form an opinion, the surveyor should flag this up immediately. Concerning the verification of compliance of the as-built situation with that stipulated on planning documents, the best course of action is to undertake a desk survey followed by a site inspection to establish any visual differences between the certified planning permission drawings and the as-built plans, sections and elevation drawings.

Attention is drawn to the tendency for as-built plans to be generated and certified by simply changing the title boxes of existing planning or execution drawings, without any cross-checking or site inspections during the acceptance process. It is therefore not unusual for the surveyor to identify some visual differences during the survey; however, in most cases the surveyor is not responsible for certifying that the as-built situation matches that permitted: this should be the responsibility of the owner/vendor through the acceptance procedures of their own design team etc. In most cases, the sales contract will/should provide a warranty from the vendor that the property is in compliance with all relevant and existing legal requirements.

Following on from the previous point, the verification and/or measurement of the lettable floor areas is not usually included in the TDD or commercial building survey. Granted, there are some multi-disciplinary surveying consultancies which can do this. However, unless otherwise stipulated, the contact instruction should make provision to state that such a measurement exercise should be done using existing as-built plans (with appropriate on-site cross-checks) *as a separate exercise* either 'in house', sub-contracted or directly contracted out by the client to a land surveyor. To do this, it will be necessary to have an electronic copy of the plans (in a format such as AutoCAD) and any measurement should be made in accordance with RICS guidance/measurement practice.

The contract instruction should detail the format for reporting the findings of the TDD. In some cases, the client will insist on a red flag report or presentation followed by a draft report (executive summary) with a schedule of condition/defects and costs and then the final report.

Probably the most important aspect of the reporting process is the assessment of costs. Generally, this will be associated with defects, non-compliance or future capital expenditure for the property. It is important to ascertain from the client their requirements concerning cost planning. In most cases, immediate costs tend to be taken into consideration as part of the sale/purchase negotiations. Depending on the economic conditions at the time of the due diligence, short-, medium- and long-term costs may be either used to reduce the price (during recession) or accepted by the purchaser in their CAPEX when the 'market' is buoyant.

The contract instruction should state what cost planning is required by the client and this may be represented as follows:

- immediate/short term (year 1)
- medium term (years 2–5)
- long term (years 6–10).

When defining the cost data used to calculate cost planning in the contract instruction, it is important to place some 'limitations' on the sourcing and nature of these. For example, and unless otherwise requested from the client it may be stated that:

> Budget estimates will be assessed relative to the limited time available to report. They are 'Day One' costs, subject to market conditions and future inflation. They are exclusive of professional fees, asbestos removal and VAT. Where prices have been estimated per m², this has been done off plans received (where applicable) or based upon approximate times required to perform tasks in accordance with published cost data information or in-house cost databases. The prices are budget estimates and are subject to a full brief and feasibility study to ascertain all relevant options and costs.

The contract instruction should mention that photographs of the property and defects will be appended to the report and these should be used to illustrate specific defects or areas of concern. While, in most cases, the language used for the report is not an issue, it is important to confirm for overseas clients or for TDD undertaken outside the UK that the report will be in English. An allowance should be made within the report to state that the fee is exclusive of any costs of translating the report from English to any other language.

Limitations

Limitations are a crucial part of the contract instruction. As much as there is a need to inform the client of the scope of works and services provided by the instruction, it is also important to place limitations on what will be included in the due diligence mission.

The contract instruction should state that the survey of the property/building will be:

- visual and no attempt will be made to open up *any* part of the building for further detailed inspection
- the upper areas of facades and other areas of construction which are difficult to access may not be inspected at close quarters
- the interiors of services, lift shafts and hidden elements (including those areas concealed by finishes) within the building will not inspected.

Furthermore, the review of the documents and site inspection *will not include*:

- verification of the concept design calculations
- confirmation of compliance with permits
- testing of:
  - materials (including fire sensitivity or fire reactivity of materials)
  - structure
  - services.

It is important to state within the contract instruction that, should such further verification or tests be subsequently considered necessary, suitable recommendations will be made by the surveyor and detailed in the reports. Where necessary, the surveyor may also arrange or facilitate appropriate specialist or further investigations.

Programme

The vendor and purchaser often agree a window of due diligence prior to appointing the relevant consultants. When the market is buoyant, this period is often very short, say between 2 and 4 weeks. The vendor is in a strong position with some assurance that, if the purchaser does not agree to the conditions, then there is likely to be another purchaser willing to step in. Likewise, during an economic downturn, the purchaser is in a position of relative strength and often the window for due diligence is significantly longer.

In both cases, the surveyor will have to run with the programme/timing of the client. They will have to mobilise staff and resources to undertake the site visit and document review with an expectation of reporting to the client in the shortest possible time. Many clients require a red flag report/presentation within days of undertaking the TDD (visit and document review) and this is the time for the surveyor to list all of the missing documents or unsatisfactory data room information. Furthermore, it is the moment when the surveyor should carefully present their initial technical findings and inform the client of possible 'deal breakers'.

Time often presents the single biggest challenge in the due diligence process and, while it is important to commit sufficient time to the document review, it is also critical to devote enough time to the site inspection. Clients and agents appear to believe that the entire site inspection can be completed in a day or less. However, depending on the complexity and size of the property, at least one day should be allocated to each 10,000 m², with another day set aside for the M&E inspection. The more time on site, the better, and it can be extremely beneficial to have a second day's visit to a property as it often yields a greater understanding of the building and can reveal additional defects.

The contract instruction should seek to clarify the execution details of the due diligence and clearly state the requirements of the client for delivery of the draft and final reports. The contract instruction should detail the 'agreed' date for delivery, subject to receiving the relevant documents, data room access, client instruction and appropriate site visit(s) or access.

Fees

Fees may be the most contentious part of the due diligence process; invariably there is always someone else willing to do a worse job for less money. Therefore, the contract instruction should seek to clarify how fees are generated in order to justify the resultant figure. While most fees for due diligence are commissioned on a lump sum or fixed fee basis, these should be calculated and agreed with the client based on the anticipated time required for the undertaking of site inspections, reviewing of available documents and preparation of draft and final reports.

It is important to state that the fees are exclusive of VAT and there should be some mention of whether the fees are inclusive or exclusive of disbursements. Furthermore, it is prudent to outline the conditions of payment and to specify the process for dealing with any dispute resolution.

Within the fee section of the contract instruction it is advantageous to include a clause governing the extra non-contracted works further to the TDD or commercial building surveying mission and this may include:

- the review of additional documents
- attending meetings with the client and legal representatives
- meeting with the vendor/purchaser/property manager or their agents to discuss technical issues
- monitoring of remedial works.

An appropriate hourly or day rate should be detailed for this. Finally, the contract instruction should be signed and dated by representatives of the client and consultant.

Case study

A client contacted a surveyor to undertake a technical due diligence on a city centre retail development (outside the UK), which included a large underground basement area. The window of due diligence was very tight and the client expressed a wish for the consultant to inspect this new build project, but only the retail areas above ground. The client expressed verbally that they were not concerned about the areas below ground and even for the consultant not to 'bother' with reviewing the documents for the project (data room) in order to save time and cost of the TDD.

The consultant produced a formal fee and services proposal based upon the requirements of the client, which explicitly stated that there would be no document review and no inspection of the basement parking area. This fee and services proposal was accepted by the client and converted into a contract instruction.

Upon commencement of the TDD the consultant received, by courier, a box containing documents for the project but did not to review these in accordance with the contract instruction. The technical due diligence inspection was completed, with numerous minor defects (essentially snagging items) being noted. In the final report, the consultant stated that the basement areas were not inspected. Furthermore, the final report also confirmed that the legal/technical documents were not reviewed as part of the due diligence procedure, as agreed with the client and stated in the contract instruction. Additionally, the surveyor made recommendations in the report that the client should ensure that the developer (vendor) complied with any legal/technical requirements.

Five years later, when the asset was sold, the client contacted the surveyor to question the original TDD findings. During the re-sale of the property a subsequent due diligence was commissioned which did review the legal/technical documents as well as undertaking a basement inspection. This revealed that there was an obligation placed by the fire authorities for a sprinkler system to be installed in the underground car park and this had not been executed.

The resultant cost was between £150,000 and £200,000 to retrofit a sprinkler installation and the client was obliged to reduce the price of the asset accordingly to compensate for this. The client was concerned that the surveyor did not visit the basement area or review the technical documents and, despite having agreed the contract instruction, was perplexed as to why the documents were not reviewed as these were received by the surveyor.

The end result was unsatisfactory for both parties; the client had to accept a reduction in the sale price and the previously good relationship that the surveyor had with the client was damaged, although no claim for negligence was brought against the surveyor.

Giving the situation due consideration, and even though explicitly not requested by the client, the surveyor could/should have foreseen the pitfalls of not inspecting the basement and not reviewing the documents. It should have been the responsibility of the surveyor to inform the client and make them fully aware of the possible risks associated with their decision to omit this from the contract instruction.

Photo courtesy of Widnell Europe.

Commercial sectors 4

Introduction

Broadly speaking, property is either classed as 'residential' or 'commercial', particularly with regard to building surveying applications. A large sector of building surveying work is concerned with undertaking valuations and surveys for mortgage lending or home acquisition and this is classed as residential surveying. All other applications of building surveying could be classed as commercial. Indeed, the RICS guidance notes concerned with building surveys are specifically divided into residential or commercial buildings.

Furthermore, when looking at the UK Building Regulations Approved Documents, there is also a defined separation to allow for 'buildings other than dwelling houses'. These volumes of the regulations are intrinsically concerned with commercial buildings.

Finally, to begin to understand the different commercial buildings, operations or sectors it is necessary to establish the definition of commercial property. What sets commercial and residential property apart is that commercial property is essentially land or buildings which are used in an act of commerce. This could be interpreted as a shop where trading occurs but at a more fundamental level it covers a wide range of activities which are encompassed under the title of business. This is where goods or services are provided for some kind of financial remuneration.

A large percentage of residential property is owner occupied and clearly this is not commercial; however, it is appropriate to consider whether the private rented sector comprises commercial properties. Buy-to-let properties are unlikely to be considered as commercial properties as they are almost always standalone investments. However, should buy-to-let owners acquire portfolios of residential properties, particularly if these are incorporated into the assets of a registered company, then these could be classed as commercial properties. A good example of commercial properties in the residential sector are the housing stocks owned and maintained by local authorities or housing associations.

In essence, commercial properties may be separated into the following principal sectors:
- residential
- retail
- offices
- industrial
- leisure
- public sector/infrastructure.

Within each sector, a sub-division can be made to further illustrate the types of properties which make up these sectors.

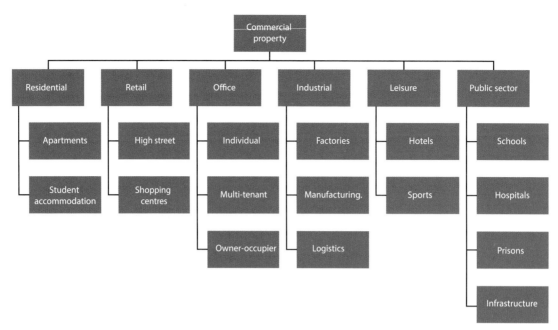

Commercial property sectors and sub-sectors.

Before looking in more detail at the individual commercial sectors, it is worth considering the reasons why organisations invest in real estate. Banks, pension funds and other large investors will have a variety of investment vehicles in which they will seek to grow capital. While stocks and shares may be considered to have potential high returns on investment, they may also be considered high risk. Property, alternatively, may be perceived as being relatively low risk and, although the returns may also be relatively low, the property or building itself is a tangible asset.

For property investment there needs to be an initial outlay of capital to purchase the property but ultimately there needs to be an end user to occupy this property and commit to a long-term lease, thus providing an income stream for the investor. As with most businesses, there will be a requirement for the investor to assess the costs or overheads required to buy, run and maintain the property. This is then offset against the return or income from the tenant or end user to derive, ideally, a profit or sometimes a loss.

Concerning general and global economic trends, there have always been periods of growth (boom) and shrinkage (bust) with the 'tipping' point between these known as recession. While it is a long-established fact that there have always been and always will be recessions, it is very hard to actually predict when these will occur. They are often triggered by global, political or economic events which are rarely foreseen, or at least foreseen to have significant consequences, at the time.

Most 'developed' economies are finely balanced and, typically, as an economy becomes 'buoyant' banks appear willing to lend ever-greater amounts at low interest rates. These loans are often made to those at the end of the investment chain, such as homeowners and small businesses. With low interest rates, high levels of borrowing or personal debt and overheating, typically, of the housing market, this results in rapid growth in property prices. There may be an overall increase in the cost of living, characterised by inflation rising faster than increases in pay. To control this, national banks raise interest rates to slow down borrowing, but for those who have over-borrowed at low rates this rate rise becomes an issue. Unable to pay back loans, consumers have to reduce overheads, such as spending on staff in the case of businesses. As a consequence, companies downscale or go out of business. The consequent vacating of premises and 'shunting' the downturn back up the chain to the investors, leaving commercial property empty, is not good for commercial investment. The commercial property market often mirrors that of the residential sector and, in particular when large banking institutions are exposed to high levels of unsecured debts arising from overvalued residential properties, economies often slide into recession.

Recession has significant effects on commercial property, with vacant tenant space affecting investment yields and overall asset values. Consequently, commercial property values drop, as do the returns paid to investors or the projected values of investment funds.

The cyclical nature of most developed economies means investors will be acquiring and disposing of assets at some point in the cycle. Ideally, investors seek to acquire property for investment when property values are low and to dispose of these properties when values are high. However, it is important to note that different investors have different investment 'cultures' when it comes to the location, type or duration of their investment strategy.

Some investors are risk averse and seek to invest in stable markets with steady but relatively low yields, such as parts of central Europe. Others look towards emerging 'markets', such as the BRICS countries (Brazil, Russia, India, China and South Africa), where the risks are greater, as are the potential returns. Some investors are 'bullish', aggressively acquiring assets, and others are 'traders' who acquire and dispose of real estate assets frequently. Some have ten-year investment plans and others five-year plans. Therefore, it is important for the surveyor to know and understand the investment culture of their client as well as their *modus operandi*.

Having discussed the different cultures and characteristics of property investors, it is important to understand the fundamental sectors of real estate investment.

Offices

Aside from where we live, where we work is probably the place where the second largest amount of our time is spent. While it is acknowledged that there are other important places of work, concerned with manufacturing, infrastructure, logistics, retail etc., offices are probably the sector which 'houses' most employees. It is no coincidence that office acquisition accounts for the majority of TDD or commercial building survey instructions. It is therefore important to understand the characteristics of this sector, as well as areas of commercial, legal and technical concern.

Primarily, investors are seeking investment properties which will generate a yield (return) on their investment. The yield and commercialisation of the investment asset form the principal advice of the investment/agency surveyors advising on the commercial due diligence. This will include the types and durations of lease contracts, price per square metre/foot as well as service charge details. All of this data is used to establish the projected return on investment, which will ultimately determine the value or price of the asset. Naturally, rental occupation and income vary significantly throughout the economic cycle therefore most investors will take this into consideration during the acquisition process and any subsequent negotiation. Of equal importance to investors acquiring office buildings are the following characteristics:

- commercial/technical considerations:
  - owner occupied
  - typically purpose built
  - single or multi-tenanted
  - governed by the Landlord and Tenant Act and the lease contract
  - service charge arrangements
  - outstanding dilapidations or lease obligations
  - mixed use development
  - flexibility

- legal/technical considerations:
    - planning and operating requirements
    - fire safety
    - deleterious materials
    - disabled access
    - easements
    - energy performance certificate (EPC)
    - test reports
- technical considerations:
    - floor plate size
    - flexibility
    - grid or module dimensions
    - floor-to-ceiling heights
    - concrete or steel framed
    - good facade insulation
    - high-quality glazing
    - high level of fire detection and means of escape provision
    - ages of construction
    - finishes to common areas
    - heating, ventilation and air conditioning (HVAC), internal climate.

Commercial and technical considerations

Office buildings are mostly purpose built for this function and, certainly post-Second World War, these are likely to resemble the office buildings that we know today. Other office accommodation may form part of more mixed-use development, which may include retail or other functions on the ground floor with offices to the upper levels. Likewise, properties dating pre-1920 and used as offices are likely to have undergone renovation or change of use from their original function. Later in this chapter the principal differences between pre- and post-1920s properties will be discussed in more detail. Changes in construction technology have significantly enhanced the ability to construct buildings, from initially low-rise to medium- and high-rise properties.

When analysing modern office buildings it may be thought somewhat old-fashioned to consider the concept of owner-occupier properties. In many cases these buildings originate from bespoke businesses where head office buildings were constructed as a sign of importance or were representative of sector dominance. Typically, such buildings were constructed for the banking sector, pharmaceutical, engineering or oil industries. Alternatively, owner-occupier office buildings were constructed by central or local government at a time when there was sufficient funding available to do this and it was seen to be both socially and culturally acceptable.

In most cases, owner-occupied buildings are constructed to a high specification, presumably due to the significance of the building to the owner's profile or to make an architectural statement portraying success, wealth or the innovative standing of the company within its sector. This may be reflected by the physical size of the building, its distinctive shape and design or by the choice of materials used for the facades and finishes. When surveying many owner-occupied buildings it is not unusual to find high-quality materials and evidence of excellent workmanship. This would simply not be replicated today due to the pressures of stakeholders' budget restrictions or public opinion being somewhat critical of the use of extravagant materials. In reality, high-cost, high-quality materials are likely to have significantly longer life cycles.

High-quality, polished natural stone internal wall linings to the reception area.

The iconic Shell Tower constructed between 1957 and 1962 epitomises the wealth and power of the oil sector in the mid-twentieth century with this London based, owner-occupied head office building. Much of the original facade detailing and internal finishes to the common areas remain.

Limestone facade cladding with polished marble recessed facade detailing.

Some office accommodation may form part of more varied mixed-use buildings and, while these may be modern and purpose-built developments, others are more localised refurbishments. This typically happens in town or city centres where older pre-1920s low-rise properties have a ground floor devoted to retail use and the upper floors as office accommodation. When instructed to undertake an acquisition on these types of properties, it is important to have an understanding of the relevant construction technology, as this is significantly different to that of post-Second World War medium- or high-rise offices. The different construction technology associated with pre- and post-1920s buildings is described in more detail later in this chapter. This construction technology knowledge is transferable between different property sectors.

Within the office sector it is important to note that the majority of buildings constructed post-Second World War were purpose built to be offices. With advances in construction technology as well as the science of materials and their applications, these became medium- and high-rise buildings. Purpose-built offices are essentially different to buildings 'converted' into offices by virtue of the fact that there will have been some design consideration given to their proposed function or end user. Central cores, shafts and emergency escape staircases should have been integrated into the initial design and not shoehorned in as a forced compromise, which is often the case with properties converted into offices.

An important consideration regarding investment in the office sector is if the building is being operated on a single or multi-tenanted basis. Sole occupancy by a single tenant can be lucrative for a landlord or investor; however, with this 'all eggs in one basket' philosophy comes the risk of significant financial adversity should the tenant leave or cease trading. Therefore, spreading the risk with multi-tenancy occupation appears a favourable solution for investors. However, to ensure the building can be marketed or effectively operated as a multi-tenant asset, it is critical to have the building infrastructure in place to support this. The floor plate sizes are important as the marketing of the space will seek to attract enterprises ranging in size from small to large and these require different amounts of square meterage. The space will need to be flexible with the provision to sub-divide individual floors, including the supply, distribution and metering of the services. Each tenancy will typically require its own secure access, which is normally provided by access controls to reception areas located off the central core and lift provision. Likewise, there is a requirement to conform to the relevant health and safety legislation as well as fire regulations regarding the provision of fire detection, firefighting and means of escape.

The actual flexibility of the office space is also an important factor in most tenant fit out requirements including partitions for offices, which could range from individual or double to open plan with meeting rooms. All tenant space will need to interface with sanitary provision as well as staff welfare facilities, such as kitchens, rest rooms or canteens.

Concerning office sector investment, and also the wider real estate investment market, the relationship between landlords and tenants is always governed by lease contracts as well as landlord and tenant legislation. Consequently, rental agreements and service charge provision are critical considerations which will allow investors to assess the relevant income streams. This information is also used to assess the amount of funding available for periodic maintenance, repair and refurbishment. Lease contracts also have specific provisions covering tenants defaulting on their rental agreement or obligations, or vacating the building. This is known as dilapidations and is a highly specialist and lucrative area for surveyors representing landlords or tenants in this process.

Legal and technical considerations

As with all of the different commercial sectors, the principal legal requirement with regard to technical implementation for office buildings is the existence of planning permission or building permits. These are often intrinsically linked to specific operating permits or licences. The legal due diligence will primarily establish the presence of any existing or outstanding planning permissions on the property. Furthermore, the technical due diligence should also be aware of this aspect. As previously discussed in *Chapter 3 – Contract instruction*, it is impossible within the scope of most TDD instructions to categorically state that the planning permissions or building permits have been 100 per cent executed. Therefore, it is important to start with the 'big picture' regarding the planning permission by establishing some basic facts, such as the number of floors, approximate facade or boundary lines and materials used. In essence, the only way to look more critically at the building's compliance with any planning requirements is to compare the approved plans (permission) with the certified as-built plans.

A requirement of most planning permissions is compliance with building regulations or building codes and, again, the only way to give the client some assurance is to check planning documents against as-built or acceptance reports. One important obligation which can be assessed during the site visit is fire safety or fire engineering at the building. Fire will be discussed in more detail in *Chapter 11 – Legal/technical*, as well as specific instances in other chapters. However, in summary, there is a requirement to understand and contextualise the fire resistance of a building, including fire compartments and the means of raising the alarm as well as the provision for a means of escape. In-situ firefighting provision, as well as any fixed firefighting infrastructure for the fire brigade, can also be established during the initial site survey.

When assessing the office sector in terms of legal and technical issues, it will be necessary to establish the presence of deleterious materials, such as asbestos or concrete elements executed using high alumina cement (HAC). The variety of potentially deleterious materials used in twentieth-century office construction is a diverse and complicated sub-specialism in its own right. Nevertheless, surveyors should be familiar with the types of materials, defect causes and effects associated with deleterious materials.

Likewise, although it is unusual to undertake an access audit as part of the TDD process, it is important to be aware of the basic principles of the relevant legislation. It is the responsibility of the surveyor undertaking the TDD to advise the client accordingly on the status of the building with respect to disabled access provision. The 'easy' solution may be for the surveyor to recommend a separate disability access audit to verify compliance; however, it only seems appropriate to do this when there is evidence to suggest that the building is non-compliant. Within the scope of a 'normal' TDD or commercial building survey instruction and the subsequent site/building inspection it is possible, relatively quickly, to establish potential issues regarding accessibility of the building to disabled users. In essence, legislation covering access provision is similar in many different parts of the 'developed' world and it is critical that the surveyor has sufficient knowledge of the existing and relevant access criteria. The principles of access provision in many countries centre around the following base criteria:

* access to the building from the surrounding area
* parking
* main entrance access
* horizontal circulation
* vertical circulation
* services provision (toilets, reception desk, counters etc.).

Specific requirements for undertaking an accessibility audit will be discussed in more detail in *Chapter 11 – Legal/technical*. However, with knowledge of the principal access criteria it is relatively straightforward for the surveyor to advise the client on any potential issues concerning access to the building.

Easements are, however, a more complex issue as these are likely to be flagged up by the legal due diligence. These concern the presence of any existing rights of adjoining owners, tenants, utility companies or the general public to pass through, under or over the property. Naturally, where such provision is granted, it is important to establish if there are any maintenance covenants or specific obligations associated with the easements. While this should normally be addressed by the legal due diligence, it is prudent for the surveyor to ask the property/asset manager, owner or occupier if they are aware of any existing easements. The findings of any such discussions should be communicated to the client and legal advisors as well as being noted in the TDD or commercial building survey report.

An important legal/technical issue is the presence of an EPC for the building; in the UK and the European Union legislation exists requiring this. An EPC is a simple one- or two-page document which illustrates an assessment of the energy consumption as well as CO_2 generated by the building on a yearly basis. This is then visually illustrated with a coloured graphic ranging from A grade to G, with A (coloured green) representing the most efficient through to G (red) which is inefficient. This has become significantly more important since April 2018 in the UK, with the introduction of the minimum energy efficiency standards (MEES) which oblige building owners to meet a minimum A to E rating before a lease can be granted or renewed for a property. This will be discussed in more detail in *Chapter 11 – Legal/technical* but it is clear that the TDD or commercial building survey will have to report on the application of this important legal requirement.

There are multiple legal/technical requirements that need to be considered during the acquisition of a commercial office building. Often these are associated with mandatory test reports and certificates concerning the following:

- heating, cooling and air conditioning
- electrical installations
- water/sanitary supply and evacuation
- fire detection, alarm and extinguishers
- lifts
- life safety equipment (lifelines, window cleaning cradles etc.).

Many of these test reports or certificates of conformity are concerned with the audit to be undertaken by the M&E consultant during the TDD or commercial building survey process. It is important to note that the actual verification of the equipment detailed in the test reports is not usually the subject of individual test or analysis during the TDD process (as discussed in *Chapter 3 – Contract instruction*). However, it is critical to review the test reports undertaken by specialist organisations to establish their validity, as well as any deficiencies or non-conformities identified in the reports. Such issues raised within test reports or certificates are likely to be highly relevant during the negotiation, with an obligation to remedy, guarantee or provide warranty included in the sale/purchase contract.

Retail

The retail sector may be essentially divided into two distinct sub-sectors:

- 'traditional' high street
- shopping centres.

The sub-sectors differ considerably in terms of location, size and construction technology, which has a bearing on the undertaking of a commercial building survey or technical due diligence.

Historically, retail properties were often located in the centre of towns or cities and were the focal point for trading. This practice dates back many centuries to the origins of a local market-place but in real terms it was the Victorian era that introduced the essence of modern-day consumerism. Many high streets and town centres today are still littered with pre-1920s retail properties. It is testimony to the quality of their construction and versatility, combined with the evolution of planning laws, which has resulted in their continued presence.

The architecture and building technology of traditional high street retail properties is essentially the same as residential or domestic construction. As a consequence of utilising shallow basements, strip foundations and loadbearing masonry walls, these are often low-rise properties, restricted to three or four storeys above ground level.

It is important when undertaking a commercial building survey or TDD on traditional high street retail properties to correctly age the property. This is highly relevant due to the age-related construction technology, material life cycles and defect analysis. The following age categories may be considered when analysing traditional high street retail properties:

- pre-1920s
- 1920–45
- 1946–79
- post-1980.

The relevant construction technology, survey techniques and pathology associated with pre-1920s to post-1980s properties will be discussed in later chapters. However, when considering traditional high street retail properties, it is important to bear in mind that the construction technology is similar to that of corresponding residential properties of the same eras. It is no coincidence that residential accommodation is often present on the upper levels of traditional high street properties.

The pre-1920s category is essentially concerned with retail properties constructed in the Victorian period (1837–1901); however, in some wealthy locations, such as Bath, Edinburgh and certain other UK cities, there is an abundance of Georgian properties which are currently being operated as retail premises. The reality is that in most 'normal' circumstances it is unlikely that instruction will be received to survey low-rise retail properties which are pre-Georgian.

Retail properties dating 1920–45 were constructed during a volatile historical period between the two World Wars and at a time of economic instability. There were significant advances in building technology with the adoption of cavity wall construction and an increased use of render to external facades. However, this period was dominated by the need to build 'homes for heroes' (Marshall *et al.*, 2014) and to address the housing crisis.

Post-1945 there began a process of urban regeneration and slum clearance, with the need to rebuild some estimated 450,000 houses destroyed or damaged during the Second World War. Limited availability of timber encouraged 'new' construction technology, with the use of solid concrete ground bearing slabs becoming more widespread. Prefabricated building components included reinforced concrete technology to provide components for structures and facades. Roofing trusses were also beginning to be prefabricated with the Timber Research and Development Association (TRADA) roofing truss being widely adopted.

Changes in architectural styles saw the reintroduction of flat roof construction and more 'modern' progressive architecture with reinforced concrete structures and facade cladding. Relatively 'lax' planning laws resulted in localised, almost random demolition and redevelopment of historic high streets, which is evident in today's streetscapes.

Post-1980 enforcement of more stringent planning and building regulations appears to have arrested the redevelopment of high street heritage architecture. However, the single biggest threat to traditional high street retail properties came in the form of purpose-built shopping centres.

Innovations in architectural style, building technology and social mobility saw a 'revolution' in retail development with the 'birth' of the shopping mall. Mirroring the social aspiration to live in mid-twentieth century high-rise residential towers, people were also intrigued to experience out-of-town, purpose-built shopping centres. Increased car ownership and consumer willingness to experience a new type of shopping resulted in the development of the first purpose-built venues. In the UK, this new approach is epitomised by the Brent Cross shopping centre in London, replicated in other European capitals with the Westland and Woluwe shopping centres in Brussels providing examples.

The progressive architectural style of the mid-twentieth century and the avid use of concrete meant that these buildings could be constructed relatively quickly. Although perhaps frowned upon architecturally by the end of the century, these large, simple but effective concrete 'boxes' provided a key change in direction to the retail sector and consumer experience. This trend has survived the test of time, with successful shopping centres proving popular with both consumers and investors alike. Ironically, home online shopping now poses the greatest threat to both high street and out-of-town retail properties.

Traditional high street shopping and purpose-built out-of-town shopping centres are, in most cases, the complete antithesis of each other. This is not just in their obvious location or siting but also in the construction attributes. With the exception of large high street department stores, most traditional retail properties comprise small buildings consisting essentially of residential type construction. In contrast, purpose-built shopping centres comprise vast surface areas with easily sub-dividable lettable units often determined by the structural grid or facade module. The loadbearing structures are largely reinforced concrete, steel or a combination of both and, generally, many of the external facade claddings and internal walls are non-loadbearing.

Retail sector characteristics

In general, both types of retail are usually low rise and rarely extend above four storeys in height but this is the limit of their common characteristics. In terms of investment, it is observed that large quantities of city centre and out-of-town shopping are owned by commercial investors. As with all commercial investment, there is always a risk associated with finding tenants to provide the relevant income streams and advice on such matters is something for the commercial due diligence. They advise potential investors and, as with property in general, the phrase 'location, location, location' could never have been more apt.

Having 'parked' the responsibility of investment risk with the investment or agency surveyors advising the vendor or purchaser, it is important to assess the technical characteristics of the retail sector.

As with office investment, the landlord/tenant relationship is governed by the applicable lease contract as well as the applicable legislation. However, the 'volatile' nature of the retail sector, where profit margins are tight and consumers have expectations of ever-lower prices, can lead to business failure and bankruptcy. While vacancy rates may be affected by economic trends and property location, in essence, most landlords or investors are concerned with the shell and core of their investment.

This is a common principle with both traditional high street retail investment and out-of-town shopping centres. As part of the TDD or commercial building survey process, it is important to understand the philosophy of the investor in terms of their investment aspirations and attitude to maintenance. Regardless of this, there will always remain a requirement for the surveyor to inspect and report on the key construction elements:

- roofs
- structure
- facades
- finishes
- services
- external areas
- legal/technical.

Roofs

The specific materials and principles of site survey/inspection associated with roofs will be discussed in detail in *Chapter 5*. However, it is important to note the fundamental difference between the roofs of traditional high street retail and purpose-built shopping centres.

As previously detailed, traditional high street shopping portfolios may comprise many small individual properties but these are essentially small-scale residential type construction. The majority of the roofs comprise pitched timber structures with a multitude of different materials used for the actual coverings.

Maintenance and replacement of the roof structure or coverings will depend upon the investment philosophy of the investor, with material choices potentially governed by the need to comply with listed building or conservation area status. However, if this is not applicable, then material choices may be driven by time, cost and quality. In essence, high-cost and high-quality materials, such as natural slate, will have a longer life cycle compared to lesser materials, such as fibre cement tiles, which are used for the same purpose. However, it should be acknowledged that, with all material choices (including natural slate), there are sub-levels of material quality depending on whether the material is sourced from the 'native' country/region or from further afield.

Ultimately, if the property is not listed or within a conservation area, the eventual material choice will be governed by the investment strategy of the building owner or investor. It is therefore almost inevitable that the low-cost option will be selected.

Progressive legislation addressing the need for buildings with increased thermal efficiency is of significance to traditional high street retail properties. In particular, the MEES in the UK oblige landlords to provide buildings with an EPC with a minimum rating of E in order to grant or extend an existing commercial lease. Therefore, the presence and validity of the EPC forms an important requirement for property management or acquisition of retail properties. The external envelope and, in particular, the roof space of pitched roof buildings may be considered an obvious place to improve thermal efficiency. Therefore, an inspection of the roof space should form part of the commercial building survey or TDD.

In contrast, purpose-built shopping centres usually have vast surface areas of roofs to maintain and replace. Most of these are flat roof construction, 'concealed' behind parapet walls, meaning that the aesthetic value of the roof covering is not usually a consideration. Therefore, material choice is driven mostly by cost, which results in material options comprising bitumen mineral felt, PVC or EPDM.

Unlike high street retail properties, purpose-built shopping centres with their large surface areas of flat roof construction are likely to be insulated (warm) decks. This is mainly due to the nature and understanding of flat roof construction technology, which has included the presence of insulation as a standard feature for many decades. The obvious point of discussion is the type and quantity of insulation, which is almost impossible to establish during the building inspection without access to the relevant as-built files. As a matter of best practice it appears reasonable to assume that any future replacement of the roof covering would include stripping off the existing waterproof layer and insulation, allowing this to be upgraded according to the current standards.

Irrespective of the type of retail property, the roof is part of the external envelope forming the shell and core of the property. It is critical to the operation of the building that this is sound and watertight, therefore it is one of the most important aspects to consider when advising potential investors or existing property owners.

Structure

When discussing structure, as will be seen in *Chapter 6*, it is important to note that this is a construction element which is sometimes difficult to inspect fully within the scope of a commercial building survey or TDD instruction. The reason for this is that much of the structure of commercial properties is often concealed by internal finishes or external claddings. However, with traditional high street retail properties the structure may be considered to be similar to residential construction, where the external facades are likely to be loadbearing, as may be some of the internal walls and roof structure. External walls are often founded on traditional foundations or single storey basements. The subsequent difficulty with such property is that tenant structural alterations, such as opening up internal areas for retail space or storage, are often largely concealed by internal finishes. It is therefore important to establish a trail of documentary evidence to understand and advise on the presence or significance of such alterations. Without such evidence the survey will have to largely rely on first principles by visually assessing the structure for any evidence which may suggest the presence of structural movement.

Purpose-built shopping centres are less likely to have been the subject of piecemeal structural alteration as these are often formed with uniform facade modules and structural grids. These can easily be opened up by removing non-loadbearing internal walls to create the necessary desirable tenant retail space. The facade materials may vary significantly and, while these may have important aesthetic requirements, they are likely to be non-structural elements. The structural components in most situations are likely to be concrete, steel or a combination of these materials. While these are mostly concealed by internal finishes to tenant retail units and the common parts, it is often the case that exposed structure is visible in tenant storage areas, technical rooms or internal emergency escape routes.

As with most commercial building survey or TDD instruction, the assessment of the structure will be visual as it is not possible to test, verify or assess the structural loadbearing characteristics within the timescales of the instruction.

Facades

Briefly touched upon when discussing the structure, the facades to traditional high street retail properties are, in most cases, loadbearing external walls. The choice of materials for repair or maintenance will again largely depend upon any existing planning constraints, which may include listing or conservation area status. In the majority of instances, loadbearing facades are likely to be in relatively good condition and there are many examples of pre-1920s high street retail properties where the facades appear largely original. Loadbearing masonry which is either solid or cavity construction has a relatively long anticipated life cycle. The notable concern with facades is associated with the glazing to the actual shop frontage. In most cases this is single glazed with a variety of materials used for the glazing frames and this raises the obvious question regarding thermal efficiency, the EPC and compliance with the MEES. Likewise, the potential upgrade of insulation to the external facades as well as the glazing poses significant technical and financial challenges for the building owner or investor.

Finishes

Aside from the finishes to the common areas of shopping centres, the majority of the internal finishes to both retail types are the responsibility of the tenant. Retail units or individual shops are almost always leased on a shell and core basis, giving the tenant both freedom and responsibility to complete the fit out. In most circumstances the commercial building survey or TDD may comment or note the internal finishes; however, these are not normally the subject of detailed inspection or analysis.

Services

The services requirements for both traditional high street retail and purpose-built shopping centres are broadly the same but the actual provision differs considerably. In essence, there is a requirement for the following:

- HVAC
- electricity
- water/sanitary supply and evacuation
- fire detection, alarm and fighting.

The provision of these services to traditional high street retail properties can again be looked upon as similar to that relating to residential property. Heating is likely to be provided by a localised boiler and heating system with static radiators as well as electrical heat curtains over the entrance. Alternatively, localised air conditioning split units may be installed with the ability to provide both heating and cooling. Ventilation to these properties is often quite basic with air extraction being provided to the sanitary rooms or staff welfare areas but often no pulsed air which has been filtered and treated.

Electrical provision to traditional high street retail properties normally comprises a low voltage installation that is metered at source and distributed from a main electrical board with lighting and power distribution circuits. Concerning the supply and evacuation of water, this is usually provided with single points for either. The incoming supply will normally be metered and distributed locally with pipework in direct or indirect distribution configuration. Water evacuation, typically from the sanitary rooms, staff welfare and kitchen areas, will be connected directly to the main town or city sewage network.

Fire detection is an important issue for investors as well as insurers for all traditional high street retail properties. Health and safety at work legislation as well as fire regulations oblige the tenant to undertake fire risk assessment but, with such a vital life safety issue, it is important to consider the need for systems to be in place. The fire detection system may comprise smoke and heat detectors that are centrally linked to ensure there is a means of raising the alarm to all areas of the retail premises. This should be mains connected with battery back-up. Firefighting provision to traditional high street properties will normally comprise fire extinguishers placed strategically in the building. It is rare to find automatic firefighting installation, such as sprinklers, in these properties.

While purpose-built shopping centres also need the provision of HVAC, this is normally centrally generated with both a primary source heating installation and chiller units. Considering the overall large surface areas that require service provision, there are normally very large or multiple technical rooms. As with the need to provide flexibility in terms of retail units and space, there is also a requirement to provide flexibility with the service installations. Tenants are normally provided with a warm water connection for heating and chilled or cold water for cooling distribution. The tenant can choose to utilise these primary source connections, which are incorporated in their fit out, with their own secondary distribution equipment. Ventilation is not usually provided to individual retail spaces but is for the common areas of a mall, and the opening of entrance doors or the open plan nature of the interface between the tenant areas and common parts may allow for some limited ventilation.

Most purpose-built shopping centres have their own high voltage installation and transformer located on site due to the high level of electrical consumption. Therefore, there will be a requirement for a main low voltage distribution board to sub-divide the electricity into individual metered connections to individual retail units. In some instances there may also be an emergency power supply (battery or diesel generated) to provide a supply to key services functions. These might include the fire detection system, sprinkler pumps, emergency lighting or lift motors. Tenants are provided with a low voltage (240V) connection and again it is their responsibility to install their own secondary distribution as part of their fit out.

Water supply and evacuation in purpose-built shopping centres is also typically provided centrally with tenants being provided with an individual metered water supply and evacuation point. These are connected centrally to the mains water connection for the shopping centre and a central sewage connection, which often includes a grease separator.

Fire detection, alarm and fighting are important requirements for purpose-built shopping centres and, with high numbers of people working in or visiting these, there is an important life safety requirement. Shopping centres are covered by a range of health and safety legal prescriptions as well as building regulations and other specific fire safety legislation. In practical terms, there is usually a central fire detection system which connects both the common parts and the individual retail units. The presence of an automatic fire detection system is also supported by manual fire alarm call points placed in common areas and in individual retail units.

The means of raising the alarm may be considered the fundamental requirement of fire engineering, followed by the need to provide safe evacuation routes from the building. In order to provide a means of escape it is necessary to ensure sufficient smoke evacuation from the building and emergency lights to illuminate escape routes. Therefore, significant responsibility for investors lies not only with the provision, maintenance and testing of fire detection systems but also smoke evacuation.

Concerning firefighting, there is often an automatic sprinkler system in place to both common areas and individual tenant occupied retail units. Supplementary to the provision of automatic firefighting installations is the requirement to provide a water connection both internally and externally for the fire brigade as well as strategically placed hose reels within the building.

External areas

Traditional high street retail properties usually interface directly with the adjacent streets and external areas may only concern rear or side access routes, possibly with courtyards for deliveries, storage or bin areas. The actual size and location of any external areas will be unique to the specific town or city centre location. There are numerous possible finishes to such external areas with in-situ concrete, paving slabs, block paving or tarmac being the most common. The commercial building survey or TDD should take note of the external areas to ascertain the type of materials, current condition and defects. It should also seek to establish the presence of surface water or drainage networks but should not constitute a drainage survey. Any areas of landscaping, including the presence, type and size of any trees in close proximity, should also be included. The inspection of the external areas should also note the presence of any outbuildings to establish if these are owned or have been installed by the landlord or tenant.

It is also not uncommon for there to be shared access routes or easements to the rear of traditional high street properties and during the inspection it is often prudent to speak to the property manager, tenant or owner regarding this issue. The findings can then be relayed or cross-referenced with the legal due diligence.

Purpose-built shopping centres are likely to have large and significant external areas, which are often devoted to access and parking. It is important as part of the pre-inspection desk survey to obtain a copy of the relevant site plan in order to establish the exact boundaries of the external areas. This should then be taken to site and used throughout the survey to navigate around the site and note areas of significance.

As the external areas are usually vast, it is again important to establish through discussion with the owner, property manager or tenant the presence of any significant reported defects. These will typically be localised areas of sunken surface, ponding during rainfall or mechanical damage caused by vehicle impact with kerbstones, bollards or perimeter fencing. It is not the responsibility of the commercial building survey or TDD to establish, calculate and verify if the number of drainage outlets or gullies is sufficient. However, any reports of surface flooding should be noted and made the subject of further discussion or investigation as part of the survey process. Likewise, the presence of external lighting should be noted and any obvious defects included in the survey. However, it is unlikely that the inspection will be undertaken at night so it is important to establish if there are any reported defects or issues with the lighting by asking the property manager to confirm this.

Legal/technical

Concerning the retail sector as a whole, the legal/technical prescriptions are largely the same for both traditional high street retail properties and purpose-built shopping centres. There is a clear requirement for both to have the necessary planning approvals, including the assignment of the relevant use class. This may include specific operating licences or permits as well as any specific conditions associated with the planning approvals. Clearly, this is an overlap between the technical and legal due diligence. It is, however, not within the remit of the TDD to verify conformity with the planning approvals (see *Chapter 3 – Contract instruction*). Information obtained during the desk and site survey may be shared between the legal and technical due diligence teams to form an opinion on the likelihood of conformity with planning approvals or to produce a dossier of evidence to establish possible infringements.

One of the most important legal/technical issues concerning the retail sector is fire engineering and conformity with fire regulations. Therefore, it is important to establish through the desk survey the presence of all relevant documents on this matter. The site survey should seek to assess the property in accordance with first principles regarding fire engineering. This will be discussed fully in *Chapter 11 – Legal/technical* but essentially is concerned with the following:

Principles of fire engineering.

High levels of occupation by staff and customers as well as large quantities of flammable materials mean that retail properties could be considered as high risk with respect to fire safety. While retail portfolios are of significant interest to property investors, fire and any potential injury or loss of life can be extremely damaging, both financially and in terms of reputation. This is, perhaps, reflected in the relatively small number of fires in the retail sector which make headline news.

Another legal/technical consideration with the potential to cause reputational damage is accessibility of retail properties to disabled persons. Therefore, part of the commercial building survey or TDD should be devoted to establishing the conformity of the asset to the relevant legislation.

As discussed above when considering the characteristics of office buildings, there are also significant legal/technical requirements concerning the presence of deleterious materials such as asbestos, concrete defects or other potentially harmful construction materials. Some of these are covered by specific legislation and others simply place a duty of care on the surveyor to notify their client accordingly.

Obligatory test reports concerned with the technical installations, including those associated with HVAC, electricity, water, fire detection, fire alarm, firefighting and lifts, should form part of the TDD or commercial building survey. However, these reports are usually the subject of the document review and it is not normal procedure to actually undertake the tests as part of the commercial building survey or TDD.

Industrial

Industrial buildings may be considered in two principal categories:
* fabrication and production
* logistics and storage.

Factories manufacturing a wide range of products are often bespoke buildings which have been designed to perform unique and specific functions. While the building fabric may be relatively simple, it is the actual fit out and plant which makes these highly unusual buildings. Many factories are owner occupied and it is unusual for private investors to acquire these with a view to generating an income or return on their investment.

Undertaking commercial building surveys or TDD on these tailor-made properties usually arises as a consequence of maintenance or asset management. This could be where defects have been identified or budget estimates are required for long-term capital expenditure forecasts. Alternatively, if the parent company of such properties needs to refinance the business or the company is the subject of a merger or takeover, it may become necessary to audit the fixed assets.

Logistics or storage buildings are quite different from bespoke factories. These are essentially large, low-quality, short life-cycle boxes, which are strategically situated on the arterial roads of the UK and Europe. While some of these buildings are owner occupied, the majority are leased by a wide range of businesses operating in the sector.

The common denominator between industrial buildings used for fabrication and those used as logistics or storage is the simplicity of their design. In terms of the principal construction elements, these are usually square in appearance with either concrete or steel frames clad externally with prefabricated concrete cladding and/or metallic sandwich panels. They need to have suitable floor-to-eaves height to accommodate and meet their functional requirements, concealed under flat roofs. Waterproof membranes to the roofs often comprise bitumen mineral felt, EPDM or PVC, largely due to their low cost and the specific characteristics of these roofing materials, which will be discussed further in *Chapter 5 – Roofs*.

In essence, industrial sector buildings are uncomplicated and epitomise the phrase 'shell and core'. However, along with the need for relatively large open spaces, which permit flexible usage, there is also a requirement for the provision of office space and staff welfare facilities. This is usually designed into the buildings with modest 'bolt on' offices, or these can be incorporated internally on mezzanine floors. The type and quality of the finishes to the office areas within industrial buildings can vary significantly but these are largely functional and often reflect the nature of the work undertaken on the site. An example of this might be the use of ceramic or vinyl floor tiles to entrances and common areas of a logistics site, where drivers have to come into the office for administrative purposes. The use of tiles, as opposed to carpet, means that the finishes are more resistant to heavy foot traffic and also easy to keep clean.

Investors, owners and landlords are principally concerned with the quality and condition of the external envelopes, as well as the structure and external areas. Often, these buildings have large surface areas and this drives the material choices for the roofs and external facades. The time required to erect 10,000 m² of industrial shed is significantly less than the time needed to build the same quantity of space to be used as an office building. This is largely due to the architectural design, choice of materials and speed of construction. This is epitomised by the time, cost, quality adage, with these relatively simple buildings being constructed in a short period of time at low cost, therefore their quality is low when compared to other building types in other commercial sectors.

The owner or occupier risk associated with these assets is often attributable to the operation of the site or buildings. Often there are high levels of combustible materials used, stored or transported in industrial buildings or around the site in which these are situated. Therefore, fire engineering is often a key consideration and it should be noted that the frequently large surface areas of individual fire compartments heightens the importance of fire detection, alarm and means of escape. Building regulations, operating licences or insurer requirements often place specific obligations on the building owner or tenant to ensure that automatic firefighting systems (sprinklers) are in place. Furthermore, smoke evacuation is a very important requirement in industrial buildings and it is common to find a significant percentage of the roof covering equipped with automatic smoke vents.

The serviceability, testing and certification of the fire engineering services are an important issue and, while it is not the specific responsibility of the commercial building survey or TDD to test these, it is critical that all relevant test reports are reviewed.

The remaining services installation to industrial buildings is often quite basic, with heating of large areas (such as warehouses) often achieved with gas-fired hot air blowers. More 'sophisticated' heating systems (typically gas-fired boilers with static radiators) may be installed to office areas, along with localised split air conditioning units. It is rare to find full centralised HVAC systems installed in the office areas and this again is an example of the relatively low quality of the buildings. Air conditioning and ventilation are also rarely seen in warehouse areas unless the space is being operated as cool storage or for other temperature-specific storage and production. Likewise, where there are multiple docking bays or loading platforms, the doors are often open throughout the day and it is not uncommon for operatives or workers to wear gloves and hats in the winter months as part of their daily work routine. In essence, industrial buildings are often hot in the summer and cold in the winter and, due to their unique operation, these do not fit the typical criteria for EPCs. Indeed, this does not appear to be a significant issue for potential investors operating in this sector.

Much like the buildings in the industrial sector, the external areas are also likely to be vast, with a combination of vehicle access and circulation routes as well as individual docking stations or loading bays. In some instances, it is not unusual to find localised fuel pumps or stations within the boundaries of an industrial site and this in itself can present specific problems or potential hazards.

These are areas of high vehicle traffic, therefore the quality and condition of the roads, surfaces and drainage are important to the function or operation of the sites. The nature of heavy goods vehicle (HGV) traffic often results in impact damage to perimeter roads, kerbstones and boundary fencing. Furthermore, blocked, defective or damaged drainage networks have the potential to cause on-site flooding, which can affect the overall operation of the site. Therefore, it is important to allocate sufficient time during the survey to inspect and assess these areas. In terms of identifying defects and repairs to large external surface areas, it can be assumed that large areas come with large costs. The external areas differ from the large surface areas of the roofs in that low-cost materials or specification are not suitable as they simply are not durable enough to cope with the heavy goods traffic.

Residential

Certain large institutionalised investors do not appear to like residential investment, as this seems to be an area of real estate investment which is just too 'personal'. Ultimately, residential investment deals with people and their everyday lives and may be considered to have a high hassle factor for the return on investment, dealing with individual problems, complaints and issues.

There is a tendency to think about individual housing properties when considering the residential investment sector. This may be the case with local authority housing portfolios or even properties owned and leased at affordable rents by housing associations. However, it is also evident that investment properties are increasingly likely to comprise large-scale apartment buildings with multiple living units. This may be the case with examples of residential investment properties, including student accommodation or retirement apartments, as well as more traditional apartments or dwelling houses.

Evolution in planning policies combined with a housing 'crisis' has resulted in increased obligations on developers to include residential units in city centre redevelopment and for some of these to be classed as 'affordable' housing. Furthermore, changes to planning laws have also encouraged the renovation of office accommodation to include change of use to residential. This has had a big impact on the amount and quality of residential property coming onto the commercial real estate investment market, which may be considered financially appealing.

For some investors there is an attraction in acquiring residential properties, with some actively setting up management companies to undertake the property maintenance; this ease of management and perceived low-risk income streams are obvious benefits. Other investors may be obliged to take residential investment properties as a 'trade off' for securing lucrative 'high end' office redevelopments. However, one observation to be made is that the tendency for increased mixed-use developments means that residential properties are likely to provide important facets of investment portfolios in the future.

Contrasted with private sector residential investment is investment in public sector social housing, which comprises a mixture of low-, medium- and high-rise tower block flats, as well as large quantities of residential dwelling houses.

Social housing is a product of the welfare state which emerged after the Second World War when there was a housing crisis and a need to rebuild an estimated 450,000 damaged homes. Post-Second World War and with the birth of the modern movement, low-rise dwelling houses made way for a host of high-rise concrete towers with multiple flats occupying a relatively small building footprint. These were often situated in clusters with the aim of maintaining existing communities. However, inadequate maintenance budgets and public sector funding, coupled with the evolution of concrete and design defects, resulted in many of these residential areas becoming run-down 'ghettos'. Coming full circle, many of these tower blocks have been heavily renovated or demolished and the sites redeveloped to include more low- to medium-rise properties.

While there has been much criticism of the 'failure' of the high-rise social housing 'project', most public sector low-rise housing portfolios comprise properties ranging from pre-1920s to post-1980s. While it may be suggested that, generally, the quality of materials used for most residential dwelling houses appears increasingly lightweight with the passage of time, it should be noted that, according to the 2016–17 survey (MHCLG, 2018), approximately 20 per cent of the English housing stock dates from pre-1920s. Although pre-1920s properties have their own specific problems and pathology attributable to their construction technology, they do appear to have been constructed with good quality materials to a relatively high standard. Consequently, significant quantities of public sector low-rise housing stock dates from pre-1920s and therefore the principal construction elements are of good quality. Indeed, government incentives promoting the 'right to buy' have encouraged home ownership for many social housing tenants. While this is a good thing, the negative aspect is that insufficient public sector housing has been constructed to replace this stock.

Irrespective of public or private sector residential development, it is logical to divide the property types into relevant categories. Naturally, these should be either low rise (dwelling houses) or medium and high rise (towers).

Low-rise dwelling houses

As discussed with low-rise retail properties, these may be divided into their relevant age categories ranging from pre-1920s to post-1980. Although there has been a slow progressive evolution in construction technology with advances in techniques and materials, the basic principles of the construction elements remain the same. Foundation depths have got progressively deeper but the concept of transferring the building load from the roof structure through the external loadbearing walls to the ground below has remained largely unchanged for centuries. Therefore, it is critical when investing in low-rise residential dwelling houses to ensure that the building envelope is sound and free from evidence of structural movement or water penetration.

The principal construction elements and components making up the building envelope are:
- structure
  - foundations
  - loadbearing masonry walls
- external walls
  - masonry (brick, block, stone, render)
  - windows and external doors
- roof
  - roof structure
  - roof coverings (waterproof membranes, insulation, coverings)
  - parapet walls
  - rainwater evacuation
  - roof lights.

The inspection principles and analysis of the materials used for the construction will be discussed in greater detail in the corresponding chapters.

Low-rise residential properties are usually leased complete with internal finishes but generally unfurnished. Tenant occupation is governed by landlord and tenant legislation as well as the tenancy agreement (lease). Therefore, the finishes are important but not critical with respect to the anticipated life cycles of other building elements and any capital investment requirements. Consequently, these can be relatively low cost for floor, wall and ceilings finishes, as these can quickly become soiled and can be replaced relatively easily. However, kitchens and bathrooms should be of sufficient quality to have a good life cycle and be relatively easy to maintain. It may be considered 'normal' to foresee renovation or replacement of the finishes to the floors, walls and ceilings every five to six years. However, it is expected that finishes to the sanitary rooms and kitchens should last 15 years. The repair, maintenance and redecoration of the internal finishes can be covered in the tenancy agreement, with tenants often obliged to redecorate, reinstate or clean the finishes to the state at the start of the lease.

Service provision to low-rise residential investment properties comprises:
- heating and hot water
- electricity
- water supply and evacuation
- fire detection.

In most cases the service provision is independent and localised to each individual property with the hot water production for heating and distribution to the kitchen and the bathrooms providing the single largest risk to the investor. As well as ensuring that the boiler is periodically serviced to extend its life it is important to ensure that boilers are safe with respect to fire prevention and to protect against carbon monoxide poisoning. Under landlord and tenant legislation it is normal that the provision of heating as well as hot water for washing is an obligation for residential occupation provision.

Services provision will be discussed in more detail in *Chapter 9 – Building services*, but it is important to note that, in many cases, the infrastructure of the services (pipework, ducting, cables etc.) is anticipated to have double the life cycle of the end user (boiler, radiators, valves etc.). This should be taken into account when undertaking a commercial building survey or TDD.

High-rise residential towers

There are many differences between low-rise and medium- to high-rise residential investment properties. The principal observation concerns the structure, which will be either reinforced concrete, steel or a combination of both materials. The foundations are likely to be piled with multiple basement levels often evident. The structural grid will have been designed to accommodate the required window/facade modules for the individual residential units. This is constructed around a central reinforced concrete core used to house lift shafts and internal emergency staircases.

The amount of overall lettable residential space may be quite high and, as this is achieved over many levels, the actual building footprints can be relatively modest and suited to tight, small plots of land. As a consequence, the areas of roof coverings also normally constitute quite small surfaces, which is good for investors as these have relatively low investment costs in terms of replacement. An important factor will be whether the roofs are pitched or flat and this will be discussed in more detail in *Chapter 5*.

The obvious area of greater concern regarding medium- to high-rise residential towers is the large quantity of surface area attributable to the facades. Therefore, it is important when advising a property owner or investor to take time to analyse the facade components and materials. Many residential towers constructed in the 1960s and 1970s were executed using exposed reinforced concrete panels and structure. This has left a legacy of defects and material failures which has had, and will continue to have, a significant impact on investment in these types of property. Latterly, it may be observed that facade cladding systems appear to have become more lightweight with increased levels of thermal insulation. However, these materials have also their own specific building pathology, which is starting to become evident. Facade materials, their inspection and defects will be discussed in more detail in *Chapter 7*.

As with low-rise residential investment properties, the finishes to medium- and high-rise properties should be considered on a similar basis, with the properties being leased essentially unfurnished. A variety of finishes to the floor, walls and ceilings may be evident, as will the specification of the fit out to the kitchen and bathroom areas.

Some central services may be provided by district heating but the common reality is that most individual residential units will be equipped with primary heat sources and secondary distribution. An emphasis on improving energy efficiency has resulted in the effective sealing of the living space to remove drafts and heat leakage, but this is only possible where individual ventilation systems with air treatment are provided.

Electrical supply and distribution is also likely to be at the individual level with separate metering and distribution panels provided per residential unit. The very nature of medium- and, in particular, high-rise residential properties means there is a duty to comply with legal requirements and regulations concerned with fire engineering. Fire compartmentation and fire resistance of the structure or facades can be provided by the selection and use of the relevant materials.

Regarding fire detection and the means of raising the alarm, there may be one system in place for the common areas, such as basement parking, entrance, stairwells and lift lobbies. This may be owned and maintained by the landlord or building owner. It is therefore important to ascertain that a duty of care is exercised to ensure that this system is serviceable. Separate fire detection and alarm systems may be installed in individual residential units and these are likely to be mains operated with battery back-up. While tenants may be obliged in their lease contracts to ensure that smoke or fire detectors are free from obstruction and batteries are changed regularly, it is up to the landlord to confirm that the principal installations are fully functional. The provision of emergency lighting may also be a legal requirement but invariably the operation and serviceability of this becomes the responsibility of the building owner.

This is perhaps where a residential real estate investment starts to become a bit more personal as the operating, maintenance and inspection of the services installations affects the everyday lives of 'normal' people. When advising a building owner, landlord or investor in medium- to high-rise residential properties it is critical to understand the extent of the services provision and establish which party (landlord or tenant) is responsible for maintaining this.

Leisure

The leisure sector comprises a large variety of potential investment properties; these may be wide ranging, from sports venues and stadia through to concert halls, restaurants, gyms or holiday complexes. The principal observation concerning properties within the leisure sector is that these are used not only to house the staff who work in them but also relatively large numbers of transient visitors. Unlike the other building sectors, the majority of people occupying or using these do so by the hour, day or week: they are rarely permanently occupied. There is an obligation on the subsequent building owners to ensure the safety of persons entering the properties and emphasis must be placed on evacuation in the event of an emergency. Leisure sector investment properties can be divided into those which are genuinely bespoke or unique, such as stadia, theme parks, visitor centres and concert venues, or more 'generic' buildings, such as hotels, pubs and cafes. However, when taking an instruction to perform a commercial building survey or TDD on these types of assets it is important to remember that, aside from the many unique features of the properties or their function, they still have the following principal construction elements:

- roofs
- structure
- facades
- finishes
- services
- external areas.

Hotels

The most common type of leisure investment properties are hotels. Often situated in town or city centres this type of investment properties serves a specific purpose and, while these are often owned and operated by organisations working within this specific industry, hotels can form part of mixed used developments. They may be owner occupied or leased and fitted out to the requirements of specific brands or chains.

Irrespective of the hotel rating or brand, and from a technical perspective, it is important to sub-divide hotels into those which are purpose built and those which have evolved through extension, renovation or conversion.

There is no doubt that many hotels are historic buildings with a wealth of important historical context as well as materials. These are certainly likely to originate from the pre-1920s or in many cases pre-Victorian or Georgian. The older the hotel, the more likely it is to be listed or located within a conservation area and these are significant factors to consider when advising owners or investors in these types of property. However, one consequence of older historic hotels is that they are likely to be low-rise properties with occupation rarely being more than four floors above ground level.

In contrast, many hotels dating post-Second World War are likely to be purpose-built leisure properties. With advances in construction technology these will inevitably be concrete or steel framed and, although also located on confined sites in town or city centres, these are often medium- or high-rise buildings. 'Modern' hotel properties encompass the same kind of structural design as office or apartment buildings, with internal separating walls which are also non-loadbearing.

As with all investment property in all sectors, owners and prospective investors in hotels need to be informed about the structural integrity of the building and whether the external envelope, comprising the roof and facades, is sound and weathertight.

The structures and building envelopes will vary considerably with the ages of construction or depending on whether these are low,

*A traditional, low-rise historic hotel (**TOP**) contrasted with (**BOTTOM**) a hotel under construction as part of a mixed use development (photo courtesy of Widnell Europe).*

medium or high rise. Individual construction elements and their characteristics will be discussed in later chapters and much of the material science, analysis or pathology is transferable between all investment sectors.

The internal finishes are important if the owner or landlord is responsible for the maintenance, repair and renovation of these. Different hotel brands appear to have different priorities when it comes to finishes; some seem to prefer high-quality, high-cost materials for entrance lobbies or receptions. Others prefer to reserve more luxurious finishes for the individual guest rooms. There are also 'budget' hotels, which are functional and low cost, therefore finishes may be considered to be more 'practicable'. One certainty is that the internal finishes to hotels are usually subject to high volumes of occupation, which reduces their life cycles. Inevitably, significant amounts of capital expenditure for properties in the leisure sector is devoted to the finishes and aesthetics of the product or experience. There are also very important considerations regarding the fit out and operation of kitchen and restaurant facilities to hotels as well as more bespoke features, such as swimming pools or gyms.

Invariably, general commercial real estate investors appear less interested in hotels or properties within the leisure sector. This is probably why most instructions for commercial building surveys or TDD are generated by organisations specialising in this sector. Where investors are obliged to take hotels as part of wider, more 'interesting' mixed use developments, these are made more attractive if such buildings or spaces are leased as shell and core with designated hotel use. In these instances, it is the hotel operator or chain that is responsible for the majority of the fit out and upkeep of the finishes.

The services provision to hotels and other investment properties in the leisure sector may vary according to the individual property but these again fall broadly into the main services categories:

- HVAC
- electricity
- water, sanitation (supply and evacuation)
- lifts
- fire detection, fire alarm, firefighting provision.

The hotel sector is customer or end user driven in terms of comfort with an important emphasis on providing individually controllable HVAC and sufficient amounts of hot water for guests. However, one of the principal concerns for commercial sectors in hotel assets is fire engineering.

Fire in a hotel can cause significant building damage and potentially loss of life; however, it is the potential reputational damage to an investor which may be perceived as a major risk. There is a relatively low incidence of hotel fires; however, when these do occur the images are usually very graphic, appearing across news channels and the printed press. Therefore, fire detection and alarm to individual rooms as well as common areas is imperative.

Although not considered one of the services, fire compartmentation is critical to hotels and should be noted, where possible, during the commercial building survey or TDD. The passage of services through fire compartments is a very important issue, which needs close attention. Certainly, with more modern hotel properties there will be vertical services shafts, which may be an 'open' fire compartment in their own right, sealed from individual floors by fireproof walls and doors. Alternatively, shafts may be vertically sealed at each floor level with any pipes, cables or ducting equipped with fire barriers or fire dampers where these pass into adjacent compartments. The integrity of the fire compartments is only as good as the fire seals, and where these have been broken or compromised by additional or new services, there is a risk of fire spread.

The provision of water supply and evacuation for hotel assets is also key to their end user/ customer operation. Purpose-built hotels are likely to have been designed and constructed encompassing sufficient pipework to accommodate this. However, in older, historic or converted properties these installations are more likely to be a bit 'ad hoc'. Dependent on any lease contracts or operating agreements, it is possible that the maintenance and repair of the pipework may be the responsibility of the landlord or building owner. Therefore, it is important to establish the condition and suitability of the provision of this element.

Lifts are also an important services feature for both investors and operators of hotel properties, but for different reasons. Lift provision is seen as a 'must have' feature by hotel operators since it is the norm for most, if not all, purpose-built modern properties. However, for investors, lifts have an element of cost risk concerning the condition and maintenance of the lift cars and motor equipment. Legal obligations require lift installations to conform with periodic safety checks, which is something routinely included in service contracts administered by the property manager. While the lift cars can and are often the subject of renovation, it is the potential replacement of the motor, winding gear and control panel which requires large amounts of capital expenditure. Furthermore, compliance of the lifts with disabled access requirements may be difficult to achieve if the existing shafts and building infrastructure are not sufficient to accommodate a car with the minimum recommended dimensions. The need to rebuild an existing shaft and install new equipment and lift cars is challenging, both technically and financially, as well as being extremely difficult to undertake with a hotel in occupation. Concerning lift provision to older pre-1920s hotels, this is likely to be even more complex, although the one reassuring factor is that the number of storeys requiring lift access is often limited to four levels above ground.

Public sector

Investment in public sector properties is either derived from tax revenues or via private investors providing investment or properties which are effectively leased back solely for public sector use. The public sector is vast and includes some of the following key sub-sectors:

- healthcare
- social welfare and housing
- education
- defence
- criminal justice
- local and national government
- infrastructure.

Within the public sector there is a wide and diverse building stock with many bespoke and historically important properties. As with all other building sectors, it is important to note that the principal construction elements (roofs, structure, facades, finishes, services, external areas) remain a constant. Therefore, when undertaking a commercial building survey or TDD of public sector properties it is important to inspect and analyse each element accordingly.

Historically, there has been a great sense of national pride concerning infrastructure projects or public sector buildings. With the Industrial Revolution and the associated advances in architectural and engineering design capabilities, high-quality construction projects were conceived and executed. There are many examples of Georgian, Victorian and early twentieth-century public sector or national infrastructure projects in existence today. These include:

- hospitals
- schools
- courts
- prisons
- town halls
- bridges
- dams
- railway stations.

Some of these are now under private sector ownership or have been renovated and transformed for changes of use, but the important point is that the building structures and envelopes are largely original and (subject to maintenance) in good condition. This is testimony to the quality of materials used as well as the skill and care of the tradesmen who executed the works.

In the aftermath of the Second World War, the birth of the welfare state in 1948 addressed the need, amongst other things, to provide homes, with an estimated 450,000 houses destroyed or damaged by the war. Large amounts of public sector investment were directed into the construction of social housing, which was a largely successful venture, up to the 1980s when government incentives granted social housing tenants the right to buy their properties. As a consequence of this and minimal construction of new social housing to replenish stocks, there has been an inevitable return to a housing crisis. Much of the current housing stock owned and operated by the public sector is aged and in need of modernisation or renovation. With local authorities having to provide accommodation to tenants at affordable rental levels there are inevitably limited amounts of capital available to maintain the properties and even less to invest in new build properties.

Funding of the public sector has historically been provided by revenues generated by taxation; however, decreased spending within the public sector means that there are large holes or deficits in these budgets. This is clearly evident in many public sector buildings, where maintenance is often reactive as opposed to planned. Low-cost, low-quality materials may also be used and invariably these have limited life cycles. Low levels of investment in public sector buildings are not confined to the UK, where income tax rates may be considered 'low', but are also found in other higher tax rate countries with developed economies.

Therefore, an increasing amount of private sector investment is being undertaken within the public sector. This is seen as a viable option for some sub-sector properties. This may be considered a win–win situation for either party with the public sector receiving investment which otherwise would not be available and private investors receiving guaranteed returns on their investment for long periods of time (up to 27 years). This process is known as private finance initiatives (PFI) or public–private partnerships (PPP).

Examples of PFI and PPP can be found typically in the healthcare (hospitals) and education (schools) sectors. In these cases, the built assets are normally to a reasonable standard and their maintenance should be assured. However, as this is essentially a commercial operation for the investor, there are inherent obligations to fulfil the public sector performance requirements or key performance indicators (KPIs) which maintain profits. Inevitably, material choices and material life cycles may not be considered to have the same level of importance as those required by an owner occupier. While there are political discussions concerning the principles and success of private investment, there have also been some technical observations. Of major concern were incidences of construction defects and failure in some private sector funded schools in Scotland. The failings were brought to light in January 2016 with the very graphic collapse of facade masonry at the Oxgangs Primary School in Edinburgh.

It was reported by the BBC on 9 February 2017 that, during a period of high winds, 9 tonnes of masonry fell down onto an adjacent playground area, which was fortunately unoccupied at the time. This prompted immediate closure of the school and a number of other schools which formed part of the same PFI project. The cause of the collapse was indicated in the BBC report to be a problem with the specification and execution of the wall ties connecting the inner and outer leaves of the external cavity walls.

This incident was the subject of political debate concerning the 'failings' of PFI as a model for private investment. It is also hard to conceive that such failure could or would occur to any of the existing schools constructed in the Victorian era. Indeed, there may be some weight to the assertion that 'we don't make things like we used to'. Such a technical failure is alarming for any private investor who has been involved in these or similar projects, as well as for consultants advising in the process. This is of significant importance with regard to professional liability and duty of care when considering undertaking commercial building surveys or TDD on these buildings.

In *Chapter 3 – Contract instruction* the limitations of the site inspection or document review were discussed. A commercial building survey or TDD is a snapshot of the building at one particular moment in time with the intention of establishing the past, present and future with respect to anticipated defects and costs. This is also undertaken in the review of technical documents or acceptance reports, but one of the principal limitations is that the survey cannot verify the structural or design calculations. Furthermore, this will typically not involve destructive tests or the opening up of areas for detailed investigation.

Therefore, and assuming there is no visual evidence to suggest failure of the cavity wall construction, it is very difficult to advise on the potential for this defect to occur when undertaking a commercial building survey of TDD. In such instances the document review and, in particular, the acceptance reports of the architect, engineer or design team, as well as building control should be reviewed to ascertain if there are any remarks, comments or infractions. However, even if this process had been rigorously undertaken, it is likely that it would not have revealed these catastrophic facade defects. Therefore, it is difficult during the commercial building survey or TDD process to legislate for design, execution or site supervision errors.

One certainty is that the coming to light of this type of problem associated with this specific PFI project, will inevitably make investors and their technical advisors cautious that this or similar defects may occur again. This may only serve to lengthen the survey or TDD period if intrusive inspection is required. Inevitably, it is also likely to place obligations on the existing owner to give a warranty to any future owner against the possibility of this event occurring again.

Schools are just one example of public sector buildings which should be considered as commercial assets, particularly for private investors buying into this sector. The principles of property management and maintenance for schools may also be considered as commercial surveying practice as with other buildings located in the public sector. Social housing is also a significant sub-sector and, although this is concerned with individual residential dwellings, the large scale of public housing portfolios, combined with the management of revenues for maintenance, gives this commercial status. Likewise, the concept of housing associations and their portfolios of residential properties should also be considered on similar terms.

When surveying public sector housing for acquisition, maintenance or investment, the same principles as those described earlier in this chapter for the residential sector should be applied. Properties should be categorised as low rise or medium and high rise as, in essence, this will normally determine the structural characteristics. Furthermore, the age of the properties is important as this has an influence on the construction technology. For example, social housing constructed pre-1920s will typically have shallow foundations with loadbearing, solid (single brick thick) external walls. For local authorities managing this type of properties, typical challenges will be improving energy efficiency to reduce fuel consumption in order to address fuel poverty and undertaking adaptive reconfiguration measures for elderly or disabled tenants.

In contrast, medium- to high-rise public sector housing will be concrete or steel framed with many being constructed in the 1960s and 1970s. Although the construction technology varies significantly, the need to upgrade this stock to improve energy efficiency is the same. Furthermore, common core central services, such as lifts, have potentially high maintenance and replacement costs.

The obvious constraints placed upon local authorities to maintain and manage these buildings is driven by limited public sector funding and high levels of demand related to the general housing crisis and the high rates of private sector rents. Often situated in prime town or city centre locations, relatively close to amenities with public transport links, the sites may be considered to have high investment value. However, with the need to keep as many tenants as possible in situ to maintain rental revenue and with limited options for relocation alternatives during works, there is a preference, or indeed a necessity, to renovate these properties with tenants in residence.

External facade renovation and upgrade of thermal efficiency to social housing in Portsmouth.

Summary

While it is important to have an understanding of the various commercial property sectors and their characteristics, they have many things in common. Primarily, all buildings will have a roof, structure, external walls, internal finishes, services provision and some external areas or interface with their surroundings.

The principal differences occur between buildings which are low rise, using 'traditional' construction methods, and those which are medium to high rise. This has a profound effect on the type of foundations, structure and some of the services provision.

It is also important to take into account the individual organisations and the investment criteria that they may have for specific investment or property sectors. Generally, all investors are seeking a return on their investment. With the exception of private investment in the public sector, this usually means buildings which have low rates of vacancy are not attractive. In order to increase take-up of lettable space, all buildings need to be structurally sound, weathertight, finished to market standards and with appropriate services provision. Therefore, when advising a property owner, occupier or investor during the acquisition process or maintenance inspection, it critical to assess these construction elements. The role of the building surveyor or technical consultant is to provide the best possible evidence-based opinion within the time constraints afforded by the client and in accordance with the information contained within the available documents.

References

Marshall, D., Worthing, D., Heath, R., Dann, N. 2014. *Understanding housing defects*, Routledge, Abingdon, Oxon.

MHCLG. 2018. *English housing survey headline report 2016–17*, Ministry of Housing, Communities and Local Government, London.

Photo courtesy of Widnell Europe

Roofs 5

Introduction

Forming part of the building envelope, roofs are one the key construction elements and have formed the basic provision of shelter since the beginning of human existence. While this may be considered a profound statement, it is important and relevant to consider the following performance criteria for roofs:

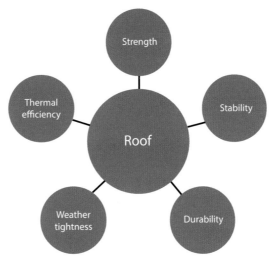

Performance criteria for all roofs.

When inspecting and analysing any roof it is essential to consider the performance criteria in the context of the type of roof and also the materials used. Throughout the passage of time there has been an evolution in roof design and the specification of materials used to execute this. However, the concept of the roof has remained largely unchanged for thousands of years. In essence, roofs fall into one of two categories:

- flat
- pitched.

These principal differences between the roof types will be analysed in more detail later in this chapter but it is also important to consider the variation in the application of roof design within the commercial buildings sectors. Not only is it important to understand the principal difference between pitched and flat roofs but also the relevance of building age; this, along with the relevant construction technology, is vital when advising an owner, occupier or investor.

*Traditional pitched roof to a pre-1920s commercial property (**LEFT**) and, in contrast, the flat roof construction over a purpose-built shopping centre (**RIGHT**) (photo courtesy of Widnell Europe).*

Strength and stability

The strength and stability of a roof is determined by the structural members which support the roofing deck. With 'traditional', residential type construction, the roof structures often comprise timber joists or rafters spanning from facade to facade with roof decks to flat roofs often also comprising timber boards supporting the actual waterproof membrane. Most commercial buildings require the roof structures to span greater distances than traditional residential properties, therefore timber structures have progressively given way to steel or concrete roofing structures. These support either concrete or steel roof decks, with the possibility of a combination of the two materials.

Durability

If protected from the external elements, roof structures can effectively have long-lasting life cycles, but this is not necessarily what is meant by the term 'durability'. In the context of roofs, durability often relates to the actual roof covering or waterproof membrane. Without this impervious layer, the life cycle of the roof structure is likely to be significantly reduced or compromised. Therefore, when undertaking an inspection of a commercial roof it is critical to identify the type and specification of the roof covering to be able to assess its quality, current condition and anticipated life cycle or durability.

Weathertightness

Closely linked to the durability of a roof is its weathertightness, as this is probably the primary requirement of this part of the building envelope. The failure of weathertightness may be directly linked to durability issues but also may be due to poor detailing and workmanship. Allowing rainwater to penetrate a roof covering may result in water damage and problems with tenant occupation as well as more sinister 'hidden' defects, such as the decay of the roof deck or structure.

Thermal efficiency

Increasingly thermal efficiency is becoming a key factor within the design of commercial buildings. With a need to reduce energy consumption, driven by legislation which has been shaped globally in response to climate change, thermal efficiency forms an important part of most commercial building surveys or TDD.

Technical due diligence process and inspection

The RICS guidance note concerning TDD and commercial building surveys recommends that an external inspection of the roof areas should be done as closely as feasible. As the roof is one of the key building elements forming part of the external envelope it is not acceptable to think that this cannot be inspected. It is not unusual for investment clients to have little or no tangible technical knowledge about the properties that they invest in, therefore, they place a great deal of faith in their technical advisors to visit and report upon an asset. Accordingly, most investors would be extremely surprised if the surveyor were unable to report upon the condition of the roof.

The conditions or limitations surrounding inspection of the roof should be stated in the contract instruction and there should always be an understanding that this can only be done where safe access is provided. In the event that safe access cannot be afforded to the roof, then this should be at least viewed from a vantage point or "cherry picker", and the use of drone technology is increasingly seen as a viable option. Using the zoom lens facility on photographic equipment, it is possible to note and record information; however, this is not the same as actually, physically being on the roof surface.

Certainly, pitched roofs pose more of an access challenge than flat roofs by virtue of the fact that these are difficult to walk on and, the steeper the pitch, the more dangerous this becomes. The RICS *Surveying safely* (RICS, 2011a) publication looks at and advises on the dangers associated with surveys, including roof inspections. However, it will come down to individual and personal on-site risk assessment to determine if a surveyor is willing to access certain roof areas. Although flat roofs appear to present fewer challenges for access than pitched roofs, without adequate safety protection they also pose a risk of falling.

Access provision to roofs varies but with modern purpose-built commercial buildings there are often specific access facilities in place. Roofs are either accessible externally from the adjacent ground level or internally via central or emergency staircases. Standalone portable ladders may be used for external access to flat roofs, but these should be securely placed and are really only sufficient for one- or two-storey heights.

Access to the roof over a single storey extension to an industrial building provided by a 'temporary' ladder. This is not an acceptable solution.

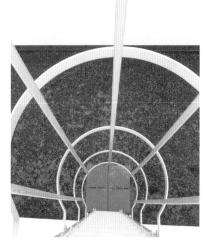

Facade-fixed 'caged' aluminium access ladder to the roof over an office building (photo courtesy of Widnell Europe).

A similar installation for an industrial building covering a height of approximately 12 m, note the provision of a 'landing' approximately three-quarters of the way up the ladder (photo courtesy of Widnell Europe).

If the flat roof is surrounded by a perimeter parapet wall or external railings with a height of, typically, 900 mm above the roof deck, then this is likely to minimise the risk of falling from the roof. However, where there are lower parapet walls the surveyor needs to be acutely aware of the risk of falling and in some cases surface-mounted lifelines or fixings are present. Where such safety provision is foreseen, it will be up to individuals to decide if they clip onto these safety fixings (provided they have the necessary personal protective equipment (PPE) to do this). However, one observation with centrally fixed lifeline systems is that the very nature of these prevents users from going too close to the roof edge, which is restrictive when assessing the roof edge detail.

The presence of a parapet wall often provides sufficient protection against falling; however, towards the ridge level of this pitched roof the wall offers no protection.

An EPDM covered section of flat roof adjacent to a curved roof finished with aluminium roof sheets. This is a very dangerous and exposed roof with no safety or lifeline fixings (photo courtesy of Widnell Europe).

Closer inspection of the parapet wall reveals parapet fixed steel brackets for the positioning of temporary rails to raise the height of protection should work be carried out on the roof (note the corrosion of the brackets, which require immediate treatment).

A static lifeline fixing point which is secured to the roof deck and finished appropriately with bitumen mineral felt upstands and detailing (photo courtesy of Widnell Europe).

Concerning pitched roof access, this is normally too dangerous to attempt but there are some instances where the provision of sufficiently high parapet walls and lifelines may permit this. Again, it is up to individual surveyors to assess the risk before performing an inspection on a pitched roof.

Inspection principles

The external inspection of the roof is necessary to establish the construction detail and condition, as well as defects in the following:
- roof covering (including roof structure, deck, insulation and waterproof membrane)
- parapet walls
- rainwater evacuation
- roof lights
- other (access, lifelines, chimneys, flues, vents, advertising, lightning conductors etc).

All of the above points will be discussed in more detail throughout this chapter.

While an external inspection will identify the condition of the roof's key construction components, an internal inspection is necessary to establish any evidence to suggest leakage or failure of these. Internal inspection should normally be in the roof space or void of pitched roofs or the top floor under flat roofs. This is not entirely straightforward as the roof areas to commercial buildings are often too large, with insufficient time allocated, to see everything or internal access may be restricted by tenant installations.

Considering the time constraints often placed upon the surveyor, it is advantageous to seek help from the building owner, property manager or tenants concerning areas of reported defects. One of the most useful assets in a commercial building survey or TDD process is knowledge. The challenge for the surveyor is to build up knowledge of the property covering the past, present and future, often with only one or two days on site. The knowledge of the building owner, property manager or tenants regarding leakage, problems or any remedial roofing works are likely to be significant. Therefore, it is important to establish this information through discussion with the relevant parties and the surveyor needs to 'befriend' the tenant or property manager. For acquisitions, this also brings to the fore the matter of disclosure and, by asking for historic information regarding roof problems, it is obliging the vendor (through their agent or asset manager) to disclose this to the purchaser. Having established the presence of reported roof leaks or defects, the surveyor will have to try to establish the potential cause of these without jumping to unsubstantiated conclusions.

Flat roofs versus pitched roofs

One of the performance criteria for roofs is weathertightness, therefore it is advantageous that roofs have the ability to shed water. It is therefore perplexing why there is such a thing as flat roof construction, which is less effective at performing this function. Flat roofs, by definition, actually have a pitch of up to 10 degrees and, if executed correctly, should provide adequate surface water run-off. This does not explain why there are so many flat roofs which are affected by standing water, known as 'ponding', and the most likely reason is that this is caused by the flat roof design and execution. Historically, the majority of roofs have been pitched and the origins of flat roof design is associated with the construction of stately homes or manor houses. Here, the presence of flat roofs was used as an architectural statement rather than a pragmatic necessity and these were often used to enhance other architectural features. Flat roof construction as a design feature remained relatively dormant until the 1920s, and was in evidence later in the 1950s as part of 'modern' architecture, but was mostly restricted to residential property.

Certainly, post Second World War, flat roof construction became more evident for commercial properties. This was both a practical measure for large span industrial properties and provided architectural features for modern medium- and high-rise office buildings.

In direct contrast to flat roof construction, pitched roofs have been ever-present in the built environment. Naturally, the main advantage is their ability to shed water, ensuring that buildings are weathertight. Pitched roofs appear to have a greater number of design features and the capacity to encompass a wide range of finishes. The roof space under the covering can be used for storage, converted into living or working space and is also used to house services installations. Typically, pitched roofs are visible to commercial properties across most building sectors, although they are less apparent on industrial buildings.

Pitched roofs are evident in all town and city centres throughout Europe and beyond, although they are perhaps less common in areas with hot dry climates, which suggests that pitched roofs primarily exist to shed water. However, there is more to their widespread use, including not only their previously mentioned flexibility but also aesthetics. Pitched roofs are also not only exclusively used for older pre-1920s properties or low-rise buildings. They are such an important feature of the skylines and townscapes of the built environment that it is often urban planners who stipulate their use. There is also an array of suitable materials for the coverings to pitched roofs, some of which are largely not specified for flat roofs. While more modern, high-profile city centre developments may utilise metallic sheets or glazing, it is often traditional tiles or slates that are chosen to blend old with new in historic town centres.

It may be argued that pitched roof construction is more complicated than flat roofs and this may be the case, with the need to cut joists, rafters, purlins or trusses when compared with the use of only horizontal members for flat roofs. However, this simplistic comparison is not truly accurate as, with large span flat roofs, there is a need to engineer trusses to provide sufficient pitch to evacuate surface water. These members must also fulfil the roofs' strength and stability performance criteria.

The real technical difference between pitched and flat roofs is the roof covering. As flat roofs cannot be seen from street level, there is no incentive to provide waterproof membranes or coverings which have aesthetic appeal. While in some cases lead, copper or zinc may be used for flat roof coverings, the reality is that this is uncommon. Certainly, the cost of metallic roof coverings may be two, three or four times that of oil-based or synthetic rolled-on sheet roofing. This is another important factor making flat roofs less complicated and more cost effective than pitched roofs. There is a direct correlation between the cost of materials used for roof coverings and their quality, and this is also intrinsically linked to the durability and life cycle. These aspects will be discussed in full later on in this chapter.

Cold deck versus warm deck

The concept of thermal efficiency as a performance criterion for roofs was discussed earlier in this chapter. In order to understand the processes involved in insulating a roof it is of paramount importance to comprehend the differences between a cold and a warm deck. Although applicable to both pitched and flat roofs, the terms cold and warm decks are usually reserved for considering flat roof construction.

A roof deck is the horizontal part of a roof directly over the occupied or useable space. This is supported by the roof structure (joists, trusses or rafters), which is the case with timber or steel decks. The roof deck may also be integrated within the structure, which is the case with cast-in-situ reinforced concrete. With a cold deck, the waterproof membrane is laid upon the surface of the deck with insulation provided below the deck. This may be considered 'old' technology, which is mostly evident on buildings constructed pre-1980. In contrast, a warm deck has the insulation placed on top of the deck but under the final waterproof layer.

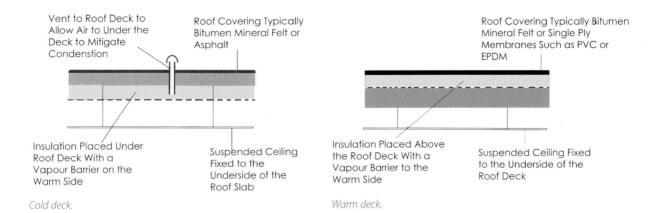

Cold deck.   *Warm deck.*

Warm deck construction usually incorporates oil-based or synthetic waterproof membranes, which are typically bitumen mineral felt, PVC or EPDM; these materials will be analysed individually in greater detail later in this chapter. Concerning the insulation to flat roofs, any insulation is better than none; however, the principal problem with cold deck construction is attributable to occupation. Warm moist air, generated due building occupation, rises up to the surface or underside of the roof. As it passes through the insulation (placed under the deck) it comes into contact with the roofing deck, which may only be a few degrees warmer than the outside air temperature. The internal building temperature is often relatively consistent throughout the year, however external temperatures are likely to vary significantly between summer and winter. In winter, much like the steam from a kettle striking cold kitchen tiles, warm air striking the underside of the cold roof deck condenses into water droplets.

This may present visually as staining to the ceiling finishes and is often misdiagnosed as a roof leak. One of the key indicators of condensation is the presence of black spots, which are associated with the growth of mould. While moisture droplets may damage internal finishes, ruining mineral wool insulation, a more sinister consequence of increased moisture content to timber roof decks and structure is the potential for rot or decay. In order to reduce the risk of condensation building up on the underside of the roof deck it is necessary to ventilate the void directly under the roof deck. To do this ventilation pipes may be placed through the roof covering and deck to allow the passage of air below. Alternatively, a waterproof membrane may also be added to the 'warm' side of the insulation to prevent moist warm air touching the roof deck. It is critical that this waterproof membrane is completely airtight to prevent condensation occurring.

Warm decks do not suffer from condensation because the insulation is placed above the roof deck, thus raising the surface temperature of the deck. However, it is still good practice to place a waterproof membrane on the warm side of the insulation to alleviate the potential for condensation to build up within the insulation itself in a process known as interstitial condensation.

One other type of roof design that may be used to insulate the roof deck is an inverted warm deck. This places the insulation above the waterproof membrane on top of the roof deck and one important requirement is that the insulation is waterproof. One example of such insulation types is high density rigid panel which, when placed above the waterproof membrane, provides an additional benefit when used in combination with bitumen-based roofing products. The placing of an impermeable insulating layer above a bitumen-based waterproof membrane protects the bitumen-based roofing product from exposure to the sun. Bitumen-based roof products are derived from oil. These are reasonably rigid materials which are quite resistant to puncturing but require heat to bond the sheets during installation. The manual process of hot bonding makes this type of roofing material softer and more malleable, hence making it easier to lay and fix. However, a negative characteristic of this type of product is that the same uncontrolled effect can happen with exposure to the sun, causing delamination. Therefore, the added presence of a layer of insulation in an inverted warm deck design serves to minimise the exposure of the roofing material to the sun and can significantly increase the life cycle of the roof covering.

Another important feature of an inverted warm deck is that there needs to be some finishing layer on top of the insulation. This is not for aesthetic purposes but to prevent the insulation being blown off the roof by wind. Rigid insulating panels are very lightweight and these are not usually fixed to the surface of the waterproof membrane but are held in place externally. Typically, this is done with either stone chippings or gravel, known as 'beaching', or with precast concrete paving slabs which may also provide a walkway for roof access or sufficient hardstanding for some services equipment.

When undertaking and reporting on the condition of a roof during a commercial building survey or TDD it is important to establish whether the roof deck is cold, warm or inverted. The survey is usually visual with no attempt to open up areas for detailed investigation; however, it is possible to determine the roof type with some basic investigation. It is important for the surveyor to use multiple senses when undertaking a building inspection and this is certainly the case when surveying a roof. In cases where access to the roof covering itself is restricted by a number of features, such as the placement of walkways, M&E installations or roof terracing, it is almost impossible to inspect the covering. For roof coverings which have been executed in oil-based materials, such as bitumen mineral felt or asphalt, solar protection is often added in the form of stone chippings or gravel beaching. This, again, significantly inhibits the ability to inspect the roof covering. Therefore, when inspecting a roof which has been finished with gravel beaching it is necessary to scrape back some of this locally in order to establish the type of roof.

The presence of 'mushroom' shaped air vents combined with a hard 'solid' feeling under foot to the waterproof membrane may indicate a cold deck. Where the roof deck is concrete (as possibly observed internally) this will feel like walking on a concrete slab with no flexibility or 'give' in the surface. The building age is important in determining possible roof construction, with such characteristics as the facade design or materials providing an indication of the age. Part of the commercial building survey or TDD process will be to review historic planning/building permits during the 'desk survey'. This again may reveal the original age of the building and with this in mind and knowledge that cold deck construction is typically 'old' technology (pre-1980s) a body of evidence may be collected to establish if the roof is a cold deck.

In contrast, most 'modern' buildings or those which have been significantly renovated are likely to have warm deck construction. There are likely to be no vents (other than those used to ventilate sanitary stacks) visible on the roof. When walking on the roof this will feel softer or 'hollow' under foot and if the roof covering feels 'spongey' it is likely that the insulation is typically mineral wool quilting. Harder but hollow-sounding material under the waterproof membrane is likely to be high density insulation boards. Where the roof covering has been covered with gravel beaching it is an indication that the waterproof membrane will be an oil-based product. The weight of these stone chippings is significant and if the insulating layer is a mineral wool quilt, which has low compressive strength, this will flatten under load. This may render the insulation unfit for purpose. Therefore, most modern roofs finished with gravel beaching are likely to comprise a reinforced concrete deck, vapour barrier and high density insulating panels finished with two layers of hot bonded bitumen mineral felt.

In order to identify an inverted warm deck, an inspection should be made of the roof covering where typically this will have some type of finishing layer, such as precast concrete paving slabs or gravel beaching. At face value this could be mistaken for a regular warm deck, and only by raking back a bit of the gravel beaching or inspecting between the concrete paving slabs will the presence of an external insulating layer be evident.

The presence of 'mushroom' shaped ventilation pipes and caps indicates a cold deck.

A large flat roof finished with gravel beaching and the presence of high levels of vegetation suggests a lack of maintenance (photo courtesy of Widnell Europe).

A typical warm deck with bitumen mineral felt roof covering and no protective gravel beaching (photo courtesy of Widnell Europe).

Closer inspection of the same roof (above) shows dislodged insulation sheets below the beaching, which confirms this to be an inverted warm deck (photo courtesy of Widnell Europe).

Roof coverings

The covering to a roof, irrespective of it being pitched or flat, essentially needs to be waterproof. In some cases this is known as the waterproof membrane or simply the roof covering. For pitched roofs there is often an aesthetic aspect to the roof covering, as these are mostly visible from ground level or other adjacent buildings. In contrast, flat roofs are mostly concealed behind parapet walls and only visible from overlooking vantage points. The typical components of a roof covering are:
- roof deck (sometimes part of the structure)
- vapour barriers
- insulation
- waterproof membrane/covering.

In all instances it is important to consider the performance criteria for roofs (strength, stability, durability, weathertightness, thermal efficiency) when analysing the roof covering.

There are typically ten common types of roof coverings associated with commercial properties:

- asphalt
- bitumen mineral felt
- PVC
- EPDM
- steel
- aluminium
- zinc
- copper
- tiles/slates
- glass.

One obvious omission from the list is lead. While this has historically been an important roof covering, it is rarely used on commercial properties as a standalone covering. Despite its high cost, its extensive life cycle does mean that it is still the material of choice for flashing details.

Each of the above-listed coverings needs to be considered in respect of its application or use within the different property sectors as well as its suitability for either flat or pitched roof construction.

The roofing materials may be simply divided into three categories:

- oil-based/synthetic
  - asphalt
  - mineral felt
  - PVC
  - EPDM
- metals
  - steel
  - aluminium
  - zinc
  - copper
- other
  - tiles/slates
  - glass.

Oil-based and synthetic roof coverings are typically used for flat roof construction; they may be considered low-cost, limited life cycle materials. Functional properties of these materials outweigh any aesthetic value.

Although some metals are used for flat roof construction, these are predominantly suited to pitched roofs. They have varying life cycles but, in essence, these are normally expected to be longer than those of oil-based materials. Having a longer life cycle invariably means that they are also higher cost materials when compared to oil-based or synthetic products. Metals have high aesthetic value and are often specified by designers for their appearance as well as their performance.

Tiles and slates are exclusively used on pitched roofs and may be considered the most 'traditional' of materials. These materials are often used for specific buildings where appearance and the type of materials used are governed by planning requirements or individual design requirements. There are variations in the life cycles of tiles and slates, with natural materials such as slate and stone having good longevity, which is also the case with clay or concrete tiles. For tiles executed in fibre cement, these are considered to be the lowest cost and subsequently have the shortest life cycle.

Glass is the 'odd one out' when considering roof coverings, this has high aesthetic value and for architects it is used primarily to introduce natural light into a building. Aesthetically it is a striking material which epitomises the antithesis of 'traditional' materials, such as natural stone or slate, and is the very essence of modern construction. It can be used for pitched or flat roofs but the drawback with introducing natural light into a building is that this often increases internal temperatures with solar gain, resulting in a need for increased cooling. Therefore, glass is often used as a material for atria or for connecting two separate building zones with a glazed walkway or lobby.

A general rule of thumb is that high-cost roof coverings have longer life cycles when compared to low-cost materials. The building type, sector and investment strategies of building owners all influence the type of roof covering that is specified for a building.

Asphalt

Fact File: ASPHALT

- ▶ Identification: dark grey colour, quite hard under foot and smooth to touch.
- ▶ Roof deck: typically reinforced concrete.
- ▶ Mostly used with cold deck configuration.
- ▶ Fine and coarse aggregates mixed with bitumen.
- ▶ 'Trowelled' hot material typically 20 mm thick in 2.5 to 3 m wide bays.
- ▶ Should be finished with non-epoxy solar reflective paint or chippings.
- ▶ In use from the early twentieth century.
- ▶ Mass use in the 1960s–1970s.
- ▶ Life cycle 20–55 years (typically 35 years).
- ▶ Used in modern construction for the covering to rooftop car parking.
- ▶ Factors affecting life cycle:
  - ▷ 'blistering' (trapped air or moisture)
  - ▷ thermal movement (material and deck)
  - ▷ pollution, damage from foot traffic or joint failure
  - ▷ very cold temperatures can make this material brittle.

TOP – Asphalt roof covering (original) to a building constructed in the early 1970s in relatively good condition with no evidence of internal leakage.

MIDDLE – Typical 'blistering' of an asphalt flat roof covering caused by solar exposure.

BOTTOM – Asphalt flat roof in excess of 50 years old – note the evidence of remedial repairs with bitumen mineral felt.

When inspecting an asphalt roof covering it is not uncommon to find that these utilise cold deck technology and this may be a factor in determining why this material is less frequently used post-1980. It is strong in compression but weak in tension and the movement of the deck below is likely to cause fracture of the covering.

Large rooftop car park situated above a shopping centre
constructed in the 1960s (photo courtesy of Widnell Europe).

Significant blistering and solar damage
(photo courtesy of Widnell Europe).

Movement joint in the reinforced concrete roof deck
finished with bitumen and prone to leakage
(photo courtesy of Widnell Europe).

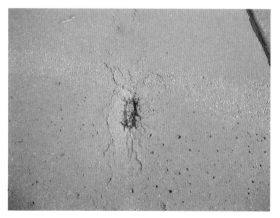

Burst blister allowing water penetration from the roof
(photo courtesy of Widnell Europe).

Irrespective of whether they are working for the buyer, seller, landlord or tenant, the surveyor undertaking a building survey or TDD should seek to work in the best interests of the building. This may appear an odd concept but it should be straightforward in the context that the condition of the building and any individual sub-construction elements is the same no matter who the surveyor is contracted to. There are instances, when instructed to undertake a vendor survey, that the client's aspirations are to receive a technical report which shows the building to be in a good state of repair with few or no defects. However, it is the surveyor's moral and professional obligation to ensure that a true representation of the building is detailed in the report.

This is particularly pertinent with large flat roofs where the roof covering may be low cost and therefore warrant a like-for-like low-cost replacement. However, the potentially vast size of flat roofs to shopping centres, car parks or industrial buildings often mean that replacement costs are significantly high.

When assessing the roof covering, it is important to establish the roof type and deck construction as well as the condition and anticipated life cycle. In the event that there is a cold deck with an asphalt covering which is approaching the end of its life cycle and, importantly, if there is any evidence of leakage, the most appropriate recommendation may be for immediate or short-term replacement. These are sometimes difficult decisions to make and could be a deal-breaker in the acquisition process. Therefore, it is critical that recommendations are evidence based and it is important to spend sufficient time on the roof to collect the relevant data.

The key principle to undertaking a commercial building survey or TDD is that, when recommending repair or replacement works, these should be done on a like-for-like basis. However, this cannot be so rigidly applied to roof coverings where there may be a requirement to 'upgrade' these to comply with relevant building regulations concerning energy efficiency or fire safety. This often leads to complex commercial negotiations between the vendor and purchaser in order to come to a compromise (financially). Irrespective of any deal negotiated, it is important for the surveyor maintain an evidence-based approach to all aspects of their initial and any subsequent advice.

Renovated 1970s car park with asphalt roof covering to a cold deck (photo courtesy of Widnell Europe).

The obvious problem with undertaking a building survey in winter, with snow concealing parts of the roof deck (photo courtesy of Widnell Europe).

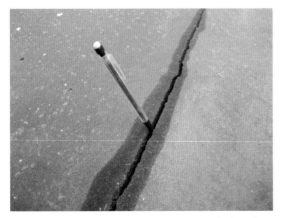

A significant crack passing through the 'new' asphalt roof covering requiring further investigation (photo courtesy of Widnell Europe).

Internal water penetration to the underside of the cast-in-situ reinforced concrete deck (photo courtesy of Widnell Europe).

Bitumen mineral felt (built-up roofing)

Probably one of the most common types of roof coverings is bitumen mineral felt. This is found widely within all building sectors on many different types of buildings. It is almost exclusively used for flat roof construction and has very little aesthetic value.

As with asphalt, it is prone to solar degradation but advances in the material's manufacture have seen the introduction of polymers resulting in improved resistance to solar degradation, hence lengthening life cycles.

Fact File: BITUMEN MINERAL FELT

- Identification: dark grey or black in colour, although quite hard under foot there does appear some flex in the material, it is rougher to touch than asphalt.
- Laid in strips, typically 900 mm wide.
- Roof deck: suitable for use on timber decks (small roof areas), steel decks and reinforced concrete.
- Used with warm and cold deck configuration but the majority of all 'new' roofs are likely to be warm deck.
- Rag fibres, asbestos fibres or glass fibres coated in bitumen.
- Can be two or three layers thick (hot bonded).
- Evidence of use before the Second World War.
- Mass use in the 1960s–1970s and still widely used.
- Life cycle 15–25 years (typically 20 years).
- Factors affecting life cycle:
  - 'blistering' (solar exposure)
  - detailing
  - loss of surface protection
  - traditionally a low-cost, low-quality roof covering.
- Note: more modern mineral felts have been/ are being developed with polymers to reduce the effects of solar exposure.

Bitumen mineral felt warm roof to a shopping centre originally constructed in the 1970s. This is the third layer of the material laid in 40 years (photo courtesy of Widnell Europe).

Evidence of solar damage with blistered exposed bitumen surface and 'rucking' of the material (photo courtesy of Widnell Europe).

Solar damage

The primary factor influencing the life cycle of a bitumen mineral felt flat roof covering is its exposure to the sun. As with almost all roofs, there is little or no protection for the roof covering from UV exposure and this has the effect of heating up the bitumen. Being an oil-based product, which relies on heat for the actual bonding process, exposure to the sun melts and delaminates the joints between the sheets of bitumen. Any air or moisture trapped under the roof covering will also heat up and the consequent expansion produces bubbles on the surface of the covering known as 'blisters'.

Throughout its life cycle, the constant heating up and cooling down of the material eventually leads to degradation. Starting initially with fine cracks on the surface of the bitumen mineral felt, this progresses to widening of the cracks and exposure of the actual mineral fibres contained within the bitumen. The felt loses its (already limited) suppleness and becomes more brittle. When walked upon it is possible to hear the material begin to crack as weight is applied to the surface.

Solar exposure causing the wrinkling or rucking of the joint between two sections of bitumen mineral felt sheets (photo courtesy of Widnell Europe).

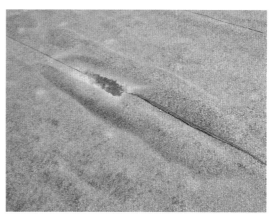

Exacerbation of solar damage showing more pronounced blistering of the material and note that in the top left of the image is evidence suggesting that rainwater collects in this vicinity (photo courtesy of Widnell Europe).

A graphic illustration showing solar exposure taken in the winter with frost persisting on the north face.

A 30 year-old bitumen mineral felt roof covering with a 'crazed' surface cracking when walked upon.

Assessing life cycles

Although bitumen mineral felt is considered a low-cost, low-quality material, there are potentially significant costs to be reported during the commercial building survey or TDD process. As a direct consequence of its cost, bitumen mineral felt is highly suited to large flat roofs, typically over logistics buildings or shopping centres. When the size of the roofs is often measured in tens of thousands of square meters, the associated replacement of the roof coverings can run into millions of pounds or euros. Therefore, it is an important recommendation that is made by any surveyor stating that a material has reached the end of its life.

As with all building surveying applications, it is critical to report upon the facts and 'say what you see'. It is important to glean as much information as possible from the building owner or property manager and the as-built documents. First, it is important to establish the date of construction or renovation and this may be obtained by analysing any planning permissions, building permits or acceptance reports. The as-built file will also be of use as this is likely to detail the specification of the roof covering. Having established the date of construction and likely age of the roof covering, this can be compared to the typical average life expectancy of the material. The on-site inspection will assess any damage or degradation of the material and during this stage of the survey it is important to ask the property manager or tenant if there are any reported leakages.

One important principle of reporting and costing defects in a commercial building survey or TDD report is that the recommendation to repair or replace can only be made if defects actually exist. While this may sound obvious, there are instances when the buyer may pressurise the surveyor to exaggerate immediate or short-term costs to seek a price reduction for the property. Likewise, when performing a vendor survey, the selling party will typically want the property to appear in good condition, therefore may request that recommendations are made for repair as opposed to replacement.

There are obvious differences between recommendations to repair or replace and there is a similar but slightly more subtle difference when using the term 'renew'. As surveyors are accountable for their advice, and remain liable for many years after giving that advice, it is 'natural' for them to be risk averse. Considering also that roofs are subjected to potentially very 'hostile' weather conditions, which cannot always be predicted, it is better to be cautious when advising on whether to repair or replace. In this circumstance, it may be appropriate to adhere to the following evidence-based process:

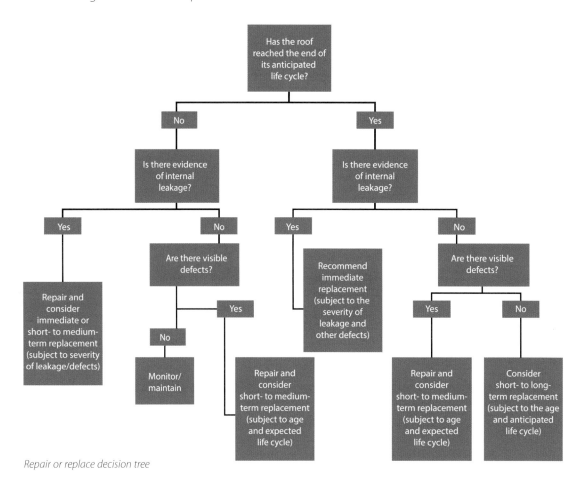

Repair or replace decision tree

The skill of the surveyor is to apply their knowledge and understanding of material life cycles to the situation observed during the site inspection. While there are recommended or suggested average life cycles for materials, bitumen in particular has a wide variation. The material life cycle is dependent upon the quality of the initial material as well as the workmanship during the execution of the works. Substandard materials or poor workmanship are likely to exacerbate the effects of solar exposure. Insufficient hot bonding of the material is likely to result in increased potential for 'blistering', as air or moisture is trapped beneath the finishing layer. In this instance, widespread and irreparable damage may occur, resulting in 'premature' replacement. Such circumstances are highly relevant to the building owner, tenant or purchaser as significant short-term investment costs are likely as a result.

Case Study: BITUMEN MINERAL FELT ROOFS

▸ Bitumen mineral felt covering to pre-cast lightweight reinforced concrete warm deck.

▸ Age ten years (typical life cycle 20 years).

▸ Internal leakage reported.

▸ Widespread external severe blistering to the complete roof covering.

▸ Recommended immediate replacement of the roof covering and upgrade of the insulation.

(Photo courtesy of Widnell Europe.)

The roof identified in the case study was replaced with an EPDM covering; this or PVC single ply membranes are possible solutions for dealing with solar damage to flat roof coverings. However, a more traditional approach is to place a layer of stone chippings or gravel beaching on the surface of the roof or to paint the bitumen with a solar-reflective paint.

Stone chippings or gravel beaching placed on a flat roof with a bitumen mineral felt roof covering. Note the centrally placed rainwater evacuation outlet with a protective metal grill. However, the grill has been dislodged from above the outlet, potentially allowing stone chippings to block the drainage outlet (photo courtesy of Widnell Europe).

Solar reflective paint applied to the upstands and covering of a parapet wall and flat roof finished with bitumen mineral felt (photo courtesy of Widnell Europe).

The placing of gravel beaching is an effective solution to the damage caused by solar exposure to a bitumen mineral felt roof as this prevents UV rays coming into contact with the roof surface. It also absorbs the heat generated by solar exposure and, as a consequence, can significantly extend the anticipated life cycle of the material. However, the principal problem with placing a layer of stone chippings on the roof covering is that it significantly hampers any investigations which may be required should a roof leak be reported.

Investigating a roof leak where the covering has been protected by a layer of stone chippings means that this must first be raked back or removed before any repair can be executed.

Gravel beaching raked back for an investigation into a roof leak; repairs were executed and the roof was left uncovered for monitoring. Failure to reinstate the stone chippings has resulted in the blockage of drainage outlets and severe ponding to the roof as well as allowing unnecessary solar exposure of the repaired bitumen mineral felt (photo courtesy of Widnell Europe).

An alternative form of solar protection is a composite bitumen mineral felt roof covering finished with a layer of aluminium foil. The foil is frequently the subject of bird damage, which can lead to localised solar exposure and damage (photo courtesy of Widnell Europe).

Single ply membranes

A low-cost alternative for flat roofs is single ply membranes. These are executed in either PVC or EPDM.

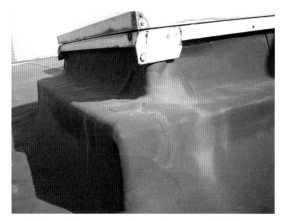

An EPDM roof covering and upstands to a rooflight housing, illustrating the flexibility of the material (photo courtesy of Widnell Europe).

PVC used as a single ply roofing membrane, which extends to form the upstands and 'dressing' of the parpapet walls.

- Single ply membranes are typically EPDM or PVC.
- EPDM was developed in the Middle East for resistance to solar degradation.
- EPDM and PVC are not susceptible to damage from solar exposure.
- Materials are glued or welded.
- The materials are prone to wind damage, puncturing or becoming torn.

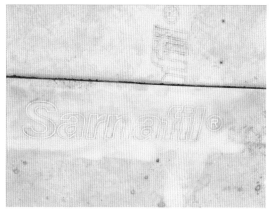

'Sarnafil' is a brand name for a PVC roof covering. During the survey it is important to record brand names to enable a subsequent desk survey to research the material and analyse its data sheets.

Ethylene propylene diene monomer (EPDM)

Fact File: EPDM

▶ Identification: dark grey, almost black in colour. It is a rubber-based product and, although used for roof coverings, similar products are used for garden pond linings.

▶ It is much softer to touch than bitumen mineral felt and has a smooth surface.

▶ Laid or glued in strips, it is a very flexible material meaning there are often seamless sections extending to form parapet linings or flashing details.

▶ Roof deck: suited for use on timber decks (small roof areas), steel decks and reinforced concrete.

▶ Used with warm and cold deck configuration but the majority of all 'new' roofs are likely to be warm deck.

▶ Developed in the 1950s and resistant to solar degradation.

▶ Average life cycle 25 years.

▶ Factors affecting life cycle:
  ▷ atmospheric pollution
  ▷ movement of the deck
  ▷ impact damage.

EPDM roof covering to a flat roof over an industrial unit; note there is no requirement for solar protection to be applied to the covering (photo courtesy of Widnell Europe).

Typical puncture to the surface of a single ply EPDM roof covering allowing potential water infiltration (photo courtesy of Widnell Europe).

EPDM roof covering in excess of 30 years old in excellent condition considering its age (photo courtesy of Widnell Europe).

EPDM roof covering less than ten years old with poor edge detailing resulting in tearing of the material, which will be susceptible to future wind damage.

Polyvinyl chloride (PVC)

Fact File: PVC

▶ Identification: light grey or even white in colour. It is an oil-based product but, unlike bitumen mineral felt, it is soft and smooth to touch with flexibility or 'stretch' in the material.

▶ Laid or glued in strips, it is a very flexible material, meaning there are often seamless sections extending to form parapet linings or flashing details.

▶ Roof deck: suited for use on timber decks (small roof areas), steel decks and reinforced concrete (flat and pitched roofs).

▶ Used with warm and cold deck configuration but the majority of all 'new' roofs are likely to be warm deck.

▶ Suitable for large areas of flat roof.

▶ Average life cycle 25 years.

▶ Factors affecting life cycle:
  ▷ trapped air/moisture
  ▷ atmospheric pollution
  ▷ movement of deck
  ▷ impact damage.

PVC roof covering with 'ornate' detailing designed and executed to look like a zinc roof covering from ground level. Executed in PVC with a reduced cost compared to zinc (photo courtesy of Widnell Europe).

Glued PVC detailing to roof covering (not noticeable from ground level) appears poor (photo courtesy of Widnell Europe).

Glued PVC warm roof deck with seamed joint to the section of PVC forming the upstand to a roof light. Note that insufficient adhesive applied to the PVC has caused air bubbles and creasing of the material locally.

Tear to the glued seam of a PVC roof covering indicating stretching or stress to the material and possible wind damage or movement of the deck (photo courtesy of Widnell Europe).

EPDM glued patch repair to a PVC roof covering; this should have been executed with a PVC patch repair to match the existing covering (photo courtesy of Widnell Europe).

Steel fixing screw left on the roof covering is evidence of poor maintenance or quality control during execution of the works. This has the potential to puncture the roof covering (photo courtesy of Widnell Europe).

Tearing of the single ply PVC roof covering to a cold deck (reinforced autoclaved aerated concrete (RAAC) planks); liable to suffer severe wind damage and infiltration in future (photo courtesy of Widnell Europe).

Defects to EPDM and PVC

During the commercial building survey or TDD process it is important to note and describe the presence of single ply EPDM or PVC roof coverings. If these have been used instead of zinc to create the appearance of zinc from ground level, this should be noted in the report. However, in the example given of this actually happening, it was only evident that the 'original' roof covering was supposed to be zinc as this was detailed on the drawings appended to a building permit. In this particular case the purchaser was informed, as was the legal due diligence to ascertain whether this had any significance with respect to the urban planning procedure.

Where there is evidence of actual material defects, which include splits, tearing or puncture of the material, it is appropriate to recommend localised repairs accordingly. However, where there is only minor localised evidence of air bubbles, indicating a lack of adhesion or bonding of the material, it appears more rational to recommend monitoring of the situation. Cutting out these small sections for repair, where there is no evidence of internal leakage, has the potential to do more harm than good. These types of repairs introduce additional areas of seams and joints, which may the site of future weakness and potential leakage.

Certainly, the placement of technical equipment on a single ply roof covering has the potential to cause punctures; likewise, areas of access or foot traffic may also lead to damage of the materials. Therefore, it is important to note this during the survey and to recommend that specific localised reinforced matting be placed in these areas to avoid future defects. Likewise, where provision has been made for such protection and this is evident during the survey, it is important to note this as evidence of good practice or maintenance.

 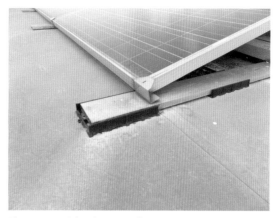

Evidence of good practice: an air conditioning compressor (split unit) has been placed on precast concrete paving slabs to absorb vibrations and these are resting on a reinforced PVC matting placed on top of the PVC roof covering. A PVC non-slip reinforced walkway gives access to the unit and protects the roof covering.

The corners of the aluminium frame supporting a solar panel installation located on a PVC roof are positioned on protective plastic supports to distribute the weight of the solar panels without puncturing the roof covering (photo courtesy of Widnell Europe).

Green roofs

Executed for either aesthetic reasons or as a planning requirement, green roofs are becoming increasingly evident in flat roof construction. There is certainly an aesthetic value for neighbouring properties that may overlook a proposed new development which includes a large flat roof. Instead of being faced with a large expanse of black or grey roofing, this can be replaced with much 'softer' low-level vegetation, which could include seasonal colours or even flowers. Furthermore, the photosynthesis of plants removes CO_2 from the atmosphere and can be used to illustrate potential reductions in the carbon footprint of a new or renovated building.

A green roof has the following construction characteristics:
*   'traditional' flat roof construction
    *   concrete deck
    *   vapour barrier
    *   insulation
    *   waterproof layer (bitumen mineral felt)
    *   root barrier
    *   drainage and planting layers
    *   vegetation.

The placing of the drainage and planting layers gives good solar protection to the bitumen mineral felt and this is likely to enhance the average life cycle significantly beyond the expected 20 years. Importantly, the choice of vegetation is crucial and only shallow-rooted but hardy plants are suitable. Typically, these may be those classed as 'alpines', which have shallow but strong roots and are accustomed to exposed mountainous habitats. They offer year-round colour and are typically evergreen, often changing colour with the seasons or flowering. A good example of these are heathers. However, negative consequences of green roofs are that any reported leakages are difficult to trace or investigate and excessive vegetation can block or conceal drainage outlets. Bamboo is completely unsuitable for green roof construction as this is a vigorously growing plant with root networks that can penetrate root barriers, waterproof membranes and even concrete. If this plant is present in green roof coverings or even externally in landscaped areas above or adjacent to basements, this should be noted in the survey report and attention drawn to any evidence of leakage (actual or reported).

More mature planting with heather. However, this is overgrown and is concealing the rainwater outlets (photo courtesy of Widnell Europe).

Green roof finished with low level 'alpine' type plants (photo courtesy of Widnell Europe).

Part green roof and roof garden with vigorously growing bamboo (photo courtesy of Widnell Europe).

Metallic roof coverings

One general observation concerning oil-based or synthetic roof coverings is that these are often placed on flat roofs. They are cost effective for large surface areas and are therefore suitable for use on industrial buildings or shopping centres. They are also suited to roofs which are not visible from ground level and have little aesthetic value. In stark contrast are the materials which make up metallic roof coverings. Primarily, these need to be functional and it is important to understand their individual characteristics with respect to the performance criteria of roofs.

When compared to synthetic or oil-based products, metallic roof coverings have generally longer life cycles, are of higher quality and, consequently, are costlier. The aesthetic value is something that cannot be discounted as, when used for pitched roofs, these metals are often visible from ground level. The form and shape of these roofs can be used to make an architectural statement.

There are differences between the individual metals concerning life cycles, quality and costs. Furthermore, the different metals are prone to both general and specific defects.

Steel

Fact File: STEEL

- ▶ Identification: steel almost always has a painted protective cover; this is primarily for functional reasons but can also be for aesthetic purposes. Roofing panels are usually galvanised steel and, at face value, it may be difficult to distinguish between steel and aluminium.
- ▶ Steel is denser than aluminium and, when tapped, has a 'heavier' sound compared to aluminium. Any scratches on the surface of the sheeting steel or fixing holes will invariably show evidence of rust. Any evidence of corrosion to the roof sheets will indicate that the material is steel.
- ▶ When walked upon, steel has less flex or movement in comparison to aluminium.
- ▶ Older panels are often a cold deck arrangement, laid in strips which are screw fixed or rivetted. New panels are likely to be sandwich panels, which include insulation.
- ▶ Average life cycle 35 years with factors affecting this attributable to failure of the protective coating leading to corrosion.
- ▶ Failure of fixings and the effects of wind load may reduce life cycle, as can atmospheric pollution or impact damage.
- ▶ Suitable for large areas of flat or pitched roofs.

Profiled painted steel roofing and cladding panels (photo courtesy of Widnell Europe).

Peeling painted finishes to the profiled steel roofing sheets of a flat roof to an industrial building. Note: the grey colour is the galvanised steel but where this is orange/brown this indicates corrosion (photo courtesy of Widnell Europe).

Screw fixed steel roofing sheets over 30 years old with evidence of cut edge corrosion (photo courtesy of Widnell Europe).

Cut edge corrosion

Steel panels used as roofing sheets or for facade claddings mainly have a painted finish; however, this is largely aesthetic. Prior to painting, the steel is invariably galvanised, which means it has been coated in a layer of zinc with the intention of protecting the steel from corrosion. Unpainted galvanised steel has a grey and silver 'mottled' pattern but is not shiny of reflective. If shiny or smooth steel is encountered during a building survey it is likely that this is stainless steel, which is completely different from galvanised steel. It is costlier, with high aesthetic appeal, and is not used for roofing sheets.

Galvanised steel railings to a rooftop car park; note that galvanised steel is easily identifiable by the matt, grey 'mottled' pattern.

In contrast, the railings in the image above are stainless steel and the recognisable characteristic is the shiny finish. However, this can sometimes be in a matt version, known as 'brushed finish'.

The principal problem with galvanised steel roofing sheets and cladding panels occurs when they are cut on site. This is normally done to fit standard sized sheets into bespoke dimensions and the sheets are cut on site using power-assisted saws. The effect of cutting the steel causes friction and the resultant high temperature damages the painted finish as well as the galvanised protection. This leaves an exposed steel 'cut edge' which is prone to corrosion. This can develop at different rates, depending on the exposure of the cut edge to atmospheric pollution. In areas of high pollution or in coastal locations, it is anticipated that cut edge corrosion will evolve relatively quickly.

The symptoms of cut edge corrosion begin with the blistering or peeling of the paint on the surface immediately adjacent to the exposed steel, which is caused by the corrosion and expansion of the steel below. The steel begins to rust and decay in a process exacerbated by time and the exposure of more steel with the increasing damage and exposure of the cut edge. It is a defect attributable to poor original execution, as cutting steel panels on site is acceptable provided a protective layer is applied to the cut edge as part of the process. It is also a defect which can be associated with poor maintenance since periodic inspection should identify this and relatively minor intervention in the early stages can resolve the issue. However, if left to manifest, cut edge corrosion will inevitably continue until it reaches the fixing points of the panels and this makes the panels susceptible to wind damage. Furthermore, cut edge corrosion can result in significant holes developing in panels, thus compromising the function of the roof or facade. Cut edge corrosion not only affects areas where panels are cut down in length but also if holes have been drilled into the steel.

With all steel roof sheets there is a certainty that corrosion will occur if the painted surface and galvanised protection is sufficiently damaged or scratched, exposing the steel below. When advising a landlord, investor or occupier during a building survey or TDD, it is important to note evidence of damage to steel roofing panels and any subsequent corrosion. It is normal to recommend in-situ repairs or even replacement of severely corroded steel panels where the fixings have been compromised and the steel is beyond repair.

It is the responsibility of the surveyor to report the presence of corroded steel. Where evidence exists that this has been allowed to evolve unchecked, it may be appropriate to comment additionally on a perceived lack of maintenance. Flat roofs in general may be a good barometer as to the maintenance regimes carried out on buildings. Out of sight, these also often appear to be out of mind, essentially meaning they are infrequently or never inspected as part of the building's periodic maintenance. In these circumstances, works to the roofs are often reactive as opposed to proactive and it may be argued that this is a costlier long-term approach overall. Owners, occupiers or property and asset managers who do not inspect the roofs of their buildings at least annually may adopt similar approaches to the other construction elements. While structures, roofs and facades can endure relatively long periods of neglect before deteriorating, the life cycles of the technical installations (M&E) are significantly reduced through a lack of maintenance in the short term.

Early onset cut edge corrosion
(photo courtesy of Widnell Europe).

Evolution of corrosion migrating towards the fixing
(photo courtesy of Widnell Europe).

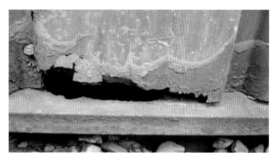

Severe cut edge corrosion with decay to the fixing point
(photo courtesy of Widnell Europe).

Aluminium

The visual appearance of aluminium roof coverings is similar to steel and it is sometimes difficult to ascertain which metal you are dealing with during a site inspection. At face value it is always appropriate to state that the roof covering is metallic as this indicates generally a higher quality material with a longer life cycle than oil-based or synthetic materials. However, it is necessary to establish and report on the actual type of metallic roof covering. Aluminium is much lighter than steel and, although it is rarely installed unfinished, any exposed cut edges or scratches to the finishes do not present corrosion or rust stains. As a direct result of being lighter than steel it has less tensile strength and is more prone to mechanical damage or deflection if overloaded or walked upon. When tapped or knocked it makes lighter, more 'tinny' sound when compared to steel.

Its relative lack of tensile strength means that it is not suited to flat roof coverings where it may be loaded with plant, technical equipment or being walked upon. Therefore, it is more likely to be used on pitched roofs and, due to the flexible characteristic of the material, it is often used to for curved coverings.

Fact File: ALUMINIUM

- ▶ Identification: aluminium usually has a painted finish but this is for aesthetic reasons.
- ▶ Aluminium is less dense than steel and when tapped has a 'lighter' sound when compared with steel. If there are any scratches to the surface of the sheeting or fixing holes these will present a shiny but non-corroded surface.
- ▶ When walked upon aluminium is likely to feel more 'springy' than steel but will readily deform if it is overloaded.
- ▶ The joints to aluminium roofing sheets may be rolled or standing seam which accommodates thermal movement or expansion.
- ▶ The average anticipated life cycle is 35 years and factors affecting this are contact with copper, brass and mild steel. This can lead to bimetallic corrosion. Likewise, rainwater running off lead will damage aluminium.
- ▶ Older panels may be in a cold deck arrangement laid in strips which are screw fixed or rivetted. New panels are likely to be sandwich panels which include insulation.

A new curved aluminium roof covering (photo courtesy of Widnell Europe).

A 40 year-old aluminium cold deck with evidence of painted bitumen repairs and deflection of the sheeting.

Screw fixed aluminium roofing sheets (over 30 years old) with patch repairs to one fixing hole and on the joint where the material has become torn (photo courtesy of Widnell Europe).

Zinc

Opinion is divided on the use of zinc as a roofing material with some believing this to be a 'poor mans' copper' and others extoling its architectural value as well as longevity. In essence, zinc is an upgrade on aluminium in terms of life cycle and it is often executed in situ with no painted finish applied. Architecturally, its rigidity presents clean straight lines which introduce shape and profile as well as shadow. It is susceptible to aging and weathering but is relatively resistive to corrosion.

Using zinc for flat roofs (as illustrated in the photo) appears somewhat extravagant as this is not visible from ground level Therefore this is not done for functional or aesthetic reasons but perhaps for the correctness of uniformity, which is typically driven by the architect and sanctioned by the owner, client or investor with few cost constraints.

Illustrating the versatility of zinc as a roofing and cladding material, it has been used to clad the gable of the pitched roof as well as forming the roof covering for the adjacent flat roofs (photo courtesy of Widnell Europe).

Zinc used as an architectural statement on a building with a curved facade; despite the straight lines of the material this creates a sweeping aesthetic of the curved pitched roof (photo courtesy of Widnell Europe).

Fact File: ZINC

- Identification: as a material zinc exudes quality, it has good tensile strength and excellent longevity. Instantly recognisable as an 'unfinished' metal it is rarely mistaken for steel or aluminium.
- Zinc is a high-cost, high-quality material but this is less costly when compared to copper and it is not uncommon for a copper roof covering to be replaced with zinc.
- When walked upon zinc appears firm and less springy than aluminium but more like steel and rarely deforms or deflects under load.
- The joints to zinc roofing sheets may be rolled or standing seam, which accommodates thermal movement or expansion.
- Zinc is often used in combination with timber boards to form the sub-deck and older zinc tends to fade or get lighter with age and exposure.
- The average anticipated life cycle is 50 years and factors affecting this are insufficient allowance for movement or soldered joints as well as leaking rolls or damaged joints.
- It is a metal with quite high value and may be the subject of theft but its placement on exposed pitched roofs appears to act as an immediate deterrent.
- Zinc can be affected by decay or corrosion which is usually associated by coming into contact with lead or copper (including rainwater run-off).
- Dampness to the timber sub-deck can cause swelling and movement of this and the subsequent differential movement with the zinc covering may cause distortion.
- Zinc is also prone to decay or corrosion from acid or alkaline attack.
- It also performs poorly during fire and has relatively poor resistance to combustion.

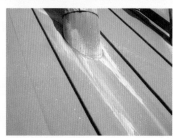

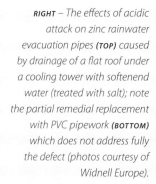

RIGHT – The effects of acidic attack on zinc rainwater evacuation pipes (TOP) caused by drainage of a flat roof under a cooling tower with softenend water (treated with salt); note the partial remedial replacement with PVC pipework (BOTTOM) which does not address fully the defect (photos courtesy of Widnell Europe).

LEFT, FROM TOP – An older zinc roof with a type of rolled/standing seam joints to accommodate movement; note the fading and discolouration of the material.

A zinc roof with evidence of bitumen patch repair suggesting this is defective and approaching end of life.

Zinc roofing laid in long strip arrangement with standing seam joints, the darker/richer colour of the zinc indicates this is relatively new (photo courtesy of Widnell Europe).

Rainwater run-off adjacent to a chimney flue on a zinc roof causing acidic attack to the material (photo courtesy of Widnell Europe).

A curved roof executed in PVC but made to look like zinc, which contradicted the planning permssion (photo courtesy of Widnell Europe).

Copper

There is evidence of copper used for roof coverings across most UK and European city centre skylines. This is a 'glamourous' high-quality, high-cost material with an excellent life cycle. As an architectural choice, it may be considered a statement material which can be used for whole roof coverings or for specific individual detailing. It is a material which appears 'living' and its colour evolves with time as it ages due to the patina forming on the surface as this is exposed to the atmosphere and pollution.

Antwerp Central Station: the copper roof in this photo is relatively new and 'brown' in colour.

The Ministry of Defence buildings in London with light-green copper roofs indicating a much older and aged material.

Further detail of the roof over the Ministry of Defence buildings in London; note the lifeline installed just below the ridge to allow safe working access.

The former City Hall building in London with copper detailing to the dormer windows.

Copper is used universally on many different types of buildings across the individual sectors but is typically not suited to industrial buildings, such as logistics sites or shopping centres where large surface areas of flat roofs make this cost inefficient.

Fact File: COPPER

A copper roof less than 20 years old. It appears brown in colour and, when viewed from a distance, could be mistaken for zinc (photo courtesy of Widnell Europe).

A copper roof between 20 and 30 years of age. This has lost its initial brown colour and is dark green (photo courtesy of Widnell Europe).

Copper roofing laid in long strip arrangement with standing seam joints. This is over 40 years old and note the fading of the green surface.

▶ Identification: copper should be instantly recognisable as a roof covering if it is aged since the surface will be green in colour. Newer copper will appear brown in colour, eventually turning a dark green before becoming lighter shades of green as it gets older.

▶ It is a high-cost, high-quality material, meaning this is usually reserved for 'statement' buildings or to create interesting architectural features.

▶ The joints to copper roofing sheets may be rolled or standing seam which accommodates thermal movement or expansion. In some cases the joints may be soldered or welded.

▶ Copper is often used in combination with timber boards, which are used to form the sub-deck.

▶ The average anticipated life cycle is 65 years and factors affecting this are insufficient allowance for movement, electrolytic action causing corrosion due to it coming into contact with lead or rainwater run-off from lead pipework or flashings.

▶ It is a metal with high value and may be the subject of theft but its placement on exposed pitched roofs appears to act as an immediate deterrent.

▶ Dampness to the timber sub-deck can cause swelling and movement of this and the subsequent differential movement with the copper covering may cause distortion.

Deflection to copper roof covering to a hotel, which appears to suggest dampness and swelling of the timber deck below.

Copper used as a facade cladding. Note the rainwater run-off from the material, which is leaving residue and causing staining to the natural stone finishes below.

Tiles and slates

Tiles and slates may be considered the more 'traditional' types of roof covering and these are almost exclusively used for pitched roofs. Their use may also be limited to relatively small surface areas as 'traditional' pitched roofs have limited spans and quite steep pitch. Therefore, tiles and slates used for roofs to commercial properties are likely to be for town centre retail and office properties or as architectural features to larger commercial properties, such as porches or mansard roofs.

Tiles and slates can be classed as natural or manmade and include the following common construction materials:

STONE

MATERIAL AND APPLICATION	CHARACTERISTICS	IMAGE
▸ A vernacular material (highly regional). ▸ Typical to older listed properties or those in a conservation area. ▸ Clients need to be advised of the consequence of listing or conservation area status, which will mean locally sourced stone and techniques will be required during repair or renovation, significantly increasing the cost. ▸ Important to note the presence of stone during any building reinstatement cost assessment.	▸ High-quality, high-cost material. ▸ Life cycle can be in excess of 100 years. ▸ Thickness typically 12–25 mm and a very heavy material. ▸ Traditional construction; fixed with wooden pegs, brass screws, galvanised or stainless steel nails. ▸ Typically, older roofs are likely to be devoid of vapour barriers (sarking felt) or insulation. ▸ Life cycle reduced by defective fixings or delamination of stone and cracking of tiles. ▸ Renovation of stone roof coverings can expect to recycle a significant amount of the tiles.	 *Stone roof coverings to a market town in the Cotswolds.*  *A cracked stone tile.*

SLATE

MATERIAL AND APPLICATION	CHARACTERISTICS	IMAGE

- A 'traditional' natural material which was popularised by the Victorians.
- Mass use of the material was aided by the expansion of train transportation engendered by the Industrial Revolution.
- Widely used material although sources for specific types are regional.
- Cheaper imports are available in Europe and also from the Far East.
- For listed buildings or conservation areas it is possible that specific slate types will be requested for repair or renovation.
- Clients should be advised of the relatively high investment cost of the material.

- High-quality, high-cost material.
- Life cycle typically 75 years but can be in excess of 100 years.
- Slate is a type of stone created under pressure and in layers; when mined, this aids the splitting and shaping into individual pieces.
- Fixed with steel nails to timber battens. Older roofs are likely to be missing vapour barriers or sarking felt.
- A typical factor affecting the life cycle is delamination (carbonates in the slate react with acid rain and calcium sulphate).
- Splitting, broken and slipped slates caused by corroded nails or wind lift.
- Water penetration via capillary action or insufficient laps.
- Renovation of slate roofs can expect to recycle a significant amount of the existing slates.
- A 'turnerised' roof involves covering a slate roof with mesh and bitumen. This is done to treat leakage. But in the long term this can do more harm than good as the slates are often not reusable afterwards.

A slate roof to a commercial building, which dates from 1902 and the roof is likely to have been renovated.

Typical damage attributable to corroded nails, cracked slates and delamination.

The use of lead strips known as 'tingles' to replace or re-fix slipped slates in situ.

A 'turnerised' slate roof.

CLAY

MATERIAL AND APPLICATION	CHARACTERISTICS	IMAGE

- ▶ The concept of clay tiles has been around for many centuries, dating back to Roman times.
- ▶ Initially considered a vernacular material as the use of clay for bricks and tiles is historically associated with the locality of this material.
- ▶ Clay tiles may be classed in appearance as 'Roman' (single or double), 'plain' or 'pan tiles' depending on their shape.
- ▶ Initially handmade but now factory pressed, these have good aesthetic value as well as being highly functional and cost effective.

- ▶ High-quality material but cost effective with varying life cycles between 60 and 100 years.
- ▶ Clay tiles incorporate a preformed 'nib' which 'hooks' onto timber battens and is secured with nails.
- ▶ Factors affecting the life cycle of clay roof tiles include delamination of the material due to frost damage, pitting or cracking associated with weather exposure or slipped tiles due to nib or nail failures.
- ▶ Clay tiles are relatively brittle and can be cracked easily by impact damage.
- ▶ Insufficient overlap to the tiles can lead to water penetration or wind lift.
- ▶ Older roofs are likely to be constructed without a vapour barrier or sarking felt.

Double Roman tiles.

Plain tiles.

Pan tiles.

Frost damage/delamination causes erosion and weakening of the nib.

CONCRETE

MATERIAL AND APPLICATION	CHARACTERISTICS	IMAGE

▶ As a material developed for roofing tiles, concrete was introduced to the built environment typically post-Second World War.

▶ Concrete tiles are pre-cast and not reinforced. They can be interlocking or plain and can be manufactured in a variety of different colours.

▶ They are versatile and can be made to look like some clay tiles.

▶ Concrete tiles are a highly cost-effective alternative to clay but are heavier and therefore not always suitable for renovation of an existing roof structure.

▶ Concrete tiles are also unlikely to be an acceptable alternative to clay, stone or slate aesthetically for a listed building or in a conservation area.

▶ Concrete roofing tiles are a low-cost material but have a relative long life cycle, with many examples still in existence 50–60 years after their initial installation.

▶ Concrete tiles are not fully proven in terms of life cycle due to their relatively 'recent' development so their overall longevity cannot be fully analysed.

▶ Concrete is very heavy and is often used to replace slate or clay tiles resulting in 'dishing' of the roof due to overloading of the existing structure.

▶ With time, there is often a loss of surface colour but this is an aesthetic 'defect'.

▶ Efflorescence can build up on the surface of the concrete. This derives from the soluble salts in the concrete and is visible as white powder on the surface.

▶ The preformed nibs to the tiles can become eroded and worn; frost damage may occur due to poor manufacture.

▶ The non-reinforced concrete tiles may become cracked or broken due to manufacturing defects, poor installation or mechanical damage.

'Dishing' caused by overloading an existing roof structure (originally tiled with clay or slate) with concrete.

Discolouration and wear to the exposed edges of these interlocking concrete roof tiles, which are in excess of 40 years old.

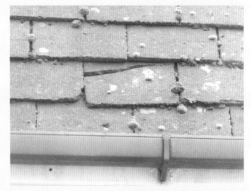

A cracked, plain concrete roof tile with general wear and erosion to the edges of the material.

FIBRE CEMENT

MATERIAL AND APPLICATION	CHARACTERISTICS	IMAGE

MATERIAL AND APPLICATION

- As an alternative to slate, fibre cement is a low-cost manmade roofing material.
- The tiles can be produced in a range of colours but these are predominately grey and are often used to replace natural slate roofs.
- They are much lighter in weight than slate and up to the end of the twentieth century there is a likelihood that these may contain asbestos.
- Fibre cement also comes in corrugated sheets. This is typically installed on industrial or agricultural buildings. Where these sheets date from pre-twenty-first century there is also a strong likelihood that they may contain asbestos.

CHARACTERISTICS

- Fibre cement tiles are typically fixed to timber battens with nails and, as these are lighter in weight than slates, they may be more prone to wind lift.
- They may be considered low cost and relatively low quality with anticipated life cycles being 15–30 years.
- As they age, they become fatigued, with the colour often fading and the material distorting around the edges.
- Fibre cement tiles are more brittle than natural slate and crack easily through poor installation or mechanical damage.
- Where asbestos is present in the fibre cement slates there often appears a high incidence of moss growth to the surface.

To the front is a roof covered with fibre cement and behind a roof covered with natural slate. The obvious visual difference is the discolouration of the fibre cement roof tiles.

Closer examination of the fibre cement tiles identifies some areas of darker tiles which suggest some localised replacement.

Fibre cement tiles and chimney flue which contain asbestos, note the high quantity of moss growth to the surface (photo courtesy of Widnell Europe).

RIGHT – Corrugated fibre cement roof sheets to an industrial storage building; the age of the building suggests that this may contain asbestos and this must be verified by reference to the asbestos inventory or by an asbestos specialist. Of significance are damaged sheets which could release asbestos fibres into the atmosphere and the report should worn of this (subject to confirmation) (photos courtesy of Widnell Europe).

Glass

Facades and roof coverings executed in glass are often striking and appear to embody the essence of high tech modern architecture. Glass is an excellent material for introducing natural light into a building; however, the regrettably by-product of this is that it also introduces solar energy too. The effect of exposure to the sun from a glass roof is to provide variable lighting levels, which can result in sun glare to occupants or users. This can be eradicated by using sun blinds, most of which are internal. Sun blinds are effective in reducing solar glare but the biggest factor of solar effects to a glazed roof is that of solar heat gain. Even in winter, solar energy passing through a glazed roof will raise the internal temperature significantly. As a consequence of increased heat gain, there is an inevitable requirement to counterbalance this with cooling to improve internal comfort levels.

The use of ventilation and air conditioning has an effect on energy consumption or energy performance of a building, therefore it may be considered less desirable to have glazed roofs to office areas for both practical and economic reasons.

While glazed roofs to office areas or occupied commercial space are relatively rare, such installations are readily used to form atriums or roof lights where there are large areas of commercial space with little or no natural light.

Glazed roofs are an effective way to introduce natural light into areas of a building with few or no windows (photos courtesy of Widnell Europe).

The technical advances in glazing have been relatively rapid compared to other standard construction materials. Initially, glazing to roofs over atria and staircases in the 1960s or 1970s comprised single glass panels. The obvious risk of breaking glass falling onto building users and occupants was mitigated by incorporating fine wires in a grid during the manufacturing process. The glass manufactured by this technique is known as Georgian wired panels (GWP); it was also seen as a method to improve the resistance of the glass to shattering during a fire. This is one of the reasons why GWP are also evident in glass vision panels in fire doors. Glass as a material performs relatively poorly in compression and even worse in tension; if glazing is subjected to tensile force it is likely to crack or break.

During the inspection of a glazed roof (where safe access can be granted) it is imperative to assess the type of glass and any evidence of safety provision. This could be the placement of internal safety film to prevent fragmentation or the use of GWP. It is of paramount importance to seek clarification from the property manager, building owner or occupier about whether there has been any recorded incidence of broken glass panels and also if remedial works, such as the placement of safety film, has been carried out.

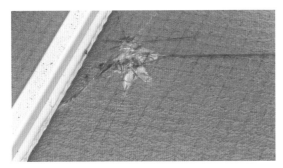

GWP with impact damage.

Public service notice warning of damaged glazing.

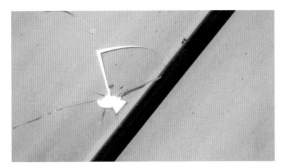

Broken glazing panel, not reinforced or laminated.

Glazed roof over the main concourse of a railway station.

While GWP glazing improved the strength and stability of roof coverings, its use has been practically negligible in terms of improving energy efficiency. Single glazing for roofs is similar to that of facades or windows, as it has high heat gain in the summer and high heat loss in the winter. Glazing innovation in the late 1960s and early 1970s resulted in the introduction of double glazing. Initially done as an internal secondary layer of glazing, this quickly became sealed units where two glass panes were sealed and a vacuum created between the glass to improve energy efficiency. Sealed unit double glazing has become the norm with early ratings of 2.5 to 3 W/m²K reducing as low as 1 to 1.5 W/m²K. Using solar reflective glass and adding inert gas to the void between glazing panels are examples of practices used to improve energy efficiency.

Improving the resistance of glass to impact damage and breakage has also advanced at a similar pace. Toughened or tempered glass has largely replaced GWP glazing. This material is created during the manufacturing process by heating and cooling the glass multiple times. Tempered glass is more resistant to impact but is not unbreakable and one of the characteristics of the glass is that it shatters into hundreds of small pieces, thus reducing exposure to large jagged edges. The addition of a layer of safety film to one side of the tempered glass has produced glazing which does not shatter easily but, when broken, it forms a panel of multiple small pieces contained in situ. Such glass in double glazing form is considered to be the 'norm' for modern glazed roofs and, as a consequence of this high specification, it is a costly material.

*A glazed entrance canopy to an office building with
one broken tempered and laminated glass panel. The
appropriate recommendation for this in the building
survey or TDD is for immediate replacement.*

The components which make up a glass roof include the supporting structure and the
covering, which includes any joints between the panels and acts as the insulation layer with
respect to the performance criteria of a roof.

Most glazed roofs are either inaccessible or dangerous to walk upon during the building
survey or TDD inspection. The surface of the glass is usually very smooth and polished; this
presents a slip hazard during wet or freezing conditions. Therefore, aside from interrogating the
building owner, property manager or tenant regarding reported problems, leakage or damage,
the roof inspection will have to be made from a vantage point. This in itself it potentially difficult
where the roof is at high level over an atrium, but the principle of the survey should always
remain the same, as this will always be a visual assessment, where safe access can be granted.

More 'traditional' glazed roofs inevitably incorporate a main roofing structure, which is often
quite slender to avoid dominating the glass. Slender structures with high tensile strength are
typically provided by steel and this is often used to support further aluminium boxed sections
in which the glass can be mounted. The joints between the glass panels require further sealing
to prevent water ingress and these need to be semi-flexible as well as resistant to solar damage.
Typically, the joints between the glazing are executed in structural silicon.

*A glass roof over and atrium connecting four separate
office buildings with a steel roof structure supporting
aluminium profiles and sealed unit double glazed roofing
panels (photo courtesy of Widnell Europe).*

*Externally it is possible to see no evidence of defects to the
glazing panels and the joints are in good visual condition.
Centrally positioned are four automatic power-assisted passive
smoke vents used to evacuate smoke from the atrium in the
event of a fire (photo courtesy of Widnell Europe).*

More architecturally sophisticated glazed roofing designs can also be embodied to make this a feature of the building. Such design includes the use of 'spider glass' where a series of stainless steel pads and fixings are placed through the glass and connected or supported by fine steel rods or cables. The joints between the glass panels require sealing with structural silicon or something similar as for more 'traditional' glazed roofs.

A visually striking glazed roof over the central core of an office building constructed using sealed unit double glazed panels connected with stainless steel fixings, tensioned with steel cables and joints. The fixings must be stainless steel to prevent corrosion and the potential weakness with this design is the hole executed through the sealed unit glazing panels; any damage to the seals may allow water ingress and any flexing or movement of the stainless steel fixings may cause cracking of the glass.

Fact File: GLASS

- Glass may be considered the odd one out of roofing materials as its use is often dictated by its aesthetic qualities as opposed to functional requirements.
- It is a high-cost material which can introduce high running costs to a building through increased levels of solar gain.
- Modern glazed roofs are likely to comprise sealed unit double glazing with toughened and laminated glass panels.
- Older glazed roofs may be single glazed with minimal or no safety glass.
- The typical life cycle may be estimated to be 45 years and factors affecting this are impact damage, worn or defective joints as well as the glazing becoming obsolete due to poor energy efficiency.

Espace Beaulieu (Brussels) by the architects Atelier d'Architecture de Genval, constructed in the early1990s, encompasses a large section of pitched roof glazing fixed to aluminium profiles and supported by a steel structure (photo courtesy of Widnell Europe).

The same roof from above, viewed from access gantry provided for maintenance inspection (photo courtesy of Widnell Europe).

A cracked internal glass panel to a roof containing sealed unit double glazing. The absence of condensation between the glass suggests that the crack does not pass completely through the panel. Such a defect internally may be a result of a defect during execution or possibly some flex or movement of the material but is unlikely to be impact damage (photo courtesy of Widnell Europe).

Parapet walls

Parapet walls are located at the interface between the facades of a building and the roof, historically they were present to high-quality commercial properties with pitched roofs. They were often constructed to obscure the visual appearance of the pitched roof to Georgian and Victorian buildings, creating aesthetically large, square, uniform areas of brickwork or stone, and giving the impression of a flat roof. On modern commercial buildings with pitched roofs parapet walls are used significantly less frequently. However, their use is more common with flat roofs and in both cases the principles of parapet construction are, in essence, the same.

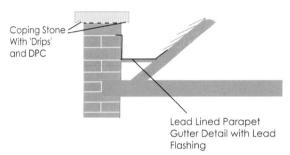

Coping Stone
With 'Drips'
and DPC

Lead Lined Parapet
Gutter Detail with Lead
Flashing

Parapet wall detail to a pitched roof.

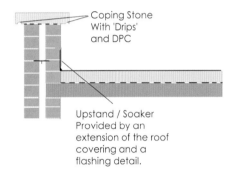

Coping Stone
With 'Drips'
and DPC

Upstand / Soaker
Provided by an
extension of the roof
covering and a
flashing detail.

Parapet wall detail to a flat roof.

Pre-1920s commercial property with a classic parapet wall detail visually concealing the pitched roof.

Parapet wall to the flat roof over a shopping centre executed in solid masonry (English bond) with bitumen mineral felt upstands and interlocking pre-cast concrete copings (photo courtesy of Widnell Europe).

Parapet walls are located to the external face or facade of a building and where these are present in the middle of flat roofs this is an indication that the building was extended or constructed in separate sections or as different buildings. This is often the case with industrial buildings where each 'shed' is effectively a separate fire zone or compartment and the external

A large industrial logistics site with several parapet walls in the middle of each of the roof areas due to the construction being completed in phases and at different historical times. Each parapet wall represents the line of an internal fire compartment to the building.

walls of these extend above roof level. This extra vertical separation, which also forms the parapet wall, is designed to prevent fire spread between the compartments at roof level.

When inspecting parapet walls during the building survey or TDD process it is important to analyse the following principal components:

- parapet structure
- lining/upstands/flashings
- copings.

Parapet structure

Irrespective of the facade materials, the external faces of the parapet walls are an extension of the facades themselves. These walls may be loadbearing for low-rise properties or non-loadbearing for medium- or high-rise buildings.

The parapet wall construction is often at the interface between the roof structure and facades where the roof structure or deck is connected to or imbedded into the facade. The obvious weakness at the junction of these two construction elements is the possibility for water infiltration.

When inspecting the structure of the parapet walls it is important to check the inside (roof side) and external face of this to ascertain whether there is any movement, damage or decay of the materials.

Linings/upstands/flashings

With flat roofs, the joint between the roof and parapet walls is typically waterproofed with an extension of the roof covering which rises vertically up the inside of the parapet wall; this is known as the 'upstand'. Alternatively, there may be more 'traditional' flashing and soaker details incorporated at this junction.

Concerning pitched roofs, the parapet wall is often used to create the rainwater drainage gutter, which also requires the relevant upstand and flashing details to provide waterproofing.

Parapet upstand waterproofing is provided by some of the materials also used for roof coverings, although these are not always exactly the same. These materials (as detailed earlier in this chapter) can be classed as oil-based or synthetic and metallic, to include the following:

- asphalt
- bitumen mineral felt
- EPDM
- PVC
- lead
- zinc
- copper
- aluminium/steel.

Concerning the life cycles and characteristics of each of these materials it is important to consider these to be the same as when used as roof coverings. However, concerning specifically asphalt or bitumen mineral felt, their vertical hot-bonded placement means that any significant solar exposure causes the materials to deform or slump in a downwards direction. The consequence of this may cause perforation or splitting of the lining, allowing potential water penetration.

Solar exposure has resulted in the hot bonded bitumen mineral felt parapet lining becoming de-bonded from the parapet wall (photo courtesy of Widnell Europe).

Right solar exposure to an asphalt upstand has resulted in blistering and slumping of the material.

Bitumen is quite a stiff and rigid material which becomes more flexible when heated and this is the principal reason why this is hot bonded and also why it can deform so significantly through solar exposure. When using bitumen as a lining or upstand, the 'correct' method is to install a 'wedge' in the corner between the parapet wall and roof covering. The purpose of this is to soften the angle created by extending the roof covering through 90 degrees into the vertical plain. Without the wedge, the 90-degree corner places stress on the bitumen which can lead to cracking or splitting of the material. This should be noted during the building survey as a possible point of defect or future infiltration. In reality, the use of the wedge is relatively rare as contractors seek to cut corners. It is also difficult to rectify this retrospectively. Therefore, unless there are defects or infiltration associated with this design or execution deficiency, it is difficult for the surveyor to justify its remedial rectification.

The use of EPDM or PVC to form upstands or parapet linings requires the material to be glued to the inside of the parapet walls. The principal observation with this is that insufficient adhesive is applied to the material, which results in air bubble or pockets between the parapet wall and the lining.

A roof renovation to a 1970s commercial building which includes a PVC single ply membrane extending to the parapet lining/upstand.

Note the wrinkles to the material caused by insufficient adhesion to the parapet wall.

The presence of air bubbles or wrinkling of PVC or EPDM linings is an execution detail and may reflect quality control issues during installation. In reality, it is difficult to repair in situ without cutting out the affected section and patch repairing, which could introduce additional joints to the lining of the surface and possible areas of future defect. Therefore, the presence of air bubbles should be noted as an area for general maintenance inspection. Clearly, areas where the material has sagged significantly need to be reported and possible patch repairs recommended but only if there is the potential for the sagged material to become damaged or torn if this is the subject of localised impact damage.

One common method of providing an upstand, lining and cover to parapet walls is to do this in one with roll-on materials such as bitumen mineral felt, PVC or EPDM. The process includes the horizontal fixing of the material to provide the roof covering and extending this vertically at 90 degrees to form the upstand then dressing this over the top of the parapet wall, which may or may not include a parapet coping. When this is done to dress over an existing coping, typically this is executed during a renovation. Dressing over the coping, as opposed to first removing the coping, dressing the wall then reinstating the coping, saves time and cost. However, this is technically a compromise which may be reflective of lower quality execution. Certainly, for bitumen mineral felt the inclusion of more than one 90 degree 'bend' in the material has the potential to cause this to become fatigued and splits may occur at the corner joints.

When inspecting such detailing it is important to analyse the horizontal and vertical fixing (hot bonded or glued) to establish whether there are any air bubbles or voids which could lead to long-term defects. Furthermore, it is necessary to inspect the outside face of the parapet wall to see how the material is secured as this is probably the most exposed edge. When dressing over parapet walls it is considered good technical practice to secure the outside face or edge of the material with a fixing profile. This is typically aluminium, screw fixed through the waterproof membrane to the parapet wall and, as this is in the vertical plane, there is little concern about rainwater collecting on the profile and penetrating the parapet as this should run off down the external face of the parapet wall. A by-product of rainwater run-off is the potential for high-level facade staining and this is one of the principal reasons why the best technical solution for parapet walls is to have a 'traditional' coping stone or sheet to alleviate this. Metallic external fixing profiles may be the subject of thermal expansion, which could potentially damage the dressing material and, if executed in steel, this may be subject to corrosion. In most cases aluminium is used to perform this function but insufficient or no fixing of the outside edge of the parapet dressing will inevitably make the material susceptible to wind lift.

EPDM dressing to a parapet wall. Note some air bubbles in the material, illustrating localised areas where there is insufficient adhesive. This appears relatively minor and should be the subject of routine maintenance inspection. In the top left of the photo is the external aluminium fixing profile (photo Courtesy of Widnell Europe).

Corroded steel parapet coping which has been dressed with EPDM. The exposed edges of the EPDM have been the subject of wind exposure and have perished. The exposed corroded steel has been the subject of corrosion and will continue to be so until this is treated (photo courtesy of Widnell Europe).

PVC used as a roof covering which extends to dress an existing low-level parapet wall.

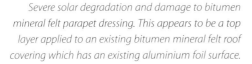

Severe solar degradation and damage to bitumen mineral felt parapet dressing. This appears to be a top layer applied to an existing bitumen mineral felt roof covering which has an existing aluminium foil surface.

Flashing details are another way of finishing the joint between the horizontal roof surface and vertical parapet wall. This still requires there to be a vertical element, which is typically bitumen mineral felt, PVC or EPDM for flat roofs.

The flashing detail to a parapet wall should be cut into the structure of the parapet wall and is typically executed in lead, although this may also be done with copper, zinc or aluminium. The flashing should allow for vertical cover or protection to the parapet wall to a minimum height of 150 mm above the adjacent roof covering. This height is recommended to prevent rainwater splashback from the roof landing on the surface of the parapet wall and potentially being absorbed into the masonry. Lead is a highly flexible material which is suitable for flashing details as it can be easily shaped and has a long life cycle. Where cut into an existing masonry parapet wall, this is usually done in a joint in the brickwork. The lead flashing is usually secured with cement-based mortar but this is sometimes done with flexible mastics or silicon-based products. With flashings secured by mortar there is a possibility of damage or wear to the joint caused by the lack of adhesion between the mortar and the surface of the lead. This may result in the lead flashing coming away from the parapet wall, allowing potential water penetration. When lead becomes fatigued or worn through exposure to the atmosphere it can become weak and typically tears or rips along the fold line. Mastic joints may also become defective but this is normally due to solar exposure causing these to become brittle and crack. In all cases it is important as part of the commercial building survey or TDD inspection to inspect the flashing details to note their condition and any evidence of defects.

Other parapet linings may include profiled steel, which is common with industrial buildings or logistics sites. Where this design is implemented there will still be a requirement for an upstand, which is often created with an extension of the roof covering, typically in bitumen mineral felt, PVC or EPDM. Profiled steel lining panels have some aesthetic value which is not always necessary for industrial flat roofs, since they are out of sight. However, the steel lining does offer good protection against solar degradation. One typical observation or defect with steel lining panels is that these are often cut to size on site and this may lead to cut edge corrosion.

A bitumen mineral felt roof covering and upstand with an aluminium screw fixed profile used to create a flashing; note the solar damage to the mastic used to seal and waterproof the joint (photo courtesy of Widnell Europe).

A lead flashing (approximately 40 years old) with an asphalt-covered roof and upstand; note that the presence of painted bitumen repair suggests there is a leakage problem.

A lead flashing cut and secured into a masonry parapet wall with cement-based mortar; note the cracking of the mortar joint requiring immediate repair.

Bitumen mineral felt flat roof covering and upstands with profiled steel lining panels.

Evidence of the onset of cut edge corrosion.

Parapet copings

The top of the parapet wall requires a coping, primarily to deflect rainwater off the surface of the wall and away from facades. Parapet copings come in an array of different materials and the most common are:

- natural stone
- concrete
- tiles
- zinc
- steel
- aluminium.

Irrespective of the type of material and in order to shed or deflect rainwater off the top of this and away from the facades, the coping should incorporate a degree of slope. It should overhang the facades sufficiently to ensure that rainwater runs or drips off away from the facades and this will require the provision of a 'drip'.

High specification buildings may be fitted with natural stone copings and low-cost versions may be precast concrete. Both types of copings require fixing and securing with the joints between sealed with either flexible mastic or cement-based mortar. The commercial building survey or TDD inspection will normally not be able to see and assess the fixings as these are typically concealed by the copings themselves. Because of their location on the building and the potentially severe consequences should they become loose, it is important to assess their stability. A sample or random test should be made along the top of the copings to establish if any are loose and also noting defective or missing joints. Such defects should be noted and reported as immediate cost recommendations. Medium- and long-term cost recommendations may be made if there is no evidence of defects but mastic joints are beginning to become fatigued or worn. There is also some justification to recommend (where there is an evident requirement) remedial cleaning of the surfaces to parapet copings if these have become soiled or stained. The reason for doing this is that external cleaning of facades and roofs should be considered as periodic maintenance. In many cases, roof areas and parapets are out of sight and, to an extent, out of mind, therefore these areas are often left to become soiled, stained and fatigued. By recommending remedial cleaning of these as part of the commercial building survey or TDD findings it proves that these areas have been inspected but also that these remedial works should not be at the cost of the new owner when they are derived from a lack of maintenance by the former; these costs may be negotiated out of the deal as part of the commercial negotiation. It is not the role or responsibility of the surveyor to perform such negotiation, but to give the options to their client to make this decision.

High specification parapet copings finished in natural stone with mastic seals to the joints (photo courtesy of Widnell Europe).

Interlocking concrete copings with defective and missing cement-based mortar pointing (photo courtesy of Widnell Europe).

Aluminium copings to a low-level parapet wall with screw fixings and defective mastic seals (photo courtesy of Widnell Europe).

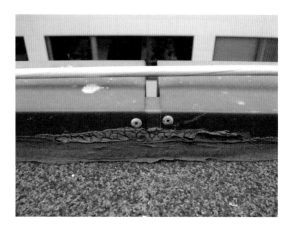

Buildings which are constructed with profiled metallic sandwich panels for facades usually have metallic copings which are screw fixed and the overlapping joints finished with mastic. This is typically the case with industrial buildings or shopping centres and it will be necessary to try to identify the coping material as well as establishing the effectiveness of the fixings and mastic joints. Metallic copings are likely to be either steel, aluminium or zinc and, regarding their characteristics, quality and life cycle, it is necessary to treat these the same as such materials used for roof coverings.

Typically, steel is subject to cut edge corrosion or localised corrosion around the fixing holes and aluminium is subject to impact or mechanical damage. Both materials are subject to wind lift or wind damage if they are poorly fixed, therefore it is necessary to randomly test these during the survey by physically lifting or pulling the material to establish any looseness.

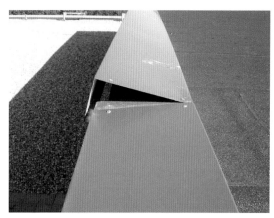

Peeling painted finish to a parapet coping reveals this to be galvanised steel. It should be recommended to repaint the coping to match the existing and also tighten existing screw fixings (photo courtesy of Widnell Europe).

Wind damage to powder-coated aluminium copings (photo courtesy of Widnell Europe).

Similar to the soiling of the parapet copings and subsequent recommendations to clean these, the typical perishing of mastic joints should be noted during the survey. If these are cracked, worn, open or missing, then the report should recommend immediate or short-term replacement (subject to their condition). Such recommendations are justifiably made as they are evidence based and should be considered as part of 'normal' routine property maintenance. Failure of metallic copings or their fixings and joints may lead to water infiltration; therefore, irrespective of the costs, which may appear relatively minor, these should be included in the findings and recommendations of the report.

Pre-cast reinforced concrete panel providing the internal fire separating wall to a large logistics building. The poor external parapet wall detail with damaged and defective steel coping (LEFT) has resulted in internal water infiltration (RIGHT) with vertical water streaks indicating this.

Copings constructed from other materials, such as zinc or tiles, are typically representative of higher quality buildings or those of more historical construction. Such materials are not presently widely used as they may be considered quite costly and, certainly with tiles, more labour intensive to execute. In essence, zinc has a good life cycle and is an attractive material; however, it is prone to fading in colour and weakening with exposure to atmospheric pollution.

Tiles are usually fixed and pointed with cement-based mortar and suffer typically from mechanical damage, including chipping or cracking. Mortar joints can become cracked or loose, requiring repointing, and in some cases tiles may also become loose. As with the inspection of all parapet walls, both zinc and tiled copings should also be randomly tested to verify that these are securely fixed.

The parapet wall to a bespoke university building constructed in the 1970s. The parapet wall detail is quite intricate with zinc coping and lead flashings to the deliberate space created in the parapet sections. Note the presence of bitumen-based paint, which is evidence to suggest historic or present water infiltration.

A combination of tiled copings and a parapet coping dressed with bitumen mineral felt. Note that the poor detailing at the junction between the bitumen and tiles is a potential source of weakness and future defect (photo courtesy of Widnell Europe).

Rainwater evacuation

Water penetration into a building has the potential to render tenantable space unlettable, as well as causing short-term damage to the internal fabric, finishes or tenants' possessions. Allied to the rendering of tenant space unfit for occupation, water can also cause longer term defects to the structure and facade. Therefore, it is important that the evacuation of rainwater is treated as a priority in both the design and the audit of an existing building.

One of the performance criteria for roofs is for them to be weathertight so, as well as utilising materials which are impervious to water penetration, it is important to ensure that rainwater is evacuated effectively from the roof surface.

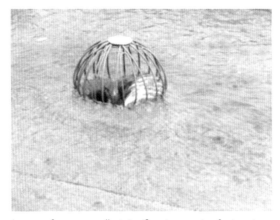

Large surface areas collect significant amounts of rainwater, which necessitates evacuation as a routine requirement (photo courtesy of Widnell Europe).

Pitched roof evacuation

The very nature of pitched roofs is that they will shed rain that lands on the surface. However, in combination with the actual pitch of the roof, it is important to collect, channel and remove rainwater water to be evacuated away from buildings. Therefore, pitched roofs generally encompass either external gutters or more intricate parapet wall drainage channels. Both of these methods of collecting and channelling rainwater require connecting to vertical downpipes to complete the task.

Parapet gutters to commercial pitched roofs are likely to be associated with older historic properties or modern properties located in town centres where there may be conservation areas requiring this type of design implementation. As previously discussed in this chapter regarding parapet walls, the location of parapet drainage gutters is at the interface between the facade and roof. Where these provide specific drainage function it is critical that the parapet is lined with a waterproof membrane. In many cases the material of choice is lead as this is very durable as well as being sufficiently malleable to form into the channel. The sub-base to the parapet gutter is usually constructed out of timber, with marine grade plywood often used. Other materials may be used to line the parapet gutter, such as zinc or bitumen mineral felt, PVC and EPDM. These are considered to have more limited life cycles but are more cost effective.

In order to evacuate and drain rainwater from the parapet channel, it is necessary to connect this to a rainwater downpipe. This is achieved by ensuring there are holes in the channel lined with pipes or lead connected to the downpipes fixed to the external face of the parapet wall and facade. This arrangement is known as a parapet 'hopper' and the top of the vertical downpipe has a splayed opening known as a 'hopper head'.

A parapet wall with parapet drainage to a shopping centre with a pitched roof.

Evidence of white surface staining to the brickwork is efflorescence, which may suggest the brickwork to the parapet is damp and is possibly caused by a defective or blocked parapet drainage channel. Note the coping or top of the parapet wall appears to be finished in lead or zinc.

Flat roof evacuation

Rainwater evacuation to flat roofs is either at the perimeter or central to the roof surface. On smaller flat roofs is it often the case that these are located at the roof edge, while the nature of large roofs, which sometimes have vast surface areas, means there are often central outlets. In some cases, there is a combination of perimeter and central rainwater outlets.

While there may be external gutters fixed to the perimeter of smaller flat roofs, most commercial properties utilise parapet walls with integrated drainage to remove surface water.

The principal requirements of a parapet drainage channel to a flat roof are not dissimilar to those of a pitched roof. However, while with pitched roofs it is necessary to provide the structure of the channel and waterproof finish (often in lead), flat roofs already have the parapet structure. In essence, by constructing a parapet wall to the roof edge this is already containing and channelling rainwater. Importantly, the flat roof should provide sufficient pitch or gradient to propagate the flow of rainwater towards the parapet. There should be sufficient drainage outlets to cope with the surface water and the waterproof lining of the channel is provided by the roof covering, which often extends to form the upstands. One important consideration is that the drainage outlets should be recessed and lower than the roof covering and failure to execute this detail will lead to localised collection of water ('ponding').

Central drainage to a large flat roof over an industrial building. The black 'spot' at the top centre of the image is the drainage outlet and insufficient pitch has resulted in ponding occurring in front of this (photo courtesy of Widnell Europe).

A parapet perimeter drainage outlet with 'wire balloon' grill, which is partially blocked with debris. Note the overflow 'slot' fitted to the parapet wall which indicates that this is a well-designed and executed drainage installation. Both drainage outlets drain internally in the building (photo courtesy of Widnell Europe).

A cast iron parapet hopper and hopper head on a telephone exchange building. Note the date on the hopper (1935), which appears to be in remarkable condition for its age. The naturally suspicious surveyor will inevitably question if this has been renovated or replaced to match the 'original'.

The drainage of parapet channels to flat roofs may be with external hoppers and hopper heads connected to vertical facade-mounted downpipes or internally within the building.

Rainwater evacuation outlets located centrally on a roof surface need to be positioned at the bottom of sloping sections and these also need to be recessed. Due to the location of the drainage outlets, the positioning of drainage pipework will be internally in the building.

Inspecting and reporting drainage provision

The RICS guidance note concerning commercial building surveys and TDD (RICS, 2011b) suggests that rainwater evacuation systems should be the subject of inspection as closely as possible or from vantage points. However, the very nature of parapet channels to pitched roofs means that these are often inaccessible areas of the roof, making this difficult to achieve. Therefore, in order to establish an evidence-based opinion concerning the condition and functioning of parapet drainage or gutters to pitched roofs, it is important to carry out the inspection with the following objectives:

- from roof lights or dormer windows, identify any evidence of blockage or standing water within the parapet channel
- internally at eaves level, seek to identify any evidence of dampness or water penetration to the internal walls, typically at ceiling height
- externally from ground level or adjacent vantage points and with optical equipment such as camera zoom lenses, seek to identify visual evidence of dampness to external brickwork or render
- visually analyse the condition of the hopper, hopper head and, specifically, the connection with the rainwater downpipe
- potentially recommend further investigation with the use of a scissor lift or cherry picker if there is evidence to suggest blockage or leakage. Alternatively, the use of drone technology allows for remote aerial inspection of inaccessible parapet drainage channels.

Concerning the inspection of parapet or central drainage outlets to flat roofs and assuming access is granted to the roof itself, then it should be relatively straightforward to visually assess these. However, with central drainage or even parapet drainage where rainwater is evacuated internally in the building, it is very difficult to assess this during the building survey or TDD visit. Such installations typically transfer rainwater via pipework located in the ceiling void under the roof to the central core where vertical pipework is positioned in shafts. For industrial buildings or sheds there are normally no internal suspended ceilings and in these cases it is possible to verify the rainwater evacuation systems both internally and externally.

In contrast, pitched roofs can be fitted with external rainwater gutters and downpipes and while these are more visible externally from ground floor level it is still difficult to establish if these are free from blockage without actually looking into the gutter.

Drainage gutters are secured to the external building facade with adjustable brackets either screw fixed to the facade cladding or to fascia boards, which are used to finish the eaves detail of the roof. Aside from blockage, defective drainage provision occurs when there is insufficient gradient to evacuate the water or damaged joints between the gutter components or connections with the downpipes.

Rainwater outlets are either gravity or syphonic type and, according to the RICS guidance note for commercial building surveys and TDD, the surveyor is expected to be able to establish what type is present on a roof. Gravity type outlets essentially comprise an open hole with a downpipe connection. A syphonic type has an inner shield known as an 'air baffle' which reduces the amount of air entering the drainage and improves the efficiency of evacuation.

A syphonic rainwater outlet with protective plastic grill to a roof finished with PVC. Note the build-up of debris which requires immediate removal.

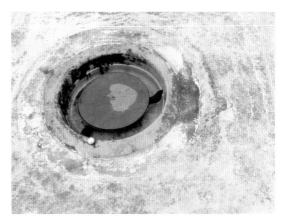

A missing protective grill reveals the inner workings of the syphonic drainage outlet with air baffle. Note the cracking to the inner plastic housing of the outlet, meaning that this requires immediate replacement.

An important requirement for all rainwater drainage outlets is for these to be fitted with protective grills. These may be manufactured from a variety of materials, which often match that of the gutters, including PVC or cast iron. There are also protective grills that can be referred to as 'wire balloons' or 'birdcages', which relates to their distinctive appearance and in some cases reflects the fact that these are executed with steel. Similar to the way an overflow pipe provides an active alert that there is a problem with a defective flow valve, blockage of a rainwater outlet grill will result in localised ponding or overflow of the gutter. This inevitably requires maintenance intervention and, when visiting site, the condition of these, including the levels of built-up debris, provide an indication of maintenance regimes at a property. The alternative to periodic maintenance of drainage outlets are blockages within the drainage pipework above or below ground level. Dead birds, tennis balls, drinks cans and other benign objects can find their way into drainage outlets. If these migrate deep into the pipework, the remedial works can be significantly more costly and inconvenient in comparison to periodic cleaning of outlet grills. For this reason it should be considered good practice during any building survey or TDD inspection to recommend the reinstatement of protective grills if these are found to be missing. Likewise, higher specification drainage pipework is often fitted with inspection covers or 'rodding eyes' to allow for the removal of blockages or as an insertion point for CCTV equipment if a drainage survey is required.

A painted zinc box-shaped rainwater downpipe with a section cut out due to a blockage (photo courtesy of Widnell Europe).

Modern cast iron downpipe with a screw fixed inspection hatch/rodding eye.

Drainage gutters and rainwater downpipes

Externally fixed drainage gutters and rainwater downpipes may be manufactured from the following materials:

- cast iron
- steel
- copper
- zinc
- aluminium
- PVC
- high density polyethylene (HDPE)
- fibre cement (asbestos).

As with the use of these materials for roof coverings, it is noticeable that these comprise those manufactured from metals and those which are synthetic or oil-based.

Cast iron may considered as a material offering one of the longest life cycles, which is typically 50 years on average. It is prone to long-term corrosion issues and, as this is a high-quality material, the cost is also high. However, it is also quite a brittle material and, when subject to impact damage, may crack or fracture. Its application to commercial buildings may be typically reserved for those which are listed or in a conservation area as all of the other materials offer a viable solution for replacement, although they lack cast iron's aesthetic appeal.

Likewise, copper and zinc may be considered high-quality materials for rainwater evacuation with high aesthetic value. Copper is rarely used for the evacuation of rainwater to commercial buildings and this is more likely to be the case with a listed building or one in a conservation area. It should also be noted that the rainwater run-off from copper pipework can be corrosive if this comes into contact with zinc. As a material used for rainwater evacuation downpipes, zinc is relatively common and, as with the use of zinc for roof coverings, this can be the subject of corrosion due to acidic or alkaline attack. Therefore, careful consideration should be given to the locating of this with respect to potential corrosive hazards, such as cooling towers or boiler flues.

Aluminium downpipes are relatively rarely used and one characteristic of their use is that these will be powder coated or painted to match the design of aluminium clad facades or roof coverings. They may also be square or box-shaped as well as the more traditional round types and, as alluded to in the section detailing roof coverings, these are difficult to distinguish from steel.

Steel downpipes should always be galvanised to resist the obvious threat of corrosion and, where used for commercial properties, these are usually powder coated or painted. It is not unusual to come across both steel and aluminium drainage downpipes with life cycles of 40 years plus, which makes them a cost-effective alternative to cast iron.

Downpipes constructed and executed on site in PVC are probably the most common to commercial buildings. These can be in a number of different colours, including white, grey, brown or black. Grey is probably the most popular colour used. PVC pipework also comes in an orangey-brown colour; however, this is a material which should be used for underground evacuation. Its use above ground and any consequent solar exposure can cause the material to weaken and become brittle. When this is observed during a commercial building survey or TDD, special attention should be paid to its inspection as well as making an allowance for its replacement in the cost planning (dependent upon the condition).

HDPE is a relatively 'modern' material used for rainwater evacuation and is easily identifiable as it has a matt black finish. Manufacturers are usually keen to champion their products and it's possible to find trade literature professing the life cycle of HDPE pipework to be 50 years. If this is the case, then this should outlast PVC and be more in line with cast iron, at a significantly lower cost.

The presence of fibre cement drainage pipes is relatively rare and, although this material can be moulded into a variety of different shapes or sizes, it is quite brittle. Its presence should be treated with a degree of caution as this is likely to contain asbestos. Where damaged, this is typically due to impact damage resulting in cracked or fractured pipes. It is easily recognisable as it is grey in colour and discolours with exposure to the elements. Concerning fibre cement pipes suspected of containing asbestos, these do appear to attract moss growth in a similar manner to asbestos roof tiles.

The inspection of rainwater downpipes should always seek to assess if there are sufficient drainage outlets and this is usually obvious when doing an inspection during wet conditions. Again, it is beneficial to ask the property owner, manager or tenant to verify if there are any reported leakages or inadequacies. While it is not the role of the building surveyor or TDD auditor to analyse or critique the design specification of a building, the location and quantity of rainwater outlets is important due to the defects associated with any deficiencies. As with most aspects of the audit process, a comparison between the actual situation and the relevant building code or regulation is always a good benchmark. However, it should be acknowledged that in most cases there is little or no obligation for retrospective enforcement of building codes or regulations, unless the building is the subject of significant renovation.

While any deficiencies should be reported to the client, it may be difficult to negotiate the cost of any associated remedial works with the building owner in the absence of any evidence that a deficiency is causing a defect. One of the principles of a commercial building survey or TDD audit is to report visible defects, therefore if it is not broken, then it is difficult to recommend repair. This, however, does not absolve the surveyor or TDD consultant from the responsibility of highlighting any area of potential defect to their client.

Having confirmed that there are sufficient drainage outlets to a roof, it is then important to establish the presence of visual evidence to suggest whether these are functional. Standing water in drainage gutters or on flat roofs may be a direct result of blocked downpipes. It may also be the result of insufficient pitch or gradient to these. It is normally possible to ascertain the difference between blocked gutters or downpipes or ponding on a roof due to poor gradient. Standing water in a gutter can usually be traced back to standing water at the top of the downpipe and, if there is no water in the downpipe but water in the gutter, this suggests localised poor gradient.

Standing water or ponding is almost inevitable to most flat roofs and while the definition of a flat roof gives this a maximum pitch of 10 degrees with a minimum of 2.5 degrees this is often difficult to execute on large flat roofs. Ponding can be easily identified given even the briefest spell of wet weather during a building survey as rainwater often settles quite quickly in the affected areas. During warm dry weather it is more difficult to assess if a flat roof may be susceptible to ponding as these large surface areas can be subject to high heat gain causing rapid evaporation. However, rainwater run-off often washes down surface dust, dirt or silt into the puddles. Following evaporation of the water, dried out surface crusting of the residual silt indicates the extent of the ponding. This should be noted during the survey and in the report as evidence suggesting the presence of ponding with any conclusions as to the effects of this established during the internal inspection or discussion with the owner, property manager or tenant.

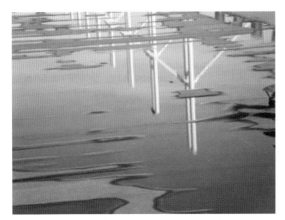 

The presence of standing water or ponding may be obvious during the survey inspection (photo courtesy of Widnell Europe).

Alternatively, it may be necessary to establish the evidence that ponding is likely to occur when wet as indicated by the dry, dusty collection on the roof (photo courtesy of Widnell Europe).

The presence of surface water can exacerbate the wearing of flat roof materials and will remain a source of potential infiltration but there is little in the way of low-cost remedial works that can be recommended as part of the survey or TDD. In essence, it is important to report evidence of ponding and, if there are no traces or reports of internal leakage, this should be monitored. However, where there is excessive localised ponding this can cause point loads to flat roofs and if this occurs on the joins between flexible decks, such as steel or aluminium, there is a potential for localised deformation or collapse of the roof deck to occur. Therefore, it is important to analyse ponding in the terms of its severity and location with respect to the construction detail of the roof deck.

Defective rainwater evacuation may occur as a result of blocked downpipes but attention should also be given to the joints between sections of pipework or the physical fixing of pipes to the external facades. Typically, corroded screw fixings, wind damage or vandalism can result in vertical pipework becoming misaligned, allowing escape of water from the joints. This is often the case with the joint between the hopper head and attached vertical downpipe, while this may be difficult to confirm from ground level or at a distance, there are some typical symptoms which may indicate this. Escaping water may infiltrate the building, in which case the owner, property manager or tenant may disclose this, or there may be visual evidence of this. Therefore, it is important to seek to establish the presence and location of vertical drainpipes and their corresponding internal position to see if there is evidence of localised dampness or staining. In most cases, the external facades of a commercial building incorporate two skins or leaves of cladding and, accordingly, it is unlikely for an external leakage of rainwater downpipes to penetrate internally due to the presence of a cavity. Therefore, external staining to facades with streaks of water or dampness running down the external surface may indicate defective external drainage downpipes. Dampness in brickwork or stone cladding often results in white staining or efflorescence to the surface of the material or, where there are mortar joints, the dampness may propagate the localised growth of moss or algae. Where possible, optical equipment such as the zoom lens of a camera should be used to investigate further and establish the presence of damaged or misaligned joints.

Vertical water staining caused by leakage of a rainwater downpipe encased in the facade (photo courtesy of Widnell Europe).

The level of humidity can suggest whether this is a 'live' or 'historic' problem (photo courtesy of Widnell Europe).

Impact damage to a steel rainwater pipe has caused this to deform and become unfit for purpose (photo courtesy of Widnell Europe).

Mechanical damage of rainwater downpipes due to vandalism or the nature and operation of the site inevitably causes rupture or blockage to occur. While cast iron downpipes are relatively brittle and crack when impacted, most other metals are quite robust and are likely to bend or deform. PVC and HDPE are oil-based plastics, which means that they are likely to crack, split or puncture when impacted.

A hybrid design between a parapet drainage channel and external guttering is a cornice drainage detail. This is typically something which may be present on older commercial buildings located in cities or town centres. This type of drainage is mostly evident with pitched roof construction and the cornice detail is often quite ornate, forming an architectural feature on the facade.

A cornice detail to a hotel which could be inspected via a 'Velux' style window. Note that it is possible to establish that this is a timber structure lined with zinc. Accumulated debris in the drainage channel and coaxial cable is not acceptable and requires addressing immediately.

Instead of a facade fixed gutter, the junction between the facade and roof edge is finished with a protruding cornice detail. This is often executed in timber or masonry to create an open box detail or channel which is lined with a waterproof layer. The lining is often lead but zinc is also quite common, as is the use of oil-based products, such as bitumen mineral felt or PVC. As with the inspection of parapet drainage channels to pitched roofs, these are normally quite inaccessible but can be viewed from dormer windows or roof lights. Likewise, it is important to view these from below at ground level to seek to identify any evidence of rainwater staining or localised decay. In most cases, the drainage of cornice channels is external and it will be possible to inspect the external rainwater downpipes, as with 'traditional' gutter systems.

Roof lights

The purpose of roof lights is to provide:
- natural light
- ventilation
- smoke evacuation.

Forming an integral part of the roof covering, it is important to consider these as part of the roof and apply the performance criteria accordingly. Concerning their strength and stability, it is important to assess how these have been constructed and if there are any signs of movement or deflection of the roof light structure.

The durability of the roof light will be linked to the type and quality of the materials used. In most cases these are finished with glass or glazing panels but low-cost alternatives may be polycarbonate or fibreglass. The condition of the material and the associated life cycle will determine whether this is watertight but additional to this is the detailing around the housing of the roof light, such as the upstands and flashing details.

Similar to the use of glazing as a roof covering, it is important to consider the thermal efficiency of a roof light and any impact that this may have on solar gain or heat loss to the occupied space below.

Roof lights comprise the following principal elements:
- structure and frame or housing
- covering
- flashings and upstands
- opening mechanisms.

As all roof lights effectively form an opening in the surface or the roof covering, it is necessary to inspect these both internally and externally during the site visit. As with all other components forming part of the roof inspection, these may be in areas that are difficult to access safely. Therefore, it may be necessary to inspect these from a distance with the use of optical equipment. It is certainly beneficial to establish through discussion with the owner, property manager or occupier any reported defects or leakages from these.

Pitched roofs and roof lights

Roof lights which are situated on pitched roofs are integrated into the roof covering and are normally placed to give natural light and the possibility of ventilation to the roof space. This usually happens when the roof area has been renovated or designed for tenant occupation as lettable floor space. It is relatively rare to find the presence of roof lights to pitched roofs where the space is solely used as storage, archives or technical rooms.

Surveyors often refer to Velux roof lights and this a trade or brand name of one of the leading manufacturers of these. There are other manufacturers of similar roof lights; however, it is not uncommon for surveyors to refer to these as 'Velux type' roof lights. In essence, the types of roof light have a frame, which can also be referred to as the housing, that forms the structure of the roof light. The frame and internal sub-frame supporting the glazing panel are often made of timber but can also be aluminium or other materials; this is usually centrally fixed with a pivot hinge. The hinge allows the roof light to be opened and for it to rotate around the hinge to about 120 degrees, which can allow for cleaning of the external glazing from inside the roof space.

Things to check from the inside are the opening mechanism, which is usually manual, and having opened the roof light it is a good opportunity check the flashing details with the roof externally. Failure of the flashings or upstands will usually result in water infiltration and it is important to check this both visually and with a moisture meter.

Aluminium framed, 'Velux' style sealed unit double glazed roof lights. The presence of external 'flashbands' to the top of the frames indicates the likelihood of water infiltration (photos courtesy of Widnell Europe).

Internal inspection has identified vertical water streaks and bubbles to the internal painted wallpaper finishes around the roof light. When tested with a moisture meter, severe levels of humidity were noted, indicating a 'live' and current problem (photos courtesy of Widnell Europe).

In some locations, roof lights may be in particularly high roof voids and operable with an extending pole or even power-assisted and mechanical. In these cases, it may be difficult to inspect these externally therefore it is important to seek visual evidence internally to suggest failure or leakage and ask the owner, property manager or occupier to sample test these. The limitations of the inspection should be noted, as should the findings in the building survey or TDD report.

In some cases, where roof lights are placed in areas of high occupation and with limited internal climate control or forced air ventilation, these may be susceptible to condensation. The internal surface of the glass is likely to be relatively colder than the surrounding ceiling finishes and the creation of warm moist air, which is typical of dense occupation, may lead to this condensing on the internal surface of the glazing resulting in localised damage of the timber frames.

'Velux' style roof lights inserted in a roof with a zinc covering forming internal meeting rooms (photo courtesy of Widnell Europe).

Internal condensation has caused some localised water damage to the internal timber frames (photo courtesy of Widnell Europe).

Condensation is likely to be exacerbated where there is the presence of single glazing to the roof lights and the use of this specification of glazing is more typical of older or historic properties. Modern, post-1980s properties are likely to encompass sealed unit double glazing panels within their roof lights.

As with glazed roof coverings that utilise sealed unit double glazing panels, these can be susceptible to impact damage or defective seals around the panes of glass, which can lead to a build-up of condensation between the glazing. When this occurs the only suitable remedy is replacement of the glazing.

Flat roofs and roof lights

When positioned on flat roofs, roof lights essentially perform similar functions to those on pitched roofs. In office areas, roof lights are mostly used to introduce natural light to areas which may have little or no glazing. Typically, they can be used to perform a similar function to glazed roofs to atriums or lightwells in the centre of buildings to facilitate natural lighting. The obvious problem with roof lights to office areas, and certainly if these are large, is the introduction of solar gain in the summer or heat loss in the winter.

A feature of office buildings, hotels and residential housing blocks is the requirement to provide internal emergency staircases. The numbers and specification of these will depend on the height or size of the building but one important feature of these is the requirement for there to be a roof light situated at roof level above the staircase. Certainly, at the highest level this provides natural light, but the primary function of this installation is to provide a means of ventilation in the event of fire and smoke entering the stairwell. These can be identified typically on the roof plan but the principal characteristic of these will be that they are relatively small, typically measuring 1 m by 1 m. As with the Velux style roof lights, they will have a frame and housing which may be manufactured in timber or aluminium and the actual covering of the roof light is often executed in a double skin of polycarbonate.

*Fractured double-skinned polycarbonate smoke vent/
roof light with bitumen mineral felt upstands. Internal
smoke vent to an emergency staircase (photo courtesy
of Widnell Europe).*

*Roof light with power-assisted opening mechanism
(photo courtesy of Widnell Europe).*

The important requirement of a roof light used for smoke evacuation in a staircase is that
this is functional in the event of a fire. In some cases, the smoke vent may be automatic and is
connected and commanded by the central fire control panel, which opens the vent when there
is a fire alarm. Alternatively, the smoke vent may have a manual control, with this often being in
the staircase on the ground floor. During the commercial building survey or TDD visit it is often
difficult to test smoke vents in the staircase, therefore it is important to seek confirmation from
the owner, property manager or occupier that these are fully functional. If necessary, it may be
advisable to obtain test reports or a certificate confirming serviceability.

Roof lights to industrial buildings or large surface areas, such as shopping centres, are
an entirely different proposition to those over office buildings. They can be used to provide
ventilation but their basic function is typically the provision of natural light to buildings with very
large surface areas, often over one or two storeys. However, the most vital function is to provide
smoke evacuation in the event of a fire.

Fire in large, high-density buildings, such as shopping centres, poses a significant risk to
life and there is normally a high degree of fire engineering provision incorporated within these
buildings. This typically includes high levels of fire and smoke detection to ensure a rapid raising
of the alarm to instigate evacuation of the building. There is also likely to be the provision of
sprinkler systems to provide immediate and localised firefighting. Fire detectors should be
installed in the apex, at the top of the rooflight, as this is the highest point where smoke will
travel.

The roof lights to shopping centres are often strategically placed over the central common
parts as well as open staircases or over escalators. Structurally, these are significantly more robust
than those used for offices or over internal emergency escape stairs. They are often constructed
using a steel or concrete structure which supports aluminium profiles and glazed panels or
polycarbonate. Externally, as these are located on large flat roofs areas, access is usually quite
good and the inspection should seek to identify any defects which may be present with the
external vertical housing and coverings. The housings may be executed in panels, glazing or
even masonry and the construction detail and analysis should seek to treat these almost as
parapet walls. The linings or upstands may be executed in an extension of the roof coverings,
with flashing details evident where necessary. The roof light covering should be examined
in much the same way as that for glazed roofs and one highly relevant piece of information
required is proof of the serviceability and functionality of the smoke extraction facility. It is not

the responsibility of the surveyor to undertake a fire audit of the property as part of the building survey unless otherwise instructed and qualified to do so. Therefore, significant time should be spent assessing and analysing any available information relating to the conformity of the smoke evacuation with the relevant regulations.

*Aluminium framed, double-skinned polycarbonate roof light with mechanical smoke extractors mounted on the roof (**LEFT**). Internally (**RIGHT**) the roof light is a series of features introducing natural light over the internal escalators to a shopping centre (photos courtesy of Widnell Europe).*

Smoke evacuation to industrial buildings should be assessed in the same manner as that to shopping centres. The occupation and operation of these buildings may be generic, in terms of some logistics buildings, or highly bespoke and specialised, with regard to factories or manufacturing. The role of the surveyor is to liaise with the legal advisors or legal due diligence to establish whether there are any specific operating permits or licences relating to the buildings. Part of this provision is likely to include fire engineering and the provision of smoke evacuation.

The building survey should seek to identify visible defects associated with the smoke evacuation provision, and this should include external inspection of the housing, upstands and coverings. Internally, it is usually difficult to inspect in detail the underside of smoke evacuation vents, as these are usually 9 to 11 m above the internal floor levels. Furthermore, the internal lighting levels in industrial buildings are often quite poor, meaning that defects are not always visually obvious.

There is a distinct difference between roof lights and smoke vents to industrial buildings, and where roof lights have a primary function of allowing natural light into the building this is not necessarily the same requirement as for smoke vents. In line with the general low-cost, low-quality concept adopted in many industrial buildings, the roof lights are never normally executed in glass. Typically, these utilise fibreglass or double-skinned polycarbonate panels as the covering and are often very large compared to roof lights to other building sectors: they can extend for many metres in length but are normally relatively narrow. As a consequence of their size, they are almost like small, independent roofs and therefore should be treated as such when undertaking the inspection.

Multiple large 'barrel' roof lights executed in fibreglass in excess of 20 m in length (photo courtesy of Widnell Europe).

Defective solar-damaged bitumen mineral felt upstands.

In order to shed rainwater, most will encompass a pitch and often these are curved and can be described as 'barrel roof lights'. In order to facilitate the curved covering of the roof lights, these are often finished with fibreglass or polycarbonate panels, as these materials are flexible and can be pre-manufactured off site. The covering is usually screw fixed to the frame, which is often constructed from aluminium with supporting steel or concrete structure incorporated into the actual roof structure. It is important to check the condition of the fixings externally as these are often prone to corrosion and if they become defective or work loose then there is potential for serious wind lift or damage of the roof light covering. Externally, it is important to check the upstands around the housing of the roof light as these are often an extension of the roof covering and a natural place where leakage could occur. Considering that the coverings to the majority of industrial buildings are likely to be bitumen mineral felt, PVC or EPDM, the same theory and life cycle analysis applies when inspecting roof light flashings as for roof coverings and parapet linings.

Although cost effective, both fibreglass and polycarbonate are prone to weathering and damage from both the sun and cold temperatures. This process can take many years but often results in exposed fibres to fibreglass with the dulling of the surface, which often reduces the amount of natural light entering the building. The fibreglass also becomes weakened and brittle, which may result in cracking of the panels and subsequent exposure to wind lift. Polycarbonate is double skinned and weathering to the external layer can make this brittle and also prone to puncture. As a result, rainwater can gain ingress and condensation can occur between the skins and lead to possible water infiltration.

In both cases it is down to the surveyor to visually assess the condition in relation to any evidence of defects and determine if the roof light is fit for purpose. This should then provide justification for recommending the relevant replacement of the covering and/or the renewal of the flashings accordingly.

Solar degradation soiling of a fibreglass roof light.

Internally there is evidence of minimal admission of daylight meaning this is not fit for purpose, requiring immediate replacement.

Multiple perforations to the external skin of a double-skinned polycarbonate roof light requiring immediate replacement.

While some smoke vents may also perform the function of roof lights, it is very important to acknowledge the primary function of these. Ultimately, the provision of smoke vents to the roofs over industrial buildings is to provide the facility for smoke to exit the building in the event of a fire. Considering the fact that industrial buildings can be used to manufacture or store some highly combustible and dangerous materials, the fire engineering in the design will be governed by the relevant regulations.

This is a typical legal/technical issue regarding a commercial building survey or TDD and it is important to acknowledge that the majority of building surveyors are not fire engineering experts. Therefore, it is important for surveyors (unless appropriately qualified) not to be drawn into confirming compliance of the building and smoke evacuation systems with the legal prescriptions. The most appropriate course of action is to undertake a document review to establish the conclusions or findings of experts who have designed, constructed or accepted the building. In some cases it may be possible to contact the architect or those involved in the design and execution process to provide evidence or to verify that the building is compliant. If there is any doubt concerning fire safety then it is appropriate to advise on the appointment of a specialist to verify this. However, this does not absolve the surveyor of the responsibility to note the number and size of smoke vents on a roof, as well as establishing from the owner, property manager or occupier confirmation of their serviceability.

Some mechanical defects with smoke vents may be obvious during this inspection and these, as well as any omissions found in the supporting document review, should be brought immediately to the attention of the owner, property manager or occupier as a life safety issue.

The actual smoke vents themselves are prefabricated and brought to site to be installed as designed on to the roof surface. Their housings may be steel or aluminium and it is important to check for any visual evidence of defects to the fixings or opening mechanisms. As these are located on the flat roofs it is often quite straightforward to access and inspect these. On some very large roofs there may be many smoke vents and, due to the time constraints placed upon the visit, it is necessary to randomly select roof lights to perform some sample analysis. This should be clearly stated and detailed within the building survey report as should a statement verifying that no tests were performed on the functioning of the roof lights.

A smoke vent to the roof over an industrial building which appears to have been installed retrospectively with flashing and upstand details executed in foil-finished bitumen mineral felt roofing.

A defective smoke vent which has external tape applied to prevent this opening with the wind and allowing water infiltration. The obvious concern is that this will fail to open in the event of a fire and this is a life/safety issue requiring immediate action.

Other

Within the inspection and reporting process dealing with roofs there is a slightly ambiguous section known as 'other'. In essence, this deals with the inspection and reporting on all other observations or features associated with roofs but not already covered by coverings, parapet walls, rainwater evacuation or roof lights.

Importantly 'other' items associated with the roofs are items or components which have a direct impact on the function or operation of the roof and can typically be categorised as:

* access
* lifelines
* technical equipment
* lightening conductors
* window cleaning installations
* ducting, flues and chimneys.

Access

Safe access to the roof should be an obligatory requirement when advising during a commercial building survey or TDD. The actual process of performing the site visit will usually identify the current access provision and this should essentially be the same regarding maintenance inspection or repairs.

Roof access is either internal, external or a combination of both with staircases or ladders providing this. Typically, large industrial buildings have external access provision with ladders screw fixed to the facades, although these can be over 10 m in height and should ideally encompass a landing at the mid-point to 'break' the climb. Once on the roof, access to other raised roofs or over parapet walls is often provided by smaller ladders or access gantries. In the absence of safe access to the roof it is almost impossible to undertake a building survey or TDD as the roof is one of the most important elements to inspect. It is therefore important to state any access requirements or assumptions in the contract instruction to ensure that provision is made, if required, to hire and use an alternative means of access. It is important to understand that most industrial buildings are limited in height to one or two storeys. However, they have quite high storey heights but are nevertheless normally considered as low-rise properties. This is why external caged ladders are a principal method of providing access but this is not an option for medium- or high-rise buildings where access ladders are not viable.

Office buildings, shopping centres or residential towers usually have access provided by internal staircases which go beyond the last occupied level to the roof. In some instances, there is ladder provision from the top floor of the staircase through a roof light to the roof.

Concerning access to pitched roofs, this is usually very limited or restricted to roof light access from the roof space. Pitched roofs are significantly more dangerous to access than flat roofs and one obvious access point may be along existing parapet gutters where the parapet wall provides a safety barrier.

If no safe access can be provided during the building survey it is important to state this in the report and limitations as well as informing the client. To an extent, it is accepted that pitched roofs are inaccessible; however, where there is no safe access to flat roofs it appears appropriate to make the necessary recommendation in the survey or TDD report. It would also appear difficult for such recommendations to be negotiated out of the report as this is a life safety issue.

Lifelines

Equally important as the provision of roof access is the presence of lifelines or safety fixings to the roof. It can be argued that it is the responsibility of operatives working on the roof or building to provide their own personal protective equipment (PPE); however, it is not their responsibility to provide the necessary fixing points or lifelines. This life safety issue is a non-negotiable area and it appears ironic that it is often the building survey or TDD inspection (without the provision of lifelines) that identifies this requirement. Therefore, it is up to individual surveyors to prepare the relevant risk assessments before deciding to access a roof without lifelines or safety fixings.

The 'safest' roofs are those with parapet walls or railings which are at a height in excess of 900 mm (typically 1100 mm) above the finished roof level. There are typically applicable building regulations which stipulate the required height of railings to roofs or parapet walls. The survey should note the existing provision and make any suitable recommendations to increase this accordingly. Irrespective of compliance with building regulations, railings in excess of 900 mm above the finished roof level provide a significant barrier to mitigate the risk of falling. However, in this instance it is more important to ensure that these are securely fixed. Therefore, it is essential during the survey to physically check the stability of railings to ensure there is no evidence of movement or potential weakness.

Original parapet railings (corroded) have been extended with galvanised steel scaffold poles and connectors.

Retrofitted parapet railings.

Having established the paramount importance of providing lifelines or safety fixings, it is necessary to consider the options. There are many different types of marketed lifeline installations and it is not the responsibility of the surveyor to comment upon the specific design or execution. It is anticipated that these are likely to have been installed during roof or renovation works and it is therefore important to cross-check any specifications held in the as-built file as well as any acceptance or test reports. It is also possible that there may be a requirement for periodic testing of the lifelines and the surveyor should establish this during the inspection. It is not the responsibility of the surveyor to perform any tests on the lifelines or fixing points unless they are suitably qualified and have been explicitly instructed to do this. These life saving fixings may have their own mandatory lifecycle rendering them obsolete over a certain age (typically 10 years). The surveyor should seek to investigate and confirm this in accordance with information made available.

Life safety fixing points are usually constructed in steel which is secured to the roof structure or deck; however, the presence of the waterproof layer usually means that the interface between these two components is not visible Externally, the fixing point will normally be presented as a stainless steel ring or 'eye' and it is considered normal practice for operatives to bring their own PPE, including harnesses, ropes or slings. Often, the fixing point will have an inventory number, which is used as identification when testing its serviceability. The obvious limitation with a fixing point and a fixed length harness connector is that this places restrictions on the distance that an operative can move from this static point. The danger is that they disconnect their lifeline to inspect or work beyond the limits, which potentially becomes very dangerous, particularly for someone who has been previously connected to a life line and forgets that this is no longer the case.

Lifelines are usually stainless steel cables which are connected in a system between fixing points. This is more advantageous as it allows operatives to move around a roof surface while remaining clipped onto the cable. In some cases cables give way to more advanced perimeter rails but these require specific compatible connectors, which should be provided by the owner or property manager.

A lifeline fixing point.

A fixing cable.

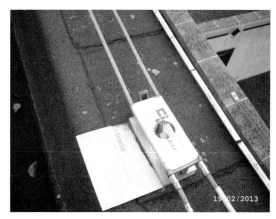

A dual cable/rail system
(photos courtesy of Widnell Europe).

Technical equipment

In most cases, the inspection and audit of technical equipment located on the roof will be undertaken by an M&E engineer. However, there is a requirement for the surveyor undertaking the commercial building survey or TDD inspection to assess the presence of any such equipment in terms of potential damage to the surface of the roof. Invariably, technical equipment is very heavy and this can impact upon the structural stability of the roof as well as potentially damaging the roof covering. This is particularly evident where there is machinery such as chiller compressors or air handling units which vibrate and this, combined with their weight, can create point loads typically to the feet of the equipment. Consequently, this load pushes down onto the roof covering and, in the case of bitumen mineral felt, PVC or EPDM, can cause puncturing of the material. To combat this, the equipment can be supported on plinths or frames, often incorporating springs. Additionally, the feet of the equipment should be placed on paving slabs or other suitable materials to distribute the load.

In some cases, lightweight concrete slabs are cast in situ to provide the support or mounting for technical equipment.

An air conditioning unit supported on pre-cast concrete paving slabs with reinforced matting to a PVC roof covering.

A lightweight concrete cast-in-situ slab on the roof of an office building supporting an air handling unit. A significant crack passes through the slab which suggest problems with the positioning or lack of reinforcement. This requires further investigation.

Lightening conductors

As with technical equipment and lifelines,
it is not the responsibility of the surveyor
to confirm conformity of any lightening
conductor installations placed on a roof
as this is a specialist installation. If the site
inspection identifies a lightning rod and
associated components it is necessary to refer
to any test reports or certificates relating to
the installation. If no test reports are present,
then the owner or property manager should
be pressed to produce evidence confirming
compliance with regulations or to provide the
purchaser with a guarantee that this is in order.

The site inspection should still note the
presence of a lightning conductor and this is
also an item which could be taken into the
M&E survey. The surveyor should note any
obvious physical damage to the installation
and, in particular, the presence of a 'bulb'
to the top of a lightening conductor, which
may suggest that this contains radioactive
material. The use of radioactive material in
lightning rods is technology which is likely to
affect some installations pre-dating 1990. The
implication of identifying such an installation
during the site inspection is that this should
be removed and disposed of safely and in
accordance with relevant regulations. This can
result in significant costs; therefore, if there
is any doubt concerning the potential for
radioactive lightning rods, it is appropriate to
seek specialist advice.

Lightening conductors can have a single
rod placed at the highest point of a building
or a series of spaced rods with earth straps
forming a grid over the facade to discharge a
lightening strike and this is known as a Faraday
cage.

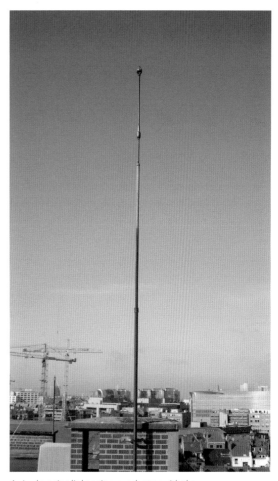

*A single point lightening conductor with the
'bulb' at the top suspected of being radioactive
(photo courtesy of Widnell Europe).*

*A facade-fixed strip forming part of a Faraday cage,
which is loose and requires immediate fixing and testing.*

Window cleaning equipment

Unless the windows to a facade are inward opening, the only way to clean these is from the outside. For buildings up to five or six storeys it is possible to clean the windows with equipment located on the ground level but for most medium- and high-rise properties glazing is cleaned from above. The most common method to undertake this is with a roof-mounted window cleaning assembly, which is effectively a crane used to lower a gondola down the facade to allow cleaning of the windows. The crane is operated from the gondola and, as well as adjusting the height, this moves around the perimeter of the building on roof-mounted rails or a concrete access way.

Facade cleaning system with roof-mounted cradle support crane.

A facade cleaning cradle suspended from the roof of an office building (photo courtesy of Widnell Europe).

A rail system to support the facade cleaning crane and allow movement around the perimeter of the building (photo courtesy of Widnell Europe).

The survey should seek to identify any visible defects with the guidance rails or access route to the roof but the serviceability of the actual window cleaning equipment should be inspected by the M&E survey and also by verification of the relevant test certificates.

An alternative method for cleaning the glazing to a facade is with a window cleaning gondola which is connected to static cradle arms fixed to the parapet walls. Again, the serviceability of this should be verified with the as-built file and any test certificates. It is not normally within the remit of the surveyor to test or certify this during the site inspection.

The building survey or TDD inspection should seek to identify evidence of visual defect associated with cradle arms or parapet-mounted brackets.

Ducting, flues and chimneys

While the M&E survey should seek to identify omission, defects or problems with ducting, flues or chimneys, their very presence on roofs means that they can be susceptible to defects or even introduce defects to the roof.

The purpose of ducting, chimneys or flues is to extract or allow the passage of toxic gases from heating systems or to allow air to be removed from or introduced into a building. As a consequence of their function they usually pass through the roof covering or form joints or junctions between the roof and facade. Therefore, it is important to inspect and assess the condition of the relevant pipework or masonry as well as any flashings or upstands.

For brickwork chimneys or shafts, it is necessary to inspect the quality of the materials as well and the condition of mortar joints and copings or chimney pots. Importantly, gases from heating installations are acidic and the effect of this, combined with excessive moisture to the masonry (from wind-driven rain), has the potential to case sulphate attack. This results in the expansion and cracking of mortar joints, which can lead to cracking and leaning of chimneys or potential collapse of masonry. The survey should seek to identify any evidence of movement or repair to brickwork chimneys or shafts.

A rebuilt section of a brickwork chimney connected to an oil-fired boiler. Note the lower brickwork has been stained by the soot from this slow-burning fossil fuel.

Localised damage to mortar joints of a brickwork chimney to a commercial building.

Stainless steel is the material of choice for the flues to modern boilers, primarily as this should be resistive to corrosion; however, these are sometimes fixed to facade or parapet walls using non stainless steel screws and, as a consequence, these corrode. Weak or defective fixings can result in potential leaning of chimney flues, compromising their function and safety. Such observations should be noted during the survey and communicated in the report.

Other flues or vents passing through the roof surface include typically soil and vent pipes (SVPs). These are used to ventilate the main vertical evacuation stacks of the sanitary installations and typically these are located on the roof in the vicinity of the sanitary rooms or main evacuation shaft. Older SVPs are manufactured from cast iron and typically have an anticipated life cycle of 50 years. More modern (post-1970) SVPs are made from PVC, and increasingly common is the use of HDPE for this purpose.

SVPs should be fitted with a cap and grill at the top to allow the pipe to ventilate but to prevent rainwater entering. Common defects include poor upstands and flashing details, allowing rainwater penetration or mechanical damage to PVC pipework, such as cracks or splits.

Other pipework passing through the roof covering may be kitchen extraction, typically from restaurants occupying the building as sole occupier or mixed use tenancy. This is usually a tenant installation and verification of compliance with all relevant regulations concerning this should be included within the lease contract. While the inspection of tenant installations is usually explicitly not included in the commercial building survey or TDD, it is important to comment, when appropriate, on issues that may affect the roof or building. One example of this is the presence of built-up cooking fat, which can catch fire if there is a fire in the kitchen extraction hood. Therefore, this appears to be the responsibility of the surveyor to notify the building owner or property manager to engage the tenant to resolve this issue. By documenting the situation in the report and notifying the owner, property manager or tenant is it hoped that any losses that might occur if a fire breaks out in the kitchen below can be mitigated.

Roofs summary

The roof is one of the key principal elements of a building forming part of the external envelope. It is one of the areas where the surveyor will spend a significant amount of time during the commercial building survey or TDD inspection.

It is important to acknowledge the performance criteria of roofs and to apply these during the inspection and subsequent report. Generalising may not always be appropriate when considering the different elements of a building, but some general observations concerning roofs are that:

- flat roof coverings usually have shorter life cycles than pitched roof coverings
- metallic roof coverings have longer life cycles than coverings made from oil-based or synthetic materials
- metallic roof coverings are significantly costlier than oil-based or synthetic but two or three times the cost does not equate to two or three times the life cycle
- the cost of replacing the roof coverings to low-cost, low-quality industrial buildings can represent significant investments by virtue of the large surface areas involved
- when making recommendations concerning immediate, short-, medium- and long-term repairs or replacement of roof coverings and their associated components, these should always be evidence based
- be risk averse when inspecting and reporting on roofs as this is an area of the building which is constantly exposed to the natural elements. Periods of freak or extreme weather can significantly influence the life cycle of components or exacerbate defects.

References

RICS. 2011a. *Surveying safely* (1st edition, guidance note, June) (available from the RICS website).

RICS. 2011b. *Technical due diligence of commercial, industrial and residential property in continental Europe* (1st edition, best practice and guidance note), London.

Photo courtesy of Widnell Europe

Structure 6

Introduction

While the building envelope shields and protects the occupants from exposure to the outside elements, this is only possible as it is supported by the structure below and above ground. When undertaking inspections of commercial properties, in almost all cases it will be not possible to view the majority of the structural components. This is a significant restriction placed upon the surveyor and it is critical that clients clearly understand the limitations of the survey in terms of the structural appraisal.

The assessment of the structure to a building will be limited to the areas where this is visible and the presence of any evidence which may suggest structural movement. After M&E, the next most commonly sub-consulted professionals during the commercial building survey or TDD process are structural engineers. While building surveyors are expected to have an understanding of structures and be able to advise on matters affecting structural integrity, there are many instances when it will be appropriate to seek the specialist advice of a structural engineer.

When preparing to undertake a commercial building survey or TDD inspection, it is normal to request copies of the as-built plans and any plans that have been attached to planning permission or building permits. The plans are typically prepared by the architect and are not specifically plans of the structure; however, they should show the positioning of columns, beams, central core and loadbearing walls. From this information, and with the dimensions annotated on the plans, it should be possible to ascertain the column grid and facade module of the building. Sections through the building are also very useful for establishing the storey heights and depth of ceiling voids and may also show some indication of the foundation design.

Another very useful piece of information concerning the structural design criteria of a building is the floor loading capacities. This information is sometimes available in the as-built files and it may be used to gauge the suitability of a building for certain operations or specific individual loading requirements. These are often stipulated or derived from building regulations or building codes which were valid at the time of construction and it is not normal to expect buildings pre-dating the latest codes to necessarily conform to their requirements. Retrospective application of building codes is not usually an obligatory requirement. If no floor loading information is available during the building survey or TDD process, this should be requested from the owner, property manager or tenant. In no circumstances is it appropriate for the surveyor to estimate the floor loading capacities unless they are suitably qualified to do this.

The assessment of the structure is a difficult and risky undertaking for the surveyor, who is often working within limited time constraints. It is therefore important to establish the best possible evidence-based opinion. In the event that evidence exists to suggest structural movement or significant water infiltration to the below-ground basement areas, then it is advisable to be cautious and seek specialist advice.

The inspection of the structure during the building survey or TDD process is essentially divided into the two following sections:

- substructure
- superstructure.

The substructure comprises all of the elements below ground and is only visible where there is a basement. In contrast, the superstructure is above ground and, although in principle it may appear easier to inspect, this is often encased by finishes and only visible in certain areas.

When inspecting the structure as part of a commercial building survey or TDD, and in particular when assessing the adequacy of this or the presence of visible defects, it is important to understand the basic construction principles. Importantly, there are fundamental differences between low-rise 'residential' type structures and the structure of medium- to high-rise buildings. It has been established that among the current commercial property stock there are large amounts of town or city centre property which is low rise. Contrasted with this are the medium- and high-rise structures associated with office and residential towers, as well as hotels. The fundamental observations or differences between low-rise and medium- or high-rise structures can be summarised as the following:

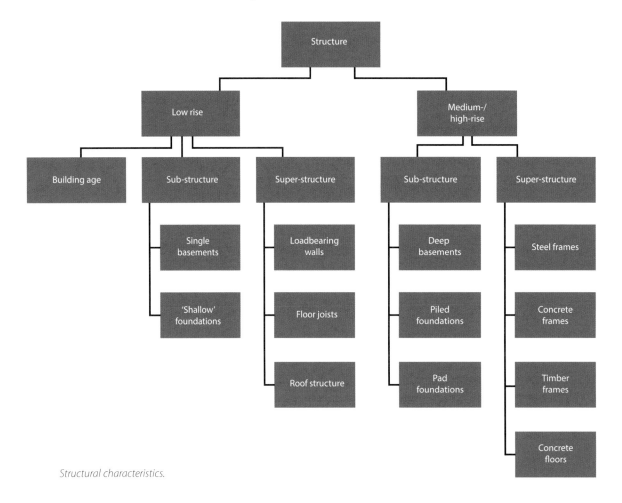

Structural characteristics.

While building age has some significance to all elements, and particularly the life cycle of materials, it has greater significance with low-rise commercial properties as these can date back to the nineteenth century, though rarely beyond this. As discussed in *Chapter 4 – Commercial sectors* there are fundamental differences between pre-1920s low-rise commercial properties and those dated post-1980s. Furthermore, there has been significant evolution in building technology between these time periods, which has had some profound effects upon the structure.

In terms of the structure of these two commercial properties, they are vastly contrasting and this is a direct result of their age category, as well as the fundamental difference between the pre-1920s property (TOP) and that constructed post-1970 (RIGHT) (photo courtesy of Widnell Europe).

Low-rise commercial buildings

Essentially, buildings which have up to four storeys above ground level should be considered as low rise. There are several different ways to categorise the ages of such buildings. One method is to divide these into broad age categories based on the key construction technology or features associated with these historical periods. For the purpose of analysing the structures associated with low-rise commercial properties, the following age categories may have been established:

- pre-1920s
- 1920–45
- 1946–79
- post-1980.

Pre-1920s

This age category spans the greatest number of years; however, in real terms, commercial property was only really established in the Georgian or Victorian eras, thus making this category span around 100 years. Importantly, there are many pre-1920s properties operating commercially today and, although these are likely to have been the subject of significant renovation or extension, their base structural components may be largely unchanged.

There are relatively rare examples of commercial properties which pre-date the Georgian era and these are typically located in older historic town or city centres. Invariably, properties of this age and type are likely to be listed and these are also commonly located within a conservation area. With commercial properties which have historic or cultural value, the building survey should be undertaken in the same manner as applies to all commercial properties. However, there are likely to be specific issues or parts of the building fabric and structure which may require more specialist knowledge. One of the principal obligations placed by the RICS on a surveyor is that they work within their own professional limits. If the surveyor has little or no knowledge of pre-Georgian construction, it would be perfectly reasonable to seek specialist advice from someone suitably qualified.

A Grade II* listed building believed to date from the fourteenth or fifteenth century, operated as a commercial property.

A pre-1920s (Victorian) property being operated as a commercial property; note the presence of a modern medium-rise commercial property to the rear, which clearly post-dates this.

One structural characteristic that most pre-1920s properties have in common is shallow or non-existent foundations. Considering specifically Georgian and Victorian properties, these are likely to have corbelled brick foundations to a typical depth of 450 mm below ground level. These may be constructed directly onto the subsoil or onto a layer of stone slabs or clinker ash. The obvious problem with shallow foundations is that they are likely to be more susceptible to movement or subsidence. Sometimes, pre-1920s commercial properties have a basement level below part or all of the property. This effectively lowers the foundation depth by up to one storey and generally results in buildings being more resistant to ground movement.

Pre-1920s basements

When undertaking a building survey or TDD of any age of property, it is always essential to inspect the basement. Lots of important construction detail can be gleaned from a basement structure and this is one of the few areas where it may be possible to see the exposed structure. Typically, the basement structure below a pre-1920s property will have loadbearing brickwork or stone external walls and it may also be possible to see evidence of corbelled foundations. The ground floor above may comprise timber joists spanning from front to back of the property. Later Victorian properties, or those of high specification, may have some metallic sections forming the floor joists or supports. Where there is evidence of concrete columns or steel structure encased in concrete, and if the basement is more than one storey below ground level, it is likely that this has been the subject of significant renovation or extension. It will therefore be necessary to consult the as-built file or technical documents to verify this.

If the space is being operated commercially as lettable floor area, the basement walls are likely to be lined or finished, as is the floor. In this case, it will be very difficult to establish the construction detail; however, defects such as water infiltration, dampness and movement should present some visual signs.

The principal purpose of the basement inspection is to establish the presence of any movement or water infiltration. With all defects, and in particular water infiltration or structural movement, it is necessary for the surveyor to establish the cause, effect and significance of this. Structural movement to a basement is likely to present as cracking to the external perimeter and internal loadbearing walls or distortion of the ground floor joists and unevenness to the basement floor. However, with pre-1920s commercial properties, these visual symptoms are almost invariably evident to a certain extent, therefore it is necessary for the surveyor to establish whether the movement of the structure is 'live' or historic. In many cases, pre-1920s commercial properties are likely to have been the subject of alteration, extension or renovation, as evidenced by the presence of steel or concrete columns and beams. The surveyor will have to establish, through visual examination, the presence of any evidence suggesting structural alteration, and also any subsequent movement or defects. This is when it may be necessary to cross-check the information in the as-built file or documentation on the building to ascertain if there are any records relating to structural works in the basement.

If the loadbearing walls have been underpinned, this will often be the case when the property has been the subject of subsidence or a basement has been created, extended or lowered beneath the building. Inevitably, all of these operations will result in some cracking to the loadbearing walls. Cracks may be visible during the inspection or there may be evidence that these have been repointed or repaired. Normally, it is possible to identify repairs to masonry joints as the colour, texture and profile is not always exactly matched. Where visible, repairs may be described in the report as evidence of 'historic' cracking, and this is particularly significant if cracking has reappeared in repaired joints. This would suggest that movement has occurred after the repair and may be 'live'. In the context of the building survey or TDD report, it would be necessary to report this and also recommend further detailed investigation. Where existing cracking is evident in walls, it is important to ascertain the size of the cracks (as per BRE Digest 251 (BRE, 1995), Table 1):

- 0 – hairline cracks 0.1 mm
- 1 – fine cracks up to 1 mm
- 2 – doors and windows may stick up to 5 mm
- 3 – service pipes may fracture 5–15 mm
- 4 – extensive repair work 15–25 mm.
- 5 – major repair if greater than 25 mm.

Crack width is an indication of the level of movement and is also used by the surveyor to form an opinion on the level of repair. Ascertaining if a crack is historic can be difficult and the surveyor will need to carry out a careful examination to establish if there are dirt deposits or dust in the crack which could have accumulated over time. Furthermore, if the internal surface of the wall has been painted after the crack has occurred this often results in a layer of paint on the inside face of the crack, thus suggesting that the crack is historic.

Crack patterns and their significance will be discussed later in this chapter and, while it may be difficult to establish the history of these and in particular if movement is live or historic, water infiltration is more straightforward.

Basement space which is being operated commercially should be dry and free from dampness. This should be achievable with 'modern' basements due primarily to advances in construction technology. However, pre-1920s basements are likely to have been constructed from solid wall masonry with minimal external tanking or waterproofing. Therefore, if there are significant levels of moisture in the surrounding ground it is possibly or even likely that this will be absorbed into the basement walls, presenting as staining or dampness to the internal wall surface. Where the walls have been painted or plastered internally, dampness often results in peeling, blistered or spalled finishes as the moisture exits the brickwork. To plain fair-faced brickwork walls, the dampness may present as efflorescence, as soluble salts are drawn through the brickwork and evaporate on the surface leaving a white powdery coating. More serious water infiltration problems may occur if there is water in the ground adjacent to a basement and this will often be transmitted through the external walls and present as 'running' infiltration or a collection of water on the basement floor. External sources of water may be the 'water table' or adjacent water courses, such as rivers, lakes, the sea or an underground water course, such as a well. Furthermore, localised flooding has the potential to cause water infiltration to basements. In essence, most water infiltration comes from above the ground due to surface water or from below by way of the water table. There are also instances where water in a basement may be attributable to damaged supply or evacuation pipework.

In some basements it is acknowledged that water infiltration is a problem and this is often collected in a drainage pit (sump) and evacuated with the aid of an automatic pump. To enable this to be done efficiently, the internal walls of the basement are often lined with masonry and include the placing of a drainage channel to direct the water into the sump. To ensure that water is fully evacuated, a drainage floor is sometimes installed, which is effectively a 'raised floor' encompassing a damp proof membrane (DPM) with grooved channels to allow the water to flow to the sump. Where this type of system is installed it is often very difficult examine either the external walls, as these are concealed by the lining, or the floor, which is also covered. Therefore, it may not be feasible to visually determine the presence of water infiltration and this is why it is important to ask the owner, property manager or tenant if they have any knowledge of water infiltration. Where a sump and pump have been installed it is vital to seek clarification as to whether this is functional and also what procedure is to be followed in the event of a power failure or defect to the pump. Most of these types of installations include an alarm integrated into the pumping equipment and a second, reserve pump which activates if the main pump fails. It is important to establish and report on the presence of water evacuation systems to a basement. Failure to do this may expose the surveyor to a potential claim if the basement floods in the future, particularly if this could have been foreseen in the survey and hence avoided. While there may be some costly repairs associated with a basement flood, it is the consequences on tenant occupation or storage that can have high cost implications.

Pre-1920s loadbearing walls

The majority of pre-1920s commercial properties are low rise with loadbearing external and internal walls. Many of these commercial properties are located in town or city centres and, as a result of this, they are often terraced properties where the external and party walls are loadbearing. The presence of solid loadbearing walls can be typically identified by the external facade, where the brickwork will be executed in Flemish bond, English bond or English garden wall bond. There is a general means of identifying the nature of the external walls as an outer leaf in stretcher bond often signifies a cavity wall, which became more prevalent post 1920. There are localised examples where the external face presents as cavity wall construction although the property is believed to date pre-1920 and this is an example of late nineteenth or early twentieth century 'experimentation'. High-quality Victorian and early Edwardian properties embraced this 'new' technology with the use of cast iron wall ties, but this practice is relatively rare and not the 'norm' with properties of this age. One way of establishing the presence of such construction is to measure the thickness of the external brickwork walls to confirm if these are one and a half bricks thick.

In most cases, the floor joists and roof trusses or rafters span from the front to the back of the properties, which is important as these play an essential structural role and tie the structure together. Typically, there are many instances of pre-1920s commercial properties being the subject of structural alteration or extension. Where this happens, this may be visible with the placement of additional beams or columns, typically executed in steel or reinforced concrete. However, in most cases the structure above ground is usually concealed by the finishes, which could be painted plaster, plasterboard or fitting out with panels etc. Aside from reviewing the as-built plans and any documentation held on the property, the surveyor will have to rely largely on their discussions with the owner, property manager or tenant to establish the existence of any historical structural works. The provision for destructive testing or opening up the property for detailed inspection is not normally covered in a standard contract instruction and would inevitably add significant time to the building survey or TDD process. Therefore, it is usually agreed that the survey will be purely visual and accordingly the surveyor should inspect internally for evidence of cracking to the walls or distortion of the floors. The survey should be continued externally with an inspection of the visible loadbearing facades. This inspection should be carried out in a similar manner to the inspection of the basement walls, with close attention paid to the presence of cracked or bulging walls.

Alteration of the roof structure to pre-1920s commercial properties is quite a common occurrence. The roofs to this age group of properties were typically constructed in situ and are known as 'cut roofs'. The significance of a cut roof as opposed to pre-formed trusses, which were typically used post 1945, is the relative ease with which these can be altered to create additional room in the roof space. Evidence of such alteration during the building survey or TDD inspection should be noted and verified in the document review with the relevant planning permission and building regulations approvals. It is important to confirm that an acceptance report or certificate has been issued for the works as it is not the responsibility of the surveyor to verify the calculations or stability of the structure. The survey is visual and will seek to identify any evidence of structural movement or defects associated with alterations such as a roof conversion. Without strengthening the structure, placing additional commercial space in the roof void may lead to overloading and movement of the existing roof or walls.

1920–45 low-rise commercial properties

This period spanned from the end of the First World War to the end of the Second World War and saw some fundamental changes to construction technology as well as differences in architectural style. It is important to acknowledge that the timeline for these changes is not exact or specific as there was no universal building code or regulations at that time. Therefore, new methods and techniques appear to have occurred locally with differences being apparent from building to building or street to street.

Typical 1920–45 buildings operated as commercial properties. The part- rendered and cavity walls, combined with singled glazed sliding sash windows and hipped roofs to bay windows, are typical characteristics of this age category.

Typically, architectural styles of this period were part facing brickwork and part render, with rooflines typically encompassing hipped details. Windows were likely to be timber framed and single glazed with leaded detailing. This period also saw the introduction of steel framed, single glazed windows. Many commercial properties dating from this period are located in towns or cities. They are typically operated as retail and offices with some residential use. It is noticeable that there is a progression or 'ripple' effect of historic development, with the oldest commercial properties being located closest to historic trading centres and newer properties emanating outwards with the passage of time. Where buildings appear to be almost randomly placed, this is usually evidence of localised or major city centre regeneration. Often, these originate from the 1960s or 1970s, when 'modern' medium- or high-rise buildings were crammed into existing town centres with little consideration for the existing townscapes. As a consequence of this, and the apparent differences between historic and current planning, many of these 'new' buildings have reached their economic end of life and a second phase of localised regeneration is occurring. The principal difference is that many of the existing older properties are deemed to have cultural historic value and are becoming listed. There is also evidence of the increased recognition and protection of 1960s or 1970s buildings as fine examples of period 'Brutalist' architecture. This places considerable restrictions on maintenance, repair or alterations and is something that needs to be noted and discussed in the commercial building survey or TDD report.

As most commercial properties dating from the period between 1920 and 1945 are likely to have undergone renovation, it is often the case that original windows, shop frontage or general glazing details have been changed. However, hipped roofs, render and brickwork facades are likely to remain and these features may be used to help identify the building age.

Structurally foundation depths were typically lowered and, although some corbelled brickwork was still in use, concrete strip foundations became increasingly common, yet were still relatively shallow when compared to today's standards. The presence of basements to commercial properties had the effect of increasing the resistance of the structure to ground movement. Above ground, the most significant advance in construction technology was the use of cavity walls. By utilising two skins of brickwork and creating a void between these, penetrating damp could be eradicated. As this is a crossover period in the switch from solid to cavity walls, there are often examples of solid brickwork walls and these should be treated the same as pre-1920s construction, although the foundation depths may be different. In essence, without excavating a trial hole it will be difficult to establish the foundation depths; however, the surveyor should always stick to the first principles of the visual survey and report any evidence suggesting the presence of structural movement.

When undertaking the inspection of a low-rise commercial property from the 1920–45 age category, it is important to ascertain any evidence of structural alteration or extension. While it is likely that any additional columns, beams or loadbearing walls will be concealed by the finishes, the inspection should seek to compare the as-built situation with any plans or documents held in the acquisition data room.

1946–79 low-rise commercial properties

Historical events influenced construction technology post 1945, with an acute shortage of raw materials, including timber, being a factor in the increased use of ground bearing concrete floor slabs. Foundation depths became progressively deeper, with these being typically up to a maximum depth of 750 mm below ground level and constructed as concrete strip.

This period saw the introduction of system building and prefabrication of construction components, but this practice was largely evident in residential properties. An increase in the use of reinforced concrete frames, particularly for low-rise retail or office construction, occurred in the decades immediately following the end of the Second World War. However, there was a widespread increase in the use of reinforced concrete in the 1960s and 1970s, as architects began to express themselves with 'modern' architecture and the birth of 'brutalism'. In this period, high-rise concrete construction became more prevalent, as well as lower rise concrete structures utilised for shopping centres, civic buildings, theatres and education establishments.

Typical mid-twentieth century low-rise commercial development.

The Urban and Regional Studies (URS) building, known as the 'Lego building', on the University of Reading's Whiteknights campus is an example of 1970s brutalist architecture.

When inspecting low-rise commercial properties from this period, it is important to acknowledge the variety and range of different structural designs and materials. For commercial properties which appear as 'residential' types with loadbearing walls and timber floor joists, it is important to assess the structure accordingly. In most cases, these properties have concrete strip foundations and loadbearing cavity walls executed from masonry with concrete blockwork inner walls. The inspection should seek to identify any evidence of structural alteration or movement, which will normally present as cracking to both external and internal walls.

Basement construction to properties post 1950 is likely to comprise reinforced concrete and it is important to inspect these to establish whether there is any evidence of water infiltration or dampness. Often, the external basement walls may be lined with brickwork and it is difficult to ascertain the condition of the concrete. In this instance, the surveyor can only report what is actually visible and, in the event that there is dampness or water infiltration rendering the space unlettable, further investigation, including the opening up of wall linings, should be recommended. The most significant risk of water infiltration or dampness to reinforced concrete external walls is the potential for the steel to corrode, causing damage and weakening of the reinforcement.

Low-rise commercial properties constructed utilising concrete framed structures are completely different from those with loadbearing masonry walls. Structural alteration to loadbearing walls appears relatively straightforward, with the insertion of lintels being the chosen method for enlarging or creating new openings. However, with concrete framed structures it is more challenging to physically remove reinforced elements, such as beams or columns, although in many instances as the frame is loadbearing, the infill walls are not. Therefore, the removal or opening up of the internal and external walls can be a straightforward exercise.

When inspecting a reinforced concrete frame it is important to establish the dimensions of the concrete elements, including the grid. This comprises the dimensions of the column and beam lines from centre to centre which, together with the dimensions of the openings (window or door modules), will give an indication of the flexibility of the space. Flexible space for investors is important as the potential to divide a floor plate often makes the space easier to let. This can be done to accommodate smaller, individual tenants or existing tenants might consider expanding into the adjacent space in a building simply by removing some internal walls. It may not be possible to inspect the elements of the concrete frame, as these may be concealed by the internal finishes or external claddings. Therefore, it will be necessary to cross-check this with any existing execution or as-built plans and, by taking some check dimensions on site, it may be possible to estimate the grid or module.

Concrete and the defects associated with this material will be discussed in more detail later on in this chapter, but it is important to ascertain during the inspection the condition of any concrete elements. There are significant differences in the quality of concrete used for structural frames to buildings executed in the 1960s or 1970s compared to today's structures. Primarily, there will always be a difference between concrete cast in situ and concrete which is pre-cast, with the latter being generally of higher quality. However, concrete is an extremely durable material, with evidence of this being used in Roman times. One observation is that, when protected internally in a building or encased by cladding for external components, it has the potential to have excellent longevity. The introduction of steel to reinforce concrete has made an overwhelmingly positive contribution to the improvement of the loading capacity and span dimension, but has also contributed to structural defects, such as carbonation.

Where visible, it is important for the surveyor to note and record the condition of the concrete frame. Particular attention should be paid to the presence of any spalling to the surface or crumbling of the material, as well as longitudinal cracking across structural members. Principally, the inspection is visual and, where evidence exists of defects to the concrete, then it is appropriate to recommend more detailed further investigation and destructive testing, including analysis of concrete samples taken in situ. This has the potential to increase or prolong the acquisition process but it is critical to advise the client on these matters. Visual inspection alone cannot provide certainty regarding the stability of the material and likelihood of future defects or problems.

Upper floor construction was sometimes executed using concrete beam and clay blocks in an arrangement known as hollow clay pot floors. It is impossible to confirm the presence of this type of construction if the underside of the floor is concealed by a suspended ceiling.

Services installations above a suspended ceiling to this 1950s, low-rise commercial property exposed the presence of a hollow clay pot floor.

The principal problems with these types of floors are their poor acoustic qualities and their potentially poor fire resistance or separation. The clay pots also have the potential to crack and loosen.

Post-1980s low-rise commercial properties

The principal characteristic of post-1980s low-rise commercial properties with loadbearing walls is the depth of the foundations, which are likely to be 1000 mm below ground level in areas of shrinkable clay sub-soils. The external walls are likely to comprise cavity construction with concrete blockwork forming the inner leaves, incorporating galvanised or, more commonly, stainless steel wall ties.

Ground floor slabs are often loadbearing reinforced concrete or pre-cast concrete beam and block. Upper floors may comprise pre-cast reinforced concrete beam and blocks or more 'traditional' timber joists but during this period these are increasingly being constructed as 'webbed' timber, encompassing stainless steel components to tie the timber members together.

Alternatively, post-1980s low-rise commercial properties may be constructed using steel frames supported by pad or pile foundations with both the external and internal walls having non-loadbearing requirements.

With most commercial properties from this age category it is difficult to establish and confirm the structure as this is often encased and concealed by finishes. Therefore, the surveyor will have to revert to any execution or as-built plans held by the owner, property manager or tenant and made available in the acquisition data room or on site. It should be expected that newer properties should have extensive as-built records as there is more capacity and a greater number of options for storing plans or information digitally when compared to paper-only documents.

It is important for the surveyor to establish the structural grid and facade module as well as the storey heights. This information is used to advise on the flexibility of the structure and the ability to sub-divide the space to enhance letting options. The purpose of the survey is also to identify any evidence of structural movement or significant water infiltration. This is achieved by accessing as much of the internal space as possible, as well as identifying, cataloguing and analysing the presence of visible cracking to internal or external walls and floors.

Structural movement of low-rise commercial properties

Structural movement is often characterised by the presence of cracks, but this can also be observed when there are no cracks, evidenced by sloping or deflecting floor joists and roof lines. Considering the fact that surveyors typically have only one day to visit a building to establish its past, present and future, regarding the condition or defects, it is important to understand why cracks appear and the potential cause and effect of these.

Structural movement may be traced back to several potential sources but typically this is from below a building with movement of the ground and foundations or movement above ground caused by structural alteration, overloading or other contributing factors, such as fire. Ground movement is probably the most complex and costly defect to remedy, therefore it is important to analyse the visual evidence during the building survey in order to establish whether this is the most likely possible cause.

Primarily, the initial evidence of potential structural movement, including ground movement below the foundations, is the presence of cracks to the external facade or internal walls. It is therefore necessary for the surveyor to undertake crack analysis.

Crack analysis

In order to assess the severity and potential causes of cracks to the external or internal walls of a low-rise commercial property, it is important to note, measure and record the position and size of these. By referring to BRE Digest 251 (BRE, 1995) the severity of the crack can be graded but it is more the shape or pattern of the crack which may suggest possible reasons for its formation. There is also a requirement for the surveyor to have geographic knowledge of the area they are working within. This is important to establish the potential presence of any known or specific ground conditions.

The external facades of low-rise commercial properties often comprise brickwork, render, natural stone (including cladding) or facade cladding systems. Certainly, brickwork or render are materials which rarely conceal the presence of structural movement. These are both rigid and have no flexible construction components, meaning that movement usually presents as cracking to the surface or joints. Facade cladding systems may have more flexibility to accommodate movement and this may mask cracks to the structural components below. Alternatively, structural movement may cause distortion or bowing of cladding panels. With all cracking observed to walls, it is important to inspect both internally and externally to see if this passes through the wall, which tends to suggest or confirm a higher degree of severity. If there is cracking to external and internal cavity walls, which are largely independent of each other, the common connecting structural components are the foundations below these walls as well as the internal floor joists or roof structure.

Cracking to external or internal walls typically presents as:

- vertical cracking
- horizontal cracking
- step cracking
- tapered cracking.

Vertical cracking

Vertical cracking is often associated with thermal movement and is typically associated with large panels of brickwork or concrete blockwork. Where this is exposed to relatively high thermal gain, such as south-facing sections of a property, the tendency may be for this to expand. Utilising the presence of a damp proof course (DPC) as a slip plane, the expanding brick or blockwork moves towards its unrestrained edge, causing vertical cracking. Where the mortar mix is stronger than the bricks, the cracking may pass through the bricks.

Lightweight plaster block laid and pointed in a cement-based mortar to an internal basement car park. The epoxy painted floor is acting as a slip plane and the mortar is stronger than the blocks, allowing the thermal movement to create cracks which pass through the blocks (photo courtesy of Widnell Europe).

(TOP) Vertical cracking through a south-facing single skin brickwork wall, which is typical of thermal movement.
(BOTTOM) This has been confirmed by the masonry overhanging the bottom two courses of bricks, where the DPC has acted as a slip plane.

It is important to contextualise the vertical cracking, as this is not always the result of thermal movement. Where there are two adjacent buildings, components or materials that are not 'keyed' into each other, there will exist the possibility of differential movement. In most cases, this is a normal occurrence and typically exists between concrete and masonry, brickwork and blockwork or masonry and plasterboard. It is not normally representative of structural movement but should be recorded and discussed in the report.

Obvious differential movement between separate sections of brickwork which are not 'keyed' together (photo courtesy of Widnell Europe).

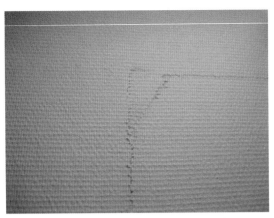

Less obvious differential movement to the corner of a window to an internal wall, which has been closed up with plasterboard and finished with glass reinforced paper. Differential movement between the gypsum-based plasterboard and the masonry opening has caused cracking, which presents as tearing to the surface of the paper (photo courtesy of Widnell Europe).

Tapered cracking

Vertical cracking can occur as a result of structural movement but this is likely to present also as 'tapered' cracking, in which one end of the crack is wider than the other. In low-rise commercial properties, tapered cracking is indicative of movement to one side or part of a building as the loadbearing structure lowers or rises relative to the adjacent brickwork. This movement is effectively rotation, and in most cases tapered cracking is a sign of structural movement.

Vertical cracking which is typical of thermal movement; however, the wall is north facing and the crack is wider at the top than the bottom. This is a tapered crack, but the mortar appears to be stronger than the brickwork, resulting in cracking through the bricks.

Tapered 'step' cracking to a panel of brickwork over a window opening, with the crack widest at the bottom. This indicates that either the brickwork to the left of the window is rising or that to the right, over the window, is dropping. By contextualising the crack as being localised over a first floor window, the evidence suggests failure of the lintel over the window, requiring immediate further investigation (photo courtesy of Widnell Europe).

Step cracking

Tapered cracking is often more commonly seen with step cracking, and where this combination is evident it usually indicates structural movement. The cracking is often localised to the area or side of the building affected by the movement and the crack pattern runs horizontally and vertically along the existing joints in the brickwork. Where the mortar is stronger than the bricks, this usually results in cracks through the bricks and therefore can sometimes deceptively present as vertical cracking. Therefore, it is important to contextualise the cracking as part of the analysis.

Horizontal cracking

The appearance of horizontal cracks to a loadbearing masonry wall can be the result of several causes. Assessing the construction detail to ascertain the age category of the property will determine the type of wall construction, with most pre-1920s properties having solid loadbearing walls. Cavity wall construction became more widely used in the 1920–45 period but, as with all construction techniques, the introduction and usage of this technique overlapped these two age categories.

Early cavity walls utilised cast iron and steel walls ties to connect the inner and outer leaves of brickwork to give collective strength to these loadbearing construction components. Corrosion of the steel wall ties leads to a reduction in the stability of the loadbearing masonry walls and can result in their distortion or even collapse. The building survey or TDD inspection will seek to identify any visible evidence of structural movement and, in the case of corroded steel wall ties, this may present as evenly spaced horizontal cracks in line with the position of the wall ties. Other visible evidence may be deflection or bowing of the external walls and in both cases it is necessary to note this and recommend further, more detailed investigation.

One of the fundamental requirements of a building survey or TDD is to advise on the presence of defects and, importantly, to determine the severity and significance of these. Where there is a defect there is a requirement to establish a course of action to investigate, treat or remedy this. Within the commercial building survey or TDD report the defects will be noted, analysed and a remedial solution proposed. The purpose of proposing a remedy is to advise on the potential cost of the defect and this is often used as leverage to achieve a price reduction or oblige the seller to repair the defect post acquisition. In the case of wall tie failure, it will be very difficult to confirm this during or after the survey inspection and it is necessary to prepare a trail of evidence to diagnose this.

Horizontal cracking to the brickwork where the roof intersects with the facade is typical of roof spread. This occurs where there may be overloading or alteration of the roof structure. Roof spread is typically associated with pre-1920s properties, as the walls are solid loadbearing structural components with the roof structure often constructed in situ. The joists and rafters are fixed to a horizontal timber wall plate integrated into the upper course of brickwork. With no cavity and any downward force exerted by the roof projecting as horizontal movement, the wall plate is forced in an outwards motion, displacing the brickwork. This causes horizontal cracking to occur in the facade at the upper level and part of the evidence-gathering exercise should be for the surveyor to establish if there has been a replacement of the roof covering or any structural alteration. This should be done through discussion with the owner, property manager or tenant, with a review of any acceptance reports or other guarantees included in the document review.

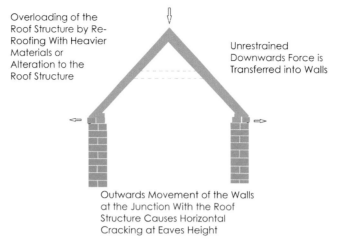

Roof spread.

While wall tie failure and roof spread have specific crack patterns, sulphate attack also causes horizontal cracking, but this is often more random and localised. Typically associated with damp or saturated brickwork, sulphate attack is a reaction between the sulphates in bricks, or from ground water, air pollution and slow-burning fuel sources and tricalcium aluminate in cement. As a consequence, there is expansion which causes cracking to mortar joints. Sulphate attack is common in areas of exposure to wind-driven rain, such as chimneys, or to retaining walls where there are deficient or defective waterproof membranes to the built-up side. With chimneys which are not lined with a flue, the sulphates produced by the burning of fossil fuels, combined with saturation or dampness from wind-driven rain associated with the exposed nature of the chimney, provide an ideal combination for the propagation of sulphate attack. Specific expansion to mortar joints of the exposed side of the chimney results in the chimney leaning away from that side. This is a potentially dangerous situation, where the chimney may be prone to collapse, particularly in storm conditions, and there is a duty of care placed upon all building owners to prevent possible injury or harm from such collapse. The visual inspection of chimneys should form part of the commercial building survey or TDD process and is usually included as part of the analysis of the roof. Where access cannot be granted to the roof to inspect chimneys in detail, this should be performed from ground level or other vantage points with the use of optical equipment, such as the zoom lens on a camera, binoculars or drone technology.

Foundation movement

While thermal movement may be considered relatively straightforward to remedy and accommodate, wall tie failure is more complex, necessitating the retrospective installation of wall ties or even reconstruction of areas of brickwork. However, foundation failure is a major structural defect as well as a source of potentially high investment risk and cost.

When assessing the relative age of the property and any subsequent structural modification or alteration, the surveyor should have some knowledge of the possible foundation type and typical depths. Furthermore, surveyors should seek to have an understanding of the geographical areas they work within and this, combined with a sense for the potential foundations depths, may indicate possible sources of ground and subsequent foundation movement.

Foundation movement can be attributed to initial building settlement or subsidence, with settlement generally occurring during the initial years after construction but usually limited to between five and ten years post construction. It occurs as the initial weight or load of the structure causes some compression of the subsoil below the foundations. Settlement cracks are usually fine or hairline and may be present around openings to the external loadbearing walls, such as doors or windows. They may also occur to internal loadbearing walls or non-loadbearing partitions.

Subsidence differs from settlement by virtue of the fact that this occurs primarily due to external factors affecting the foundations or the stability of the ground below these. It is therefore important for the surveyor to have or obtain knowledge of the ground conditions if there is a suspicion of subsidence. There are a wide variety of subsoils below low-rise commercial properties and understanding the evolution of town or city centres is important to begin to assess why subsidence and structural movement may have occurred.

Historically, most commercial town or city centres were initially established due to their geographic importance. Whether this was due to strategic trading posts at the intersection of trading routes or hilltop towns where natural topography provided a means of defence, this invariably contributes to the subsoil and bearing capacity for foundations. The majority of town or city centres have historic centres and often it is possible to see or plot through the changes in architecture the ripple effect of expansion throughout the subsequent years. Typically, commercial centres constructed on high ground for defensive purposes are on stable or solid subsoils, which can include rock formations, and cohesive soils such as clays. Those commercial centres constructed adjacent to major waterways or in valleys or by the coast, were historically wealthy due to their trading importance but, as a consequence of their location, the subsoils are usually less stable and are non-cohesive. This may include granular silts, sands or gravels, which are associated with low-lying river or coastal beds, and, typical to these locations, the water table is often relatively high, as is the potential for flooding. There are, of course, many variations of subsoil and it is not unusual to find pockets of different soil types interspersed with those that may be considered the 'norm' for a certain location.

Surveyors should have knowledge of the geographic areas in which they are working and part of the commercial building survey or TDD process will involve a document review or desk survey. Information concerning subsoils and ground conditions can be found on geological maps; however, these are only indicative of what might be expected and do not take into account local, random dispersions or pockets of different soil types. This information also does not take into account previous construction activities, such as backfilling ground or underground cavities created by services or, in some cases, landfill.

Knowledge of the principal differences between cohesive and non-cohesive soils is essential for surveyors assessing and contextualising evidence of structural movement which may be attributable to subsidence.

Cohesive soils

These are essentially clay soils and are cohesive as the microscopic particles stick together. Importantly, clays contain a degree of moisture and they have a degree of elasticity, meaning that they can swell or shrink depending on the amounts of moisture contained within them. There are many different types of clay soils, including relatively soft, sandy clay, which is almost 'buttery' in texture and has low compressive strength. There are other, stiffer clay soils and boulder clay, which are significantly stronger but more susceptible to shrinkage and swelling.

Subsidence of foundations on shrinkable clay soils occurs as moisture levels in the soil reduce to the extent where there is volumetric change under the foundations, causing these to drop. Significant amounts of moisture need to be removed from shrinkable clay soils to propagate subsidence and this is not something that happens quickly. Typically, clay shrinkage occurs during periods of exceptionally dry and hot weather, but it is mainly the action of tree roots extracting moisture from shrinkable clay soil that can cause subsidence.

Non-cohesive soils

Granular soils, such as sands, silts and gravels, are non-cohesive and do not stick together. They are typically susceptible to erosion or being washed away by water and this is graphically evident in coastal areas where there are high-profile examples of this. Foundations on non-cohesive soils are not generally more likely to suffer subsidence or movement compared to cohesive soils and, when contained in situ, these soils have good loadbearing characteristics. However, rapid erosion of the soils by storm water, defective drainage connection below ground or fractured water supply pipes can cause large cavities to occur under foundations. This can lead to rapid and sudden foundation failure or collapse, which often presents as sinkholes.

The definition of a sinkhole is the sudden appearance of a cavity or opening up of the ground. While this may be caused by the erosion of non-cohesive subsoils, it is often more associated with the erosion of specific subterranean rock strata, such as gypsum and chalk. These are not soils but rocks, which can be eroded by the passage of underground watercourses, creating large cavities under buildings. Similar rapid erosion and subsequent potential collapse of properties can occur on landfill or made ground, as well as in areas of historical mining activity.

Non-cohesive soils are also more susceptible to volumetric change as a result of vibration and this often occurs when an external source, such as the heavy movement of trains or traffic adjacent to or under a property, causes the shift and compaction of fines in the soil. The most graphic example of this is with an earthquake and, while this is less likely to be present in or affect northern and central Europe, it is of real concern to southern Europe and many other parts of the world. Therefore, surveyors working in these areas should be aware of the potential for this to occur.

An example of the effect that an earthquake can have on non-cohesive subsoils occurred with the 1985 Mexico City earthquake. Although the epicentre of this was on the West coast of the country, major damage and loss of life occurred in Mexico City, some 220 miles (350 km) to the east. Built largely on a historic lake bed, the non-cohesive soil was significantly more susceptible to extreme volumetric change than that closer to the epicentre.

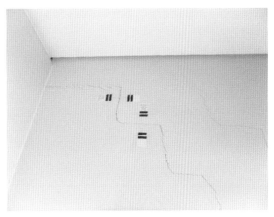

Severe cracking to the facade of a school, discovered during a private finance initiative survey (non-UK). Internal examination revealed no evidence of the crack but further desk studies and discussion with the owner identified that this 'historic' damage was attributable to an earthquake that had occurred some 30 years previously (photo courtesy of Widnell Europe).

Step cracking to the internal wall and column line of a commercial building identified during an acquisition survey (photo courtesy of Widnell Europe).

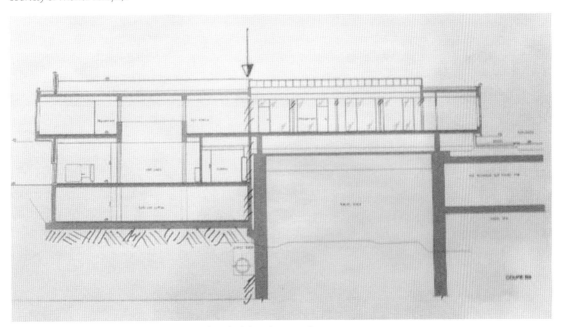

Further investigation and document reviews identified that this part of the building was cantilevered over two underground railway lines. The consequent vibrations were deemed to have caused the cracking (image courtesy of Widnell Europe).

Ground engineering and subsidence

Ground engineering and subsidence is a specialism in itself and surveyors who do not routinely work in this field should avoid jumping to conclusions if structural movement or subsidence is suspected during the building survey or TDD. Foundation failure or structural movement of low-rise commercial properties associated with ground movement may be attributable to the following causes:

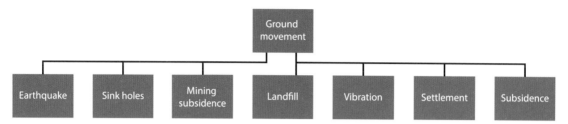

Principle causes of ground movement.

In most cases, this will present as cracking or distortion to the facade or loadbearing walls and floors. It is unlikely that the surveyor will be able to diagnose the cause of this through visual evidence alone and it will inevitably need to be reported to the client as a red flag issue. The commercial building survey or TDD process should avoid leaving open points or issues in the findings and where there is an observation or defect, there should also be a recommendation. Clearly, with structural movement or possible foundation failure it is necessary to recommend immediate further investigation. This may need to be done by a 'specialist' building surveyor or structural engineer and usually involves a selection of the following processes:

- desk survey, geographic/geological analysis
- crack analysis
- trial holes
- soil analysis
- tree root analysis
- camera survey of drains
- level survey
- crack monitoring.

The investigation process can probably be completed within a relative short period but the most important requirement prior to proposing repair is stabilisation of any movement. This can only usually be determined by crack monitoring over a minimum 12-month period to establish whether the movement has ceased. With all building defects, it is paramount to treat or address the cause prior to repairs being executed. Therefore, structural movement or foundation failure can potentially be a deal-breaker with respect to the acquisition process.

Having identified cracking or distortion to the walls, floors or roof of a commercial building, it is important to notify the client but also to begin to form an opinion as to whether this is 'significant', 'live' or 'historic'. This is an area of potential risk for the surveyor, who is liable for their actions in reporting this issue, therefore it is of paramount importance to form an evidence-based opinion. If there is insufficient evidence to validate this opinion, then it may be prudent to recommend further investigation.

Cracking to non-loadbearing elements of low-rise commercial properties may be considered to have less significance in terms of structural integrity but will inevitably have an effect. This could be aesthetic or more serious, allowing water penetration or compromising a fire compartment, and it is therefore still very important to report on the significance of cracks.

It is always necessary to analyse and report on the cause and effect of cracking or suspected structural movement. Being risk averse and giving an opinion is not always straightforward and the surveyor will have to contextualise and do this for each and every situation.

Medium- and high-rise commercial properties

The fundamental difference between low-rise and medium- or high-rise commercial properties is the height of the buildings, with medium-rise considered to have more than four storeys above ground level. The majority of medium- and high-rise commercial properties are likely to be dedicated as offices, hotels or residential use; however, properties can often comprise a mixture of all of these, plus retail in city or town centres.

Generally, it was not until the post-Second World War period that medium- and eventually high-rise properties were adopted as part of the commercial townscape in the UK and Europe. They were initially confined to the capital or larger cities, with specifically high-rise skyscrapers representing status, power and wealth.

Loadbearing external walls need to be extremely thick to support structures above four storeys and it was evident with Victorian architecture how the thickness of walls tapered as the height of the building increased. Therefore, in order to increase the height of the properties and widen clear internal spans of commercial space, loadbearing brickwork gave way initially to iron and steel structures, which have been superseded by steel and reinforced concrete.

Logically, with city centre land and space a limited resource, the most efficient way to maximise property value and investment is to increase the lettable floor area for the building footprint. The way to achieve this is to increase the numbers of floors above and below ground, but this requires specific construction technology which invariably places costs, restrictions and legal prescriptions on the use or occupation of the property.

Generally, lettable floor space to commercial properties below ground level is usually dedicated to storage, archives, parking and technical rooms. However, it should be noted that for some commercial properties, such as retail or hotels, this space may be used for customer or client activity. The absence of windows and natural light tends to make basement space unsuitable for office or residential occupation and, accordingly, the lettable value of the space is less than that above ground.

One common factor regarding all buildings is a requirement for them to have foundations, and one consequence of increasing the number of floors below ground level is a requirement to undertake deep basement construction with increased foundation depths.

Foundation and basement inspections

When undertaking a commercial building survey or TDD of a medium- or high-rise property, one of the few places where it is possible to see the exposed structure is in the basement. It is not the responsibility of the surveyor to analyse the foundation design or suitability, as this will have been done by the design engineer. The control or conformity of the foundation design should be signed off and accepted by the design team, local authority and possibly external control organisations. It is, however, within the remit of the surveyor to review these documents and comment on any remarks or snagging that may be contained within the reports.

In essence, the role of the surveyor is to undertake a visual inspection of the basement areas and this will be limited to those where access is granted. In some instances, certain basement rooms may not be accessible during the visit or may form private tenant archives or storage areas with key access only. In these cases, it is important to note and detail in the report all areas or rooms where access was not granted.

It is usually impossible to inspect the foundations under a medium- or high-rise commercial building as these are below the lowest basement level and are incarcerated or concealed by the basement floor and walls. However, there may be some instances where sub-basement voids may be accessible during a survey and, where this is the case, a specific risk assessment should be made before entering these areas.

Due to the relatively 'modern' design and execution of medium- and high-rise commercial properties, it is common for the foundations to comprise reinforced concrete deep piles below the building. However, the only way to understand the detail of the foundation construction will be to identify this on the as-built documents, plans or sectional drawings.

The top of a pile and preparation of the pile cap for a new build commercial development.

A pile cap and reinforced concrete foundation which was accessible via a ventilated void under a shopping centre. Note the floodwater erosion of the non-cohesive soil around the top of the pile (photo courtesy of Widnell Europe).

The basement perimeter walls are sometimes non-loadbearing for the support of the building above but they may be integrated and connect to the structure below ground. The basement walls do act as retaining walls to the external ground surrounding the property and therefore these are mainly constructed in reinforced concrete and comprise one or more of the following:

- diaphragm walls
- secant piles.

The building survey or TDD will seek to establish the type of basement wall construction and note any visual defects. In some cases, the walls are lined with masonry, therefore obscuring or even masking potential defects, and this should be noted in the survey report. Considering that there can be many basement levels below a building, and particularly when the space is used to accommodate parking, the basement levels can extend to a significant depth. It is necessary to inspect as much of the perimeter basement walls as possible, subject always to access availability.

Water infiltration

Typically, the biggest issue regarding basement construction and operation is the presence of water infiltration. Part of the desk study or knowledge of the surveyor will concern the anticipated ground conditions and factors affecting the water table. Clearly, in coastal locations or sites situated adjacent to significant water courses, rivers or lakes there is a potential for water penetration to occur in the basement. If the lettable space in the basement is being operated as a commercial space, then it is normal to expect that this is free from any infiltration or dampness. In essence, the space has to be dry and normally as a consequence of commercial operation this will include full air supply and extraction. The only exception where dampness may be accepted is in underground basement parking; however, running water of any description is not acceptable and this should be noted as a red flag issue in the report.

Water infiltration in basements comes from three possible directions:

- above
- side
- below.

Typically, water infiltration from above is due to insufficient waterproofing or tanking of the basement at ground floor level and in many ways this is not dissimilar to the function of a roof. Often basements extend beyond the footprint of the building above and it is these areas which are most at risk from water infiltration. However, the principal difference between such basements and roofs is that, in almost all cases, the external face of the basement is concealed by paving, roads or landscaping. As a result of this, it is very difficult to investigate and treat such infiltration. The commercial building survey or TDD will not actively undertake any destructive testing or opening up of areas for detailed further investigation, and infiltration from above typically shows as damp patches, drips or even stalactites forming on the underside of the ground floor slab. Where there is actual formation of stalactites, then this is conclusive evidence that the infiltration has been a long-standing issue. Likewise, the provision of drainage channels or water collectors under the slab addresses the symptom but not the cause.

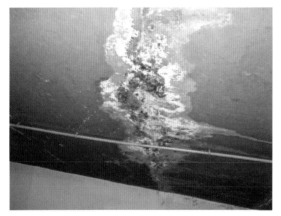

Calcification or the forming of limestone or stalactites is an indication of a long-standing water infiltration problem. Downward streaks (as shown on the right of the image) or water to the concrete blockwork and corrosion of steel cable trays indicate infiltration from above (photo courtesy of Widnell Europe).

The presence of a corrugated polycarbonate 'roof' over electrical boards is treating the symptom and not the cause; this is unacceptable (photo courtesy of Widnell Europe).

The commercial building survey or TDD should note the presence of water infiltration and its potential to cause damage to any finishes, the concrete (spalling), tenant occupation or even dripping onto parked cars, which can damage paintwork. The recommendations concerning water infiltration may be for further specialist investigation or to undertake repairs externally by re-tanking the basement 'roof'.

(TOP & TOP RIGHT) An office building under construction, note the incomplete placement of bitumen mineral felt tanking to the 'roof' of the basement with the presence of surface water causing infiltration in the basement (photos courtesy of Widnell Europe).

Remedial works to the basement of an office building which extends beyond the office footprint; external concrete paving slabs have been removed and the bitumen mineral felt tanking replaced.

Water infiltrating from above in a basement will travel downwards due to gravity and, if there is insufficient sealing between floors, this has the potential to pass to different levels. Although leaks may be initially difficult to trace, treatment can be relatively straightforward.

In contrast, water infiltration from the side or below a basement can rarely be investigated externally and can prove to be a lifelong persistent defect. The cause of such infiltration is usually the presence of the water table and the constant pressure this exerts on the basement walls and floor. During the construction of a basement, where ground water is an issue, it is normal to undertake a de-watering process to prevent the excavation from flooding. Once the basement is completed, the dewatering measures are usually removed, allowing the ground water to effectively press against the external walls and floor. Ground water pressure is generally high and this has the potential to infiltrate through any fine holes, joints or weak spots in the basement walls or floor as the water finds the path of least resistance. This is often treated by injecting the areas with products to seal any voids but these measures can result in water infiltration occurring in other adjacent areas of subsequent relative weakness.

In some buildings where the water table may be particularly high or where, historically, it has not been possible to fully prevent water infiltration into a basement, systems are put in place to treat the symptom in order to make the basement functional. By placing drainage channels to divert the water to a sump at the lowest point in the basement it is possible to pump and evacuate accumulated water. To guarantee that the basement will never flood, it is essential that there is a back-up pump and also that the power supply for the pumps is connected to an emergency or no-break power supply. Alarms should be fitted to the pump and sump installation and monitored constantly to ensure that a rapid response is made to any failure of the system. There have been cases where basement car parks have been flooded by the failure of sump and pump systems, resulting in claims for negligence against the building owner or car park operator. Therefore, it is very important to report the presence of such an installation in the building survey or TDD report and to highlight the risks of potential failure of the equipment. Assessment of the pumps, electrical supply and alarm installations should all be undertaken as part of the M&E survey, which is attached or integrated into the final building survey or TDD report.

There are some instances where remedial drainage channels are concealed behind internal masonry lining walls and impossible to inspect. Remedial recommendations should include the placement of inspection points or even opening up of these channels if there is evidence of system failure.

Placement of a drainage channel inside a basement wall to a 1970s office building; note the vertical pipe which is used to evacuate water from the drainage channel on the floor above (photo courtesy of Widnell Europe).

A drainage sump and pump used to evacuate water from a basement (photo courtesy of Widnell Europe).

A concrete blockwork wall constructed to conceal a basement drainage channel; note the overflowing of the channel has caused water to collect at the base of the wall (photo courtesy of Widnell Europe).

Basement floors

The building survey or TDD report needs to comment on the construction detail and condition of the floors to the basement. Where the space is being operated as car parking or technical rooms it is normally possible to note the detail and also to check for defects as these are not concealed. However, if the space is being operated as retail or customer space, such as in hotels, there will normally be floor finishes which will prevent inspection. In all cases, the type and thickness of the basement floors should be checked or verified from as-built plans and documents.

Basement floors to medium- and high-rise commercial buildings are always constructed from reinforced concrete and these are either pre-cast, cast-in-situ or a combination of both. The floors are mostly finished with floor screed and for parking areas painted finishes are sometimes applied, as well as car parking and circulation markings. Floor screed is usually cast in bays with movement joints around each bay and also adjacent to the basement columns. These are used to accommodate movement and shrinkage of the screed post construction and are evidence of well-designed and competently executed works. Sometimes, where pre-cast concrete hollow rib slabs are used to form basement floors, there is a need to place steel reinforcing mesh in the floor screed, meaning that it is not possible to cut movement joints.

The consequence of shrinkage cracks to the screed is not solely aesthetic as these can be wide enough to allow the passage of water from a car parking floor into the concrete below. This is significant, as water brought into car parks on the wheels of cars can be contaminated by pollution from the roads or salts attributable to the gritting of roads in the winter. If this is allowed to pass into cracks in the floor screed it may be absorbed into the concrete floor slabs and result in carbonation. This will be discussed later in this chapter but, in essence, is the corrosion of the steel reinforcement contained within the concrete, leading to cracking or spalling of the concrete cover.

Painted floor screed to a basement car park with movement joints cut per column line; note the painted finishes are worn and should be reinstated immediately (photo courtesy of Widnell Europe).

Movement joints cut into the floor screed extend around the base of the columns (photo courtesy of Widnell Europe)

Worn painted finishes to concrete floor screed in a basement car park allowing collected water to pass into the cast-in-situ reinforced concrete floor slab below (photo courtesy of Widnell Europe).

Water absorbed into the concrete has caused corrosion of the steel reinforcement (carbonation) causing the concrete to spall (photo courtesy of Widnell Europe).

The most effective way to prevent water penetrating cracks in a basement car park floor is to finish the floor with epoxy paint. This will remove the cause of carbonation, allowing the treatment of the defective concrete; however, having potentially recommended to paint the basement floors, there will be a requirement to re-do this every five to six years (depending on wear) and this should be foreseen in the medium- to long-term cost forecast.

Movement joints are not restricted to the floor screed to basement car parks but are an important consideration for the whole structure of a building, both above and below ground level. Movement joints to the structure of a basement pass through the structural members, such as the beams and columns, and are used to effectively separate different parts of a building's structure. Sometimes they are left open to column and beams but to floors they should be filled with flexible sealants. Where movement joints are bridged by masonry or filled with non-flexible sealants these will be prone to cracking. The site survey, along with cross-checking of any available as-built plans, is useful in determining if movement has occurred in line with existing movement joints. Relatively minor levels of movement may be considered 'normal' when occurring in the location of movement joints and this should be noted in the survey report.

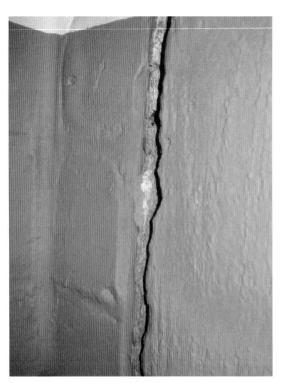

Severe cracking through a basement floor where the screed appears to have been placed over a movement joint in the structure; the cracking also passes vertically through painted blockwork lining walls. The current orange and previous yellow painted finishes have left residual paint on the inside of the crack, suggesting that the crack pre-dates the painting. Estimating that painting occurs every five to ten years, then it may not be unreasonable to believe that the crack is at least five to ten years old, or even older. This is an example where evidence-based judgement should be further explored by questioning the 'history' with the owner, property manager or tenant (photos courtesy of Widnell Europe).

Concrete structure

The use of concrete as a construction material is not a new concept as there is evidence of this being used in Roman times. However, concerning its use in the UK and Europe for commercial buildings, this became of significance in the twentieth century but more prevalent post Second World War. The basic properties of concrete are that it is strong in compression but weak in tension; a good example of this being concrete paving slabs, which can be walked or driven upon but can be easily 'snapped' when tensile force is applied. Therefore, in order to improve tensile strength, steel reinforcement is added. The steel reinforcement comprises bars which run the length of the concrete and stirrups which are horizontal and used to construct a 'cage' or the external corners of the concrete elements. Some concrete structural elements are pre- or post-tensioned and essentially this means they contain steel cables which are placed under tension to improve the loading capacity of the concrete. In most cases it is not possible to ascertain if a structure is pre- or post-tensioned during a commercial building survey or TDD inspection. This is usually visible only with buildings under construction, otherwise the structure and evidence of post-tensioning is likely to be concealed and hidden by the internal finishes or external cladding. Often there are no visible markings to indicate that the structure is pre- or post-tensioned, although such information may be contained within the as-built files. It is not the role of the surveyor to analyse or critique the structural design but to seek to identify any evidence suggesting structural defects. The design engineer and architect are responsible for the structure and conformity with the relevant regulations. Therefore, the surveyor should seek to establish that this has been approved and signed off at the completion or acceptance of the building.

Vertical steel reinforcement bars with horizontal 'stirrups' forming a cage of reinforcement.

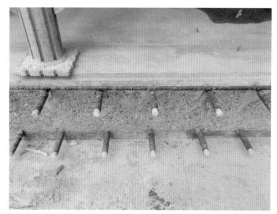

A section through a reinforced concrete floor slab with the exposed steel forming the 'starter bars' for the adjacent section of floor.

Cast-in-situ core walls can also include 'twin wall' construction, which utilises two thin pre-cast reinforced concrete panels which are linked by steel reinforcement and held together with concrete poured and sandwiched between the panels. This is a hybrid of pre-cast and cast-in-situ concrete and a relatively modern method of construction.

Concrete used for structural components in commercial properties is either pre-cast, cast-in-situ or a mixture of both. There are fundamental differences between pre-cast and cast-in-situ concrete; these have an effect on the performance of the material and can affect the prevalence of certain concrete defects. Therefore, it is important during the building survey or TDD process to establish the type of concrete used.

Concrete is a material that is widely used in almost all of the built environment and comprises the following key elements:

- aggregate
- cement
- water
- reinforcement.

The aggregate usually comprises coarse materials, such as gravel, combined with fine aggregates, like sand, mixed in defined ratios with cement and water. There are many different ratios concerning the aggregate, cement and water and it is not usually within the scope of the building survey or TDD instruction to critique these. Essentially, the concrete specification will be designed by the concrete contractor or structural engineer and will be prepared or supplied by the concrete manufacturer. Whether the concrete is pre-cast or cast-in-situ, this is a very flexible material which may be considered a 'liquid' prior to setting with support from formwork to give create the shape. It is a low-cost and very versatile material with excellent fire resistance, which makes it one of the most widely used materials for structural components.

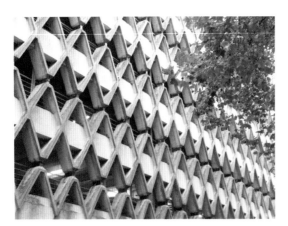

*Examples of the versatile nature of concrete which can be used to form almost unlimited
structural shapes. There is no other structural material which can be used to do this in
the same cost-effective manner (photos courtesy of Widnell Europe).*

Concrete used for structural frames is flexible in design; however, once executed, there is
little possibility to open up or widen the span of the grid or alter storey heights. Therefore, when
considering the life cycle of a concrete framed building, the structural grid, facade module and
storey or floor-to-ceiling heights are an important consideration. Flexibility of lettable floor space
is one of the most important factors for commercial buildings, therefore concrete frames with
wide spans and minimal internal columns to accommodate evenly spaced or 'regular' facade
modules are preferable. Storey heights need to be sufficient to allow clear floor-to-ceiling heights
in excess of typically 2.50 m, with the necessary provision of space in the ceiling voids for the
distribution of HVAC and other services. Therefore, the inflexibility of existing concrete framed
buildings usually results in these being either heavily renovated or redeveloped. There are,
however, some rare examples of buildings with concrete frames being extended in storey height.

Other factors affecting the life cycle of concrete framed buildings are:

- loading capacity
- the mix specification
- protection of steel reinforcement
- permeability
- workmanship/quality
- exposure and air quality.

When undertaking a commercial building survey or TDD it is not possible to verify the
specification of the mix or the positioning of the steel reinforcement. The assessment of the
structure is purely visual and reference should be made to any relevant documents held in
the as-built files or TDD data room. In most cases, much of the structure is not visible and is
concealed by the finishes or cladding. From a technical perspective, this should be considered
as a positive characteristic as most concrete defects occur when the concrete frame is exposed
externally.

The process of forming concrete includes the mixing of the water, aggregate and cement.
When this is poured microscopic air bubbles form and can be exacerbated as moisture
evaporates from the concrete during the curing process. The presence of air bubbles or
microscopic voids results in the concrete having a degree of porosity; this is not usually a
problem when the frame is concealed internally. However, externally moisture can be absorbed
into the concrete, which has the potential to cause carbonation.

Carbonation

As a material, concrete is strongly alkaline and when poured onto steel reinforcement this prevents the steel from corroding. Exposed concrete absorbs moisture and where this is in areas of poor air quality, this acidic pollution is also absorbed into the concrete. In a process which often takes tens of years, the concrete becomes more acidic and, when this comes into contact with the steel reinforcement, corrosion occurs. As the steel corrodes, it expands and causes the existing concrete cover to crack and work loose; this is referred to as spalling. The whole process of this type of concrete decay is known as carbonation.

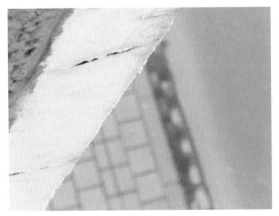

Exposed cast-in-situ reinforced concrete frame with evenly spaced horizontal cracking to a vertical column in line with the steel reinforcement stirrups (photo courtesy of Widnell Europe).

Corrosion of the stirrups and vertical steel bars has caused severe spalling of the concrete (photo courtesy of Widnell Europe).

Cracking along a pre-cast reinforced concrete window cill (LEFT) caused by carbonation and corroded steel reinforcing bar (RIGHT) (photos courtesy of Widnell Europe).

Carbonation is typically more pronounced or likely to occur when the concrete has high porosity or the steel reinforcement is relatively shallow. These may be seen as issues which concern the quality of the mix or may be the result of poor workmanship and quality control. The exposure of the concrete is also relevant as carbonation is often exacerbated in areas of high pollution and moisture, typically in city or town centres where there are high levels of CO_2 and to the exposed upper areas of the building. The potential for carbonation also increases in coastal locations where there are also high levels of moisture with a salt content which is acidic.

While reinforced concrete was used in the early twentieth century it was not until the late 1950s and early 1960s that this became more common for medium- and high-rise construction.

From the 1970s onwards, there was a relative boom in the development of high-rise structures for both residential towers and office buildings. This coincided with the brutalist movement, which embodied an architectural design and style in which the concrete structure, in its 'raw' form, was used for both functional and aesthetic purposes. Typically, concrete used for structural frames in these relative early periods either had little quality control or the future evolution of concrete defects was simply not foreseen or understood. With improved understanding of concrete technology and increased building regulations or building codes, the depths of reinforcement and thickness of the concrete cover has increased. This should, in theory, increase the ability of reinforced concrete to resist the formation of carbonation. However, the principal method to avoid carbonation is to protect the surface of the concrete and this is either done with a painted finish, or more commonly and aesthetically acceptable, by cladding over the concrete.

The perfect conditions for the propagation of carbonation appear therefore to be exposed concrete frames of buildings constructed in the 1960s and 1970s in city centre or coastal locations. All of this information can be gathered or assessed during the commercial building survey or TDD inspection. It is not possible to verify the porosity of the concrete but one important factor to establish is whether the concrete is cast-in-situ or pre-cast.

Pre-cast versus cast-in-situ

The very nature of pre-cast concrete is that this is manufactured under factory conditions, the mix is uniform and the depth of the reinforcement is controlled. In order to reduce the air bubbles and voids in the concrete, the material is vibrated during manufacture and the quality of the formwork often results in smooth construction elements with few or no surface blemishes. Structurally, the concrete has relatively lower porosity when compared to cast-in-situ concrete and is therefore less susceptible to carbonation.

'Real Life' pre-cast reinforced concrete 'Lego' type blocks used as security barriers outside an embassy building. The smooth concrete and perfectly shaped features with no blemishes is achieved in a factory environment. There is always a defined joint between pre-cast concrete components.

Cast-in-situ reinforced concrete is identifiable by the imprint of the formwork (lines in the vertical section), the less smooth surface with fine pores or air bubbles present despite the painted finish. Note there is no discernible joint between the column and beam.

Cast-in-situ concrete is effectively the complete opposite of pre-cast, with formwork for the structure prepared on site and the placement of the steel reinforcement also fixed in situ. Concrete is delivered and poured by a specialist supplier or sub-contractor and this is vibrated with the use of mobile hand-held equipment. Once the concrete has set or cured, the formwork is removed and this often leaves an imprint or pattern on the surface of the concrete, which can be used as an identification feature for the surveyor. The process of casting concrete in situ has less quality control compared to factory manufacture. Accordingly, cast-in-situ concrete is prone to being more porous with greater variability in the location of the steel reinforcement, making the concrete more susceptible to carbonation.

A cast-in-situ reinforced concrete floor slab being executed as viewed from above.

Formwork and temporary support for a cast-in-situ reinforced concrete floor as seen from below.

Post-tensioning cables inside a cast-in-situ reinforced concrete floor slab; these are usually concealed and covered up after construction and not normally visible during a building survey or TDD inspection. The placement, position or design characteristics of these are not normally analysed as part of the building survey or TDD.

The underside of a 1970s cast-in-situ reinforced concrete floor slab; the presence of the formwork imprint is a key indicator that this is cast in situ. This is only visible as the building is under renovation and the suspended ceilings have been removed (photo courtesy of Widnell Europe).

Investigating carbonation

Ruston (2006) and Marshall *et al.* (2014) fully detail a range of concrete defects, including the science behind these, dealing with carbonation, the use of high alumina cement (HAC) and alkali silica reaction (ASR). In essence, carbonation is more likely to be an issue with cast-in-situ reinforced concrete but does also occur in pre-cast reinforced concrete. The commercial building survey or TDD is to advise on the current condition of a building and possible future defects, based upon visual evidence. Therefore, any signs of carbonation need to be assessed in terms of their future evolution or occurrence. The most basic form of repair for carbonation is to break off all of the loose and spalled concrete around the steel reinforcement and then to paint the steel to treat the corrosion. The cover over the steel then requires reinstating with a finish to match the existing in colour, texture and profile. Finally, the surface of the concrete should be painted with a protective coating to mitigate the possibility of a recurrence of the carbonation. While this may appear straightforward, localised repairs to the concrete often fail to match in colour, texture or profile. Furthermore, poorly executed repairs may be prone to recurrence of this defect. Having undertaken a repair locally, there is the possibility for carbonation to occur in other parts of the exposed concrete structure, therefore it is often a requirement to manage the repair and treatment of this defect as a proactive maintenance concern. In order to advise a client on the likelihood of the spread or evolution of carbonation it may be necessary to recommend more detailed further investigation of this issue. Such investigation is not the responsibility of the surveyor but is usually a sub-specialism for a concrete engineer, contractor or surveyor specialising in this area.

The exposed edge of a cast-in-situ reinforced concrete roof slab which has been painted as part of a renovation; the rust stains and longitudinal cracking suggest carbonation (photo courtesy of Widnell Europe)

Repair of carbonation to a pre-cast reinforced concrete facade panel does not match in colour, texture or profile.

In order to investigate carbonation, it will be necessary to undertake some destructive testing and this is usually done under the guidance of a specialist. Several samples need to be taken of the exposed reinforced concrete elements in the area where carbonation is evident and also in other unaffected or less affected locations. This often requires a temporary scaffold or cherry picker to gain access to exposed areas and samples are extracted with a drill operated by a specialist contractor.

Once removed, the concrete sample is sprayed with a phenolphthalein solution, which turns the surface of the sample pink except in those areas affected by carbonation. This will indicate the depth of the carbonation and its proximity to the steel reinforcement. The samples can then be sent off for further laboratory analysis to ascertain the compressive strength of the concrete.

Drill hole from a pre-cast reinforced concrete panel identifying the depth of the steel reinforcement at 28 mm building (photo courtesy of SECO Belgium S.A/N.V).

Spraying of the subsequent sample with phenolphthalein indicated carbonation to a depth of 3 mm, making these sections less susceptible to damage (photo courtesy of SECO Belgium S.A/N.V).

Carbonation to cast-in-situ concrete (LEFT) on the same 1970s building with sample analysis (RIGHT) identifying the depth of the reinforcement at 9 mm and the carbonation at 15 mm, meaning that the management of carbonation is a significant issue for this part of the building (photos courtesy of SECO Belgium S.A/N.V).

The results of the sample analysis will enable the surveyor to make recommendations on the immediate repairs and likelihood for future occurrence of carbonation. This inevitably takes time to complete and is not normally within the scope of the initial commercial building survey or TDD contract instruction. Therefore, if there is any visual evidence of spalled concrete or corrosion spots to the structure, which may indicate corroded steel reinforcement, it is imperative that this is raised as a red flag issue.

Alkali silica reaction (ASR)

Some aggregates in concrete can react with highly alkaline cement and this chemical reaction can cause expansion in the concrete resulting in 'map' cracking to the surface or localised 'popping' (similar to spalling). It can cause decay of the concrete, which may affect the strength of the material, but only in severe cases.

It is not possible to verify ASR through the site inspection alone and, as with carbonation, it will be necessary to take samples of this for laboratory analysis. If the surveyor observes map cracking and spalling of the concrete during the commercial building survey or TDD visit, they should recommend specialist further investigation.

High alumina cement (HAC) concrete

Considered a deleterious material, concrete using HAC was a material developed in the early to mid-twentieth century. Designed for rapid curing, HAC concrete was used in significant quantities in the 1950s and 1960s but ceased to be used the UK after 1976 due to two sudden collapses of structures. This is a potential issue when undertaking a building survey or TDD on commercial properties constructed up to the mid to late 1970s. The application of HAC concrete structure was typically in pre-cast, pre-stressed X or I beams (Rushton, 2006). Other features of this concrete include a dark grey or brown colour. HAC concrete has relatively high porosity and, as a consequence, is prone to rapid carbonation leading to sudden failure of the material as pre-stressed reinforcement cables snap. Investors or building owners should be advised on the potential for HAC concrete in the structure of a building based upon the age and type of construction as well as any visual evidence which may suggest its presence. In the event that there is evidence to suggest the use of HAC concrete in the structure, it is necessary to recommend a further and detailed investigation.

Pre-cast reinforced concrete floors

In addition to cast-in-situ reinforced concrete floor slabs which can be post-tensioned, there is also the possibility to use pre-cast reinforced concrete planks. These often comprise 'hollow rib' construction and are manufactured with hollow voids running through the centre of the slabs. Plastic caps are placed over the holes to the end of each slab for storage on site. However, one consequence of the central voids in the slabs is that these can collect water. Where exposed on site, it is often possible to observe small holes drilled to the underside of these concrete planks to allow water to be drained from the core. This does not affect the strength or stability of the slabs. However, if the presence significant longitudinal or cross cracking to the slabs is observed, this should be the subject of further investigation.

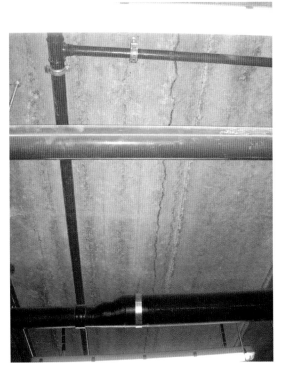

'Hollow rib' precast reinforced concrete floor slab with a 'significant' longitudinal crack (photo courtesy of Widnell Europe).

TOP & BOTTOM – Holes drilled into the slabs have been made to allow water to drain from the slabs (photos courtesy of Widnell Europe).

Cast in situ reinforced concrete slabs – industrial buildings

Typically, those industrial buildings used for light manufacturing and logistics effectively have one storey, although there may be mezzanine or 'bolt on' office accommodation of more than one storey. The ground floor slabs in the actual warehouses or 'sheds' are typically raised above the external ground floor level to ensure there is no change of level for the docking stations. There should be zero level change between the internal shed floor height and the docking lorries to allow goods to be transferred with forklift trucks. To ensure this, there is often a manual or power-assisted dock levelling system which enables palleted goods to be transferred from the warehouse to the lorry trailers.

Forklift trucks are used to move goods internally around industrial building as well as for the transfer of incoming or outgoing goods. They have particularly sensitive wheels which exert very high point loads due to their relatively small surface areas and the high loads carried. The very minimal changes in level or damage to the edges of reinforced concrete floor slabs can cause damage to the wheels, which are costly to repair.

In situ repairs to the floor slabs can involve localised patch repairs but these are not always successful in repairing the edges fully. Therefore, more drastic action may require the cutting of a 'sine wave' movement joint, which is finished with a steel edge.

The floor slabs are often raised above the existing external level and comprise cast-in-situ reinforced concrete laid in bays with movement joints. There is a tendency for these to be the subject of localised settlement, which can result in uneven surfaces or loosening of the slabs which can sometimes physically move when forklifts trucks drive over them. This can be remedied by a number of different methods, including pressure injecting grout or liquid polymers beneath the slab to raise its level.

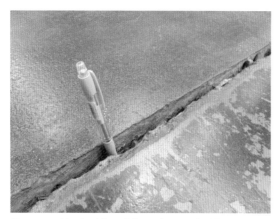

Uneven and 'sunken' floor slab to a logistics building with an unacceptable threshold between the slabs causing damage to forklift truck wheels (photo courtesy of Widnell Europe).

An in-situ polished concrete repair with 'sine wave' movement joint (photo courtesy of Widnell Europe).

Steel structure

Metallic structures have been in existence since the Victorian era with cast and wrought iron being used for significant landmark projects or infrastructure. Examples of these include railway stations such as Paddington Station in London or the Eiffel Tower in Paris. The use of steel as framework for medium- and high-rise buildings occurred towards the end of the nineteenth century and extensively throughout the twentieth and twenty-first centuries. Steel-framed construction is evident in all commercial sectors to many different building types including low-, medium- and high-rise structures.

Steel frames differ from concrete by virtue of the fact that they can span wide distances yet remain relatively slender in dimension. They are comparatively quick to erect with structural sections bolted together or, as with wrought iron or early steel structures, rivetted.

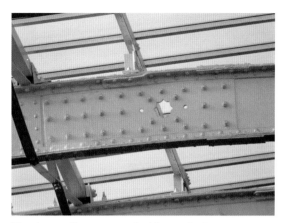 

Historic metallic structure at Paddington Station in London (LEFT), a 1960s steel structure office in London (RIGHT) and a steel structure under construction in 2017 (BOTTOM). Evidence that the worth of metallic structures has been proven throughout many decades and their versatility ensures that they remain a viable option for commercial buildings.

Steel structural frames are often combined with reinforced concrete floor slabs and, where possible, it is necessary for the surveyor to inspect the frame to ascertain any visual evidence of defects. The life cycle of steel structures is similar to that of reinforced concrete as this is only likely to become obsolete if it is not deemed fit for purpose. Examples in which this may be the case concern the grid, module or storey heights, and factors such as the loadbearing capacity or quality of the steel or detailing may also be important. However, these are difficult to assess during a survey as the structure is likely to be concealed by finishes or cladding. In this case it will be necessary for the surveyor to rely upon any reports or documents in the data room or as-built file which detail any acceptance, control or investigations undertaken on the structure. Determining whether a building is steel framed from a site visit alone can be difficult when the structure is not exposed. It is important to understand the culture of construction; for example, in the UK there has historically been a preference for steel-framed construction whereas in other countries in Europe, such as Belgium, reinforced concrete is more common. Also, the age of construction is important as the use of reinforced concrete for medium- and high-rise buildings is a practice which typically occurs in the mid to late twentieth century in both the UK and Europe. Therefore, medium- to high-rise structures constructed from the end of the nineteenth century to the mid-twentieth century may be more likely to encompass metallic structural frames.

Ideally, it would be possible in all building surveys or TDD instructions to have access to execution or as-built plans or documents. These are often extremely valuable in ascertaining the structural nature and characteristics of a building. However, this information is rarely available so it is necessary to rely on visual observations. Inspection of the internal and external column lines may provide evidence indicating the type of structure, with reinforced concrete typically having wider columns and deeper beams to achieve similar spans in comparison to steel-framed construction.

A building survey of a Belgian hotel (TOP) renovated in the late twentieth century discovered photographs on the wall describing the renovation (BOTTOM) which detail the encapsulation of the existing early nineteenth-century cast and wrought iron structure in reinforced concrete. A further desk study identified that the building was built in by a John Cockerill, a British-born industrialist (photos courtesy of Widnell Europe).

TOP & BOTTOM – An office building in central Brussels which was renovated in the early 2000s retaining the original 1950s steel-framed structure, which was encapsulated in concrete. A desk study confirmed that the original building was known as 'The House of Steel' in a country where there is a widespread preference for reinforced concrete (photos courtesy of Widnell Europe).

As with reinforced concrete frames, if there are any potential concerns or issues regarding the structure, the surveyor should seek the advice of a structural engineer unless they are suitably qualified to deal with the issue. Steel frames can be the subject of movement, dependent on their exposure to variations in temperature. They are also susceptible to corrosion, which is why the majority of these are concealed and protected. It is relatively rare to find exposed steel structure to commercial office buildings and where this is used there is inevitably a requirement to treat any corrosion and apply protective painted finishes. This should typically be foreseen in any cost recommendations or maintenance planning attributed to the short, medium and long terms.

One principal defect associated with corroding steel frames is 'Regent Street syndrome', which occurs when steel frames, clad with brickwork, stone or concrete, corrode. It was initially associated with steel-framed structures in Regent Street, London, hence the association with this in the name. In reality, this defect can affect most steel frames, should the appropriate conditions prevail. Similar to carbonation in concrete, where the corroding steel reinforcement causes cracking and spalling of the cover, corroding steel frames cause expansion which results in cracking or movement to the cladding. It is a serious defect and can prove to be very costly to repair. Therefore, it is important for the surveyor to ascertain if the structure is steel, the age of this and any evidence of cracking to the column or beam lines. This may be evident internally or externally and age is a likely determining factor as it should be anticipated that modern commercial properties constructed post 1980 will have incorporated protection against corrosion, although this may not always be the case.

Vertical cracking along the column line of a suspected steel-framed pre-1920s commercial property.

TOP & BOTTOM – Opening up of cracked faience cladding to a commercial property revealed the corroded steel frame (photos courtesy of Tom Goodhand/Paragon).

Steel is susceptible to loss of tensile strength in the event of a fire, therefore one of the key requirements and observations concerning a steel structure should be the presence of fire protection. Historical metallic or steel-framed buildings were constructed with little evidence of specific fire protection, with masonry or stone cladding often offering the only form of protection. It was not until the early to mid-twentieth century that asbestos was used to provide fire protection. The health risks posed by asbestos are well documented but it should be noted that, for several decades, it was the fire protection material of choice. Asbestos comes in several forms but the most common were fibre cement panels used to box around the steel or flock-sprayed asbestos, where sticky fibres were blown onto the steel.

*Flock-sprayed asbestos fire protection to a 1970s office
building (photo courtesy of Widnell Europe).*

Steel frames encased in concrete can have good fire resistance, depending on the depth of
the cover, and it is not uncommon for the steel sections used in such an arrangement to also be
covered with asbestos insulating panels. The visual survey should seek to identify the presence
of a steel frame and, having established this, it will be necessary to determine the existence of
fire protection. It is not the role of the surveyor (unless suitably qualified) to verify compliance of
the fire protection with the relevant regulations, as it is not possible to undertake sample analysis
during the survey. The surveyor should seek to establish the presence of as-built files or data
sheets relating to the protection of the steel and, if these are absent, then the owner, property
manager or tenant should be pressed to deliver this information. Naturally, the surveyor should
be risk averse and, in the event of discovering flock-sprayed coatings or fibre cement boards,
these should be cross-checked and verified against the asbestos inventory or register held for the
building.

Asbestos was banned in the UK and Europe in 1999 and fire protection of steel structures is
now provided by intumescent paint coatings or 'modern' fire protection boards. As-built files or
data sheets relating to the conformity of these materials to the legal prescriptions should be held
in the data room. It would appear unwise for the surveyor (unless suitably qualified) to assess,
gauge or verify the suitability of these materials without referring to manufacturers' information,
including British or European standards.

A steel structure with little evidence of fire protection. The steel has a painted finish which could be intumescent; however, this has been damaged locally with evidence of corrosion to the steel. Note the stainless fixing bolt, which is not fire protected and appears to be an obvious weak point (photo courtesy of Widnell Europe).

A glulam timber roofing member is connected to the concrete vertical structure. The presence of a steel fixing bracket is correctly concealed by fire protection boards (photo courtesy of Widnell Europe).

Timber structure

Compared to reinforced concrete and steel structures for commercial buildings, timber is relatively rarely used. As discussed earlier in this chapter, some timber-framed pre-Georgian commercial properties are in existence but these are usually limited to historic town or city centres. They are usually listed and form part of wider classified conservation areas. The surveying of these is a specialism which many commercial practices rarely get involved in. Timber frames are also used for residential housing but this is not necessarily classed as commercial property other than large portfolios of housing owned and operated by housing associations or local authorities.

For medium- and high-rise construction, the use of timber structural components is usually limited to roofing members; however, it is used in some cases to form structural columns and beams. In order to achieve structural strength as well as design flexibility, glued and laminated timber components are used in a system known as glulam. One of the key features of this type of material is its aesthetic appeal. For some designers, glulam represents a 'natural' material (albeit fabricated) in a manufactured world. It is interesting to look at and can be curved to create striking and unique facade or roof lines.

Portsmouth City Council Somers Town community hub has a glulam structure used to create a unique and visually striking building.

Most buildings which utilise a glulam structure often have this exposed due to the architectural 'feature' that this presents. Therefore, this makes the structure more readily available to inspect than either reinforced concrete or steel.

The Henley Business School at the University of Reading is an example of the aesthetic and functional values of a glulam structure.

While the individual timber strips are glued using 'finger' joints, larger composite columns or beams are usually tied or connected at junctions in the structure with steel brackets. The glulam typically forms the structural support for the eternal facades or roof, but there is often a requirement for this to be complemented with steel or concrete structural elements for the internal floor loading areas. Therefore, it is important to establish the complete structural design while undertaking the commercial building survey or TDD to ascertain any visual evidence of defects.

Typical structural fixing joints between glulam sections. The survey inspection should seek to identify that these appear in good condition with no visual evidence of damage to the joint, fixings or the adjoining timber.

When inspecting a glulam structure, it is important to seek to identify any evidence of structural movement or deflection. This may be difficult where there is an intentional curve in the structure. However, when under excessive load stress, glulam splits along the grain or, in more serious cases, can be prone to sheer cracking against the grain. Any evidence of cracks or splits to the timber sections should be noted and reported as a red flag issue. Accordingly, it may be necessary to seek specialist advice with respect to the severity of the cracking and any specific in-situ repairs. Where repairs have been executed to the glulam, it is also important to ascertain the history of these from the building owner, property manager or tenant. It should be expected that any repairs which have been made will have been done following further investigations, therefore copies of any reports or advice on this issue should be requested for review. There have been recorded cases of sudden glulam failure, but this is usually associated with unexpected overloading of the structure due to adverse weather and snow loads.

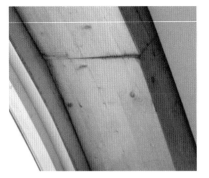

Cracking along the grain of a glulam beam (LEFT), more severe cracking across the grain (MIDDLE) and in-situ bolt repair to cracking along the grain (RIGHT) (photos courtesy of Widnell Europe).

Glulam can be used as a material externally, as the glue should be waterproof and, as with all timber products, this should be treated with wood stain or preservative to ensure that it does not degrade. When undertaking an inspection of external glulam structures, it is important to note the presence of any timber decay or deterioration as well as recommending periodic treatment of the structure with an appropriate treatment.

External glulam structure has been bleached by exposure to the sun, this is an aesthetic detail but it is important to ensure that the timber is treated for protection.

Exposed glulam structure in a coastal location, which does not appear to be detrimental to the functioning of the structure.

Vertical glulam structure to a low-rise shopping centre with evidence of good maintenance and painted timber.

Structure summary

When undertaking a commercial building survey or TDD inspection, it is important to assess the structure as either a low-rise 'residential' type building, where this usually comprises load bearing masonry, or medium- and high-rise buildings, where there is a structural frame. Particularly with framed buildings, the majority of the structure is likely to be concealed during the inspection with this often only visible in the basement, roof areas or in shafts.

Within the scope of the commercial building survey or TDD contract instruction it is not common practice for the surveyor to verify the structural calculations of a building but there is a requirement to seek and record visual evidence which may indicate the presence of structural movement. Some key characteristics or requirements of the survey relating to the structure are summarised below:

- It is important for the surveyor analysing low-rise construction to have an understanding of building age and the likely construction technology relative to each age category.
- It is rare to have access to or to expose the foundations during most building surveys or TDD inspections and there is a requirement to rely on as-built plans, where available.
- Document review and local knowledge are important factors in forming an opinion on the existing structure and any apparent defects, particularly acceptance or completion reports for a building, which may make reference to structural defects, investigation or repair.
- The majority of structural movement initially presents as cracking or deformation to structural members or parts of the building supported by the structure. Undertaking contextualised crack analysis is a key duty of the surveyor.
- Ground engineering and subsidence is a specific sub-specialism undertaken by surveyors or engineers, therefore it is important not to jump to conclusions regarding possible subsidence. Surveyors should have an understanding and local knowledge of the areas they are working in, and this will include the anticipated ground conditions, water courses and historical subsidence issues. While it may be possible to offer advice on the potential causes of structural movement or subsidence to a commercial property, this will inevitably require further detailed investigation, which is usually beyond the standard commercial building survey or TDD instruction.
- Structurally basements should be free from water infiltration and localised dampness is only usually acceptable to parking areas. Where permanent solutions are in place to drain and evacuate ground water, it is important for the surveyor to understand and report upon this provision as well as any potential consequences of a system failure.
- Above ground, the structural storey heights, column grid and facade modules are important structural features which should be assessed. These affect the operation of the building, including the flexibility or sub-division of space as well as the ability to install services, raised floors and suspended ceilings.
- The floor loading capacities of the building are important relative to the building use and this information may be available from design criteria or as-built information. In no circumstances should the surveyor attempt to calculate or verify this unless they are suitably qualified.
- Structural frames are either made from reinforced concrete, steel, timber or a combination of these materials. It is important to establish the type of frame and to advise on the limitations or potential defects which may be associated with this.
- Concrete and steel frames have the potential to last for many decades or even centuries provided they are given adequate protection from the external elements as well as from fire. Therefore, it is important to assess the frames in situ and to comment on the presence and suitability of any protection.

- Concerning specifically reinforced concrete, the surveyor should ascertain the relevant types and locations of concrete used in the construction of a building. There are fundamental differences between pre-cast and cast-in-situ reinforced concrete, and these have an effect of performance, particularly if the concrete is exposed.

- Any evidence of cracked, spalled or loose concrete should be reported, with the location and severity noted. In areas of significant or widespread spalling it may be necessary to seek further specialist advice. Spalled and loose concrete poses an imminent risk to the building users and general public if this is allowed to fall from a building. This is a red flag issue and the owner, property manager or tenant should be informed immediately.

- Steel frames characteristically achieve long spans with relatively slender columns and, with the exception of industrial buildings, the frame is usually concealed or boxed by the facade or cladding.

- Steel is susceptible to thermal movement but the principal defects concern corrosion and insufficient fire protection. Corrosion of masonry-, stone- or concrete-clad steel frames can result in substantial damage, with this defect known as Regent Street syndrome. This presents as cracking to the cladding in line with the structural frame.

- The provision of fire protection is an important requirement for steel-framed buildings and this was typically achieved using asbestos-based materials up to the end of the twentieth century. It should be expected that this information will be contained within the asbestos register or inventory for the building. In the event that there is no information available concerning the presence of asbestos, and based upon the known or anticipated age of the building, it is appropriate to advise on the potential presence of asbestos in the fire protection.

References

BRE. 1995. *Assessment of damage in low-rise buildings* (BRE Digest 251), Building Research Establishment, Garston, Watford.

Marshall, D., Worthing, D., Heath, R., Dann, N. 2014. *Understanding housing defects*, Routledge, Abingdon, Oxon.

Rushton, T. 2006. *Investigating hazardous and deleterious building materials*, RICS Books, London.

Photo courtesy of Widnell Europe

Facades   7

Introduction

In direct contrast to the structure of a commercial building, which is mostly concealed, the facades are generally visible when undertaking a commercial building survey or TDD. However, the visible part of a facade or cladding often only relates to the aesthetic, and while this has historically and continues to be one of the primary purposes, there are other, unseen functional requirements. The wide variety of building types within the different commercial sectors and the age span, which can be in excess of 150 years for 'normal' commercial properties, mean that the surveyor needs extensive knowledge of construction technology to advise on the past, present and future.

The facades to these two commercial properties are vastly contrasting and this is as a direct result of their age category as well as the fundamental difference between the pre-1920s property (LEFT) and that constructed and renovated post-1970 (RIGHT) (photo courtesy of Widnell Europe).

The RICS guidance note (2010) concerning commercial building surveys and TDD recommends the visual inspection of the exposed elements of all walls with attention paid to detailing such as flashings, copings, cornices and similar components. Specifically, when inspecting external cladding components, the guidance note recommends that the surveyor should assess any degradation, movement or cold bridging as well as defective sealants, coatings, fire stopping, cracking or corrosion.

Practically, and dependent upon the size and function of the commercial building, the facade can equate to a large surface area. Concerning the visual inspection, it should be noted that, while it may be possible to assess this externally with few access constraints, internally the facade detailing is likely to be concealed by the finishes. Also likely to be concealed are the fixings, insulation and other important detailing. Therefore, this may result in the surveyor having to form an opinion and make subsequent recommendations based solely on visual evidence and some relatively basic tests. Consequently, this can result in potentially high-risk judgements with large cost implications, particularly with larger commercial buildings. It is therefore important to inspect the facade and give an evidence-based opinion while taking into consideration the functional requirements of the facade.

Functional requirements

Irrespective of the aesthetic value or appearance, facades form part of the building envelope and, similar to roofs, their primary function is to be weathertight. A facade which is not weathertight is simply not fit for purpose for most commercial buildings. The assessment of a facade should also take into account the following key functional requirements:

- strength and stability
- durability and freedom from maintenance
- fire safety
- admission of daylight
- provision of ventilation
- resistance to the passage of sound
- resistance to the passage of heat.

Strength and stability

Every facade has materials which require an element of strength; this is either compressive or tensile and reflects the ability of the material to resist the load of its own weight as well as live loads derived from building occupation or wind and snow. Loadbearing facades also need to act as structural members and this is typical to low-rise commercial properties where the walls are often constructed from masonry. The inspection and in-situ analysis of this type of facade will be intrinsically linked to that of the structural assessment. Where there are structural problems, such as cracked or displaced walls, this may lead to water penetration or damaged internal finishes and this should be separately identified, discussed and reported to the client. Cracked and damaged mortar joints may be the symptom of structural movement, but they can also be the cause of facade defects which present as internal symptoms such as dampness. There are inevitably trails of evidence which link many different building defects and construction elements. It is important to address these for each building element in a logical, systematic manner.

Linked to the strength of facade materials is the facade's overall stability and, while problems with loadbearing facades often result in or arise due to structural movement, non-loadbearing facade stability is a different matter. Invariably, this is related to the facade fixings, which can vary significantly between materials and can result in facade degradation or wind damage. There is primarily a public liability issue with defective facade cladding and building owners have a duty of care to ensure that safety of the building occupiers and general public. Sections of unstable facade may fall from the building and pose a serious risk to life as well as to the reputation of the owner or occupant. Therefore, building owners, property managers and tenants seek to mitigate their risk by instructing surveyors to assess the facade stability. Inspecting the facade may be relatively straightforward for a low-rise commercial property and the visual inspection is usually undertaken from ground level or vantage points, such as adjacent flat roofs or windows. However, inspection of medium- or high-rise facades is much more complicated to achieve within the relatively short timescales afforded with normally limited access equipment.

It is therefore important to look for visual evidence of facade movement and this is typically characterised by cracked or deformed materials as well as misaligned or missing components. There may also be some indication internally of water penetration associated with the facade movement, which should be used to establish a trail of evidence. If there are any concerns about the stability of a facade and the potential for this to loosen or fall from a building, it may be prudent to recommend immediate further detailed inspection.

Durability and freedom from maintenance

One of the purposes of a building survey or TDD is to establish the facade type as well as the construction detail in order to advise upon the relative condition, anticipated life cycle and possible maintenance implications. Ideally, facades should be constructed with materials which have good durability; typically, materials with good durability have long life cycles and often high investment costs. Materials with high aesthetic appeal can also be expensive but do not always have long life cycles or durability.

Associated with the durability of facade materials are their maintenance implications and it is not always the case that a facade with good durability requires little maintenance. There is often a link between the physical location of a building and its facades in respect of exposure to the weather or pollution. It is important when assessing a facade to note the effects of the location on the evolution of any defects past, present and future. Natural stone is a good example of a high-quality facade material which has good longevity; this is clearly evident in the range of commercial properties which have an age of 100 to 150 years. However, this material is susceptible to staining and soiling due to rainwater run-off and pollution. As a consequence, there is a maintenance requirement to clean, treat and protect the stone, which can be a repetitive and costly process.

The soiling and staining of facades is not solely restricted to natural stone but is also common to brick and concrete, as these materials have surfaces which are not smooth or polished. One interesting observation is that with metal or other manmade facade cladding panels, such as those used for curtain walling, facade staining is less of an issue.

 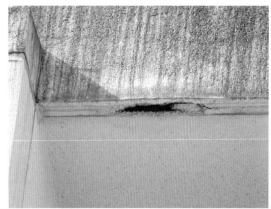

*Rainwater run-off and staining to a pre-cast reinforced concrete facade (**LEFT**) which requires immediate and periodic cleaning. The added potential consequence of this is concrete defects such as carbonation (**RIGHT**) and the porous nature of concrete, brick or stone often means that this is more susceptible to facade staining when compared to metallic cladding panels, curtain walling or glazing.*

Fire safety

Facades are expected to have a degree of fire resistance and this should be deemed sufficient to allow building occupants enough time to evacuate a building. However, one of the most important fire safety requirements of a facade is the ability to separate vertical sections or fire compartments. Fire spread internally in a building is typically through internal walls, ceiling voids or doors. This is often mitigated by constructing internal walls from certified fire resistant materials and the placement of fire doors. Furthermore, ceiling voids can fitted with adequately spaced fire barriers. Despite this, fire will spread if fire doors are left open or walls are 'punctured' by cables or pipes. However, one of the biggest risks of fire spread is externally, via the facades, and it is therefore critical during the design and execution of these to ensure that there is sufficient vertical separation, and this extends to the facade materials, insulation and fixings as well as openings, such as windows. Similar to the internal placement of fire doors, any substandard or open windows may also result in vertical fire compartments being compromised and this may exacerbate fire spread.

Annually, there are global examples of fires to commercial properties which are either uncontained or spread via the facade. Typically, when these are either high-profile buildings or prominent medium- and high-rise properties or where there may have been a fatality, this attracts media coverage. Such fires include the Windsor Tower in Madrid in 2007, The Torch in the UAE (in 2015 and 2017) and the Grenfell Tower in 2017.

The Windsor Tower was a commercial office building from the 1970s and this was engulfed in fire on 12 February 2005. Fire spread vertically via the facade and this was a similar situation with the Marina Torch Tower, which was completed in 2011 and was the subject of two fires in 2015 and 2017. The principal difference between the fires at these two different building is that the Windsor Tower fire effectively destroyed the building. But, in the case of The Torch, both fires were brought under control and the damages contained. This perhaps suggests that modern buildings are more resistant to fire spread or have better firefighting capability with the implementation of more stringent fire regulations. Globally, fire regulations appear to have evolved progressively to encompass increased amounts of medium- and high-rise construction. It is therefore not unreasonable for surveyors believe that new or renovated buildings, which have the requisite planning permission or building permits, have been designed to comply with the applicable fire regulations.

This, however, does not appear to have been the case with the Grenfell Tower fire and the tragic loss of life which occurred on 14 June 2017. This was a local authority owned residential tower located in West London, which had been the subject of renovation with all of the relevant works appearing to have been approved by the design team and building control. These works included a facade renovation, which was an over-clad of an existing concrete facade with insulated aluminium cladding panels. The fire was caused by a faulty electrical appliance in one of the flats and spread rapidly up the facade to engulf the building. Clearly, there were failures with the integrity of the fire resistance to the facades, which were one element among other fire engineering provision that had been deemed acceptable to meet fire the relevant regulations. This raises many important issues concerning the design and execution of facade cladding panels but, from a commercial building survey or TDD perspective, this also impacts upon the audit process.

One of the key principles of a commercial building survey or TDD is that the survey is visual with no opening up of any part of the building or testing of materials. This is particularly relevant when inspecting the facade or anything associated with the integrity or sealing of fire compartments. Unless otherwise suitably qualified, most surveyors are not fire engineering experts. The majority of contract instructions include provision for the building survey or TDD to analyse and report upon legal/technical documents and this is done to verify the compliance of the building with fire regulations. Unless the surveyor was explicitly involved in the design, execution and acceptance of the building, they clearly have insufficient evidence to confirm or certify compliance of the building with all relevant regulations. While this may appear a logical position, necessary to limit the exposure of the surveyor to potential negligence claims, it is likely, in the light of the Grenfell Tower fire, to provide insufficient safeguards to the owner, property manager or tenant. The principal fact is that, unless the surveyor can confirm 100 per cent that the entire facade complies with all relevant regulations then they should rely on the acceptance of other qualified professions. By verifying compliance with fire regulation without having seen the works, the surveyor is effectively providing incorrect information, which is inappropriate from both the client's position and also regarding the mitigation of any potential claim for any subsequent negligence in this matter. The surveyor should, however, review the as-built documents, plans and data sheets, which should include the fire rating or certification of materials and comment on any part of this which poses a potential fire risk.

Concerning fire engineering and, in particular, the vertical fire separation between fire compartments afforded by the facade, individual developed countries have their own regulations and standards and it is critical for the surveyor to have knowledge of these. Typically, vertical fire separation between fire compartments is stipulated to be in the region of 1 meter (depending upon the regulations relevant to the applicable country or state), as this has historically been the perception of the height that flames can reach between floor levels. This has to be taken in context with the knowledge that any subsequent 'normal' glazing above the 1 meter barrier glass has little or no resistance to fire, fracturing with the intense heat of the fire and allowing potential further spread of fire. To achieve fire separation, the facade materials must have sufficient fire resistance, including fire barriers to cladding, and this should be certified in the as-built plans and documents, such as technical data sheets.

Clearly, with the fires affecting the Torch and Grenfell towers there appears to be fundamental issues regarding both the height of the separation and also the resistance of the materials. Both fires spread vertically up or through the facade with visual evidence suggesting that a vertical fire separation of 1 meter would appear to have been insufficient to prevent this. When undertaking a building survey or TDD, the surveyor should pay particular attention to vertical fire separation of the facades and accordingly they should seek to measure and verify the amount of separation, where possible. If it is not feasible to do this on site due to lack of safe access, the measurements should be taken off the as-built plans. Ultimate conformity should be verified by reviewing the acceptance reports of the architect, fire officer, authorities and any building control organisation (private or public sector).

Certain materials have naturally better fire resistance than others, with reinforced concrete or masonry being examples of this. There is almost an honesty quality to mass solid reinforced concrete facades or cladding executed in the 1960s and 1970s. While these may be perceived to have low aesthetic value and be prone to concrete defects, they do have relatively good fire resistance. Aesthetically upgrading the facades by over-cladding these and also improving thermal efficiency has the potential to compromise the fire separation. This may be the case if the insulation or cladding panels have insufficient fire resistance or there are inadequate fire barriers in the cavity created by the over-cladding. The surveyor should note and make reference to this as part of their survey. However, unless they have been present during the execution and acceptance of the works, they cannot be in a position to verify the integrity of the fire barriers. Once more, they must report upon the relevant acceptance afforded by the design team and control organisations. Concerning the fire at the Grenfell Tower, this appears to have been a reinforced concrete clad building constructed in the 1970s, which was renovated and over-clad in 2016. While the fire devastated the facades, spreading rapidly across the new cladding panels, the concrete structure and 'original' facades beneath appear to have remained largely intact. There are also important wider issues concerning the internal spread of the fire at the Grenfell tower and other high-profile buildings. These include the placement of fire doors and integrity of fire compartments; the principles of this will be discussed in *Chapter 11 – Legal/technical*.

Vertical Fire Separation is Provided by the Conrete Slab and the Masonry Cavity Walls to This Facade With Infill Panels. The Fire Risk and Difficulty Comes With Other Materials Such as Glazing and Cladding Panels

The Vertical Separation Varies in Different Countries as per Regulations But Typically This is 900-1000mm

Vertical fire separation (section).

1970s pre-cast reinforced concrete facade cladding.

Curtain glazing (photo courtesy of Widnell Europe).

In situ renovation of the same facade (LEFT), upgrading the glazing while the building appears occupied.

The Grenfell Tower, and despite the failure of the vertical fire separation and evident charred facade, the original base facade constructed in cast-in-situ concrete appears intact and probably would have offered significantly greater resistance to the vertical spread of the fire up the facade in its non-renovated state.

The issue of vertical fire separation is further complicated where the building has curtain glazing with little evidence of vertical fire separation. Curtain glazing is technology typically associated with medium- and high-rise buildings constructed in the 1970s. The technology has evolved but the principles have largely remained the same since this period, with the facade panels or glazing being non-loadbearing. Curtain glazing is essentially fixed or 'hung' from the facade, with the frame secured to the outer edge of the concrete floor slabs or upper structure. Where there are concrete floor slabs, these provide a degree of fire resistance from floor to floor. However, the relatively lightweight nature of curtain walling or glazing potentially allows the passage of fire rapidly up the facades and it is critical that this offers sufficient vertical fire resistance. In the 1970s and 1980s, curtain glazing often had an opaque lower glazing panel and, to facilitate vertical fire separation, an asbestos panel was often placed behind this. This was a fairly standard mode of design for this age of medium- and high-rise buildings.

Typical 1970s medium-rise office building with curtain glazing; note the low-level opaque panels below the windows which may contain asbestos.

1970s aluminium-framed casement window; note the presence of fibre cement (asbestos) panels used to provide fire separation. The internal window cill has been executed in a type of marble, which is evidence that the 'original' building was built to a high specification (photo courtesy of Widnell Europe).

While it is not possible to confirm the presence of asbestos through a visual inspection alone, the surveyor should be risk adverse and assume this until otherwise verified, if visual evidence suggests this. The placement of asbestos panels in a curtain glazed facade is difficult and costly to remove, with these works often only occurring if there is a renovation of the building, including replacement of the facade. In many cases such asbestos, if concealed and not openly accessible to the tenants' demise, is considered to be 'safe' and not posing a risk to the occupants of the building. This is then placed on the building's asbestos register and subject to management as part of the applicable asbestos regulations. This means that the material will be the subject of routine inspection and there may also be a requirement to encapsulate or seal it to prevent any friable fibres entering the atmosphere. With respect to fire safety provision to curtain glazing, it may be observed that for buildings from the 1970s and 1980s it is better to have asbestos panels, managed and controlled in a safe manner, than nothing at all.

The surveyor should seek to measure and verify the height of any vertical fire separation and, while this may be difficult to achieve on site with potentially limited access, it can also be done off any as-built plans.

Post 1999 and the widespread prohibiting of asbestos, other fire resisting panels have been developed for use in curtain glazing or walling.

One fundamental design change has been the use of full-length glazing, effectively creating floor-to-ceiling windows. This inevitably removes the presence of low-level fire resisting panels and one way to reinstate this element is to use fire resisting glass. This has the potential to increase construction costs significantly as the manufacture of this type of glazing is expensive and the surface areas to the external envelope are large. Therefore, architects and designers have become 'creative' in providing vertical fire separation and one way to achieve this is to provide it internally.

Traditional aluminium-framed curtain walling (London) with low-level panels appearing to provide vertical fire protection.

Modern post-2000 office building (Brussels) with floor-to-ceiling glazing; vertical fire separation is achieved with the cantilevered floor plates providing a flame 'barrier'. This is an example of 'creative' design to comply with regulations (photo courtesy of Widnell Europe).

It is not the responsibility of the surveyor to provide an acceptance of such visible solutions but it is their responsibility to review and report upon any as-built documents and acceptance procedures from the architect, engineer, control organisations, fire officer and local authority. These will stipulate whether the design and execution of the facade is in conformity with the relevant regulations. There will always remain the potential for vertical fire spread to exceed the heights stipulated in regulations and this is something that the surveyor cannot dwell upon too much when advising a client. The heights are supposed to have been derived from a raft of scientific testing, typically undertaken by fire engineering specialists, and there is always the possibility for the building owner, property manager or tenant to seek supplementary specialist advise on this matter. Concerning the minimum requirement for the building survey or TDD, this is that the facade is in conformity with the relevant fire engineering regulations and has been accepted accordingly.

Admission of daylight

With the exception of logistics sheds or other industrial buildings, the facades provide an important function in allowing the admission of daylight. There are very few commercial spaces above ground which have no access to the provision of daylight and, in these instances, the lettable value is often reduced. Such space is usually restricted to storage areas or important unseen support activities, such as IT server rooms. Concerning office space, natural light appears to maintain or enhance the well-being of the occupants, irrespective of the level or quality of the artificial lighting.

Large amounts of natural light are not always welcome since they can cause solar glare to computer screens as well as high levels of solar gain, causing discomfort to the end user or occupants. This is not always evident during the commercial building survey or TDD inspection, particularly when the external weather conditions are overcast. One tell-tale sign that solar glare or gain may be an issue is the presence of internal sunscreens or blinds. This is typically a problem to the south-facing facade of buildings located in the northern hemisphere and the opposite for those in the southern hemisphere. The consequence of high levels of solar gain is potentially high costs attributable to increased amounts of cooling required to provide internal comfort. This can have an adverse effect on the energy performance certificate (EPC) or energy rating or grading of the building and is something that should be taken into account when cross-checking the relevant legal/technical documents.

The admission of daylight is invariably achieved with the presence of windows which occupy a percentage of the facade. However, where there is curtain glazing, there are inevitably significantly higher amounts of natural daylight. This can exacerbate the problems associated with solar glare and gain. Advances in glazing technology have resulted in glass panels with improved heat isolation as well as anti-glare function. The incorporation of such technology will typically be evident on the as-built data sheets for the glazing or possibly discussed in the EPC. This information, where available, should be discussed in the findings of the report. It is not the responsibility of the surveyor to perform any on-site tests into the lighting levels, unless they are suitably qualified to do so and this is stipulated in the contract instruction.

In buildings with large floor plates where limited natural daylight would be experienced to those areas furthest from the facades, an acceptable way of introducing this is the provision of lightwells in the design. These can vary in size from relatively small 'shafts' of light to the very large, and often they travel vertically over several floors. The commercial trade-off of incorporating lightwells to introduce natural daylight is a reduction in potential lettable floor area for the plot or building footprint. This is largely the concern of the investment or valuation surveyor and, technically, the facades of lightwells should be treated the same as the rest of the building with respect to the performance criteria. Lightwells generally suffer less than full facades in terms of solar glare and gain as they often have dimensions which are greater in depth than width; this invariably creates areas where the building surrounding the lightwell partially shades the internal facades from direct sun light.

A large central lightwell with landscaping and decorative sculptures (photo courtesy of Widnell Europe).

A much smaller lightwell which permits relatively limited amounts of natural light but presents no problems regarding solar glare or gain (photo courtesy of Widnell Europe).

One of the principal methods of establishing if there is an issue or problem with solar glare or gain at a property is to make specific enquiries of the owner, property manager or tenants during the inspection. When asking tenants to give an opinion on such matters, the surveyor should accept that any opinion offered is subjective, and should also take into account the potentially fractious relationships that can exist between landlords and tenants. This may result in tenants exaggerating problems or issues with the building and landlords understating these.

'Natural' ventilation

Historically, the placement of windows in a facade was to allow for the provision of natural daylight as well as ventilation. This principle is largely the same with low-rise commercial properties which are constructed with 'residential' type construction technology. For larger commercial medium- and high-rise properties, such as offices, hotels or residential towers, there has been a shift in construction technology for the provision of fixed glazing to reduce heat loss in the winter and avoid introducing drafts into the building. The aim of this is to improve the overall EPC of a building but this does require the provision of a mechanical ventilation system to introduce fresh air and create an ambient internal climate. This can work well for offices where the occupation is typically during the working hours of a day and work etiquette often means that occupants come into the office for the day, bringing with them minimal belongings and undertaking relatively low levels of physical activity. Often, the ventilation systems are centralised, with individuals having little or no control over the flow rates or air speed to control temperature.

However, for modern or renovated residential buildings and occupation this is completely different, with 24-hour occupation and activities such as washing, cooking and drying clothes being common. This introduces large amounts of moisture into the internal atmosphere, necessitating ventilation to avoid condensation and the typical associated mould growth. Therefore, ventilation systems are required to have 24-hour operation, but these are often individually controllable per residential unit, meaning that tenants can adjust settings or even switch these off. Such actions can cause imbalances in the air extraction and supply, resulting in inadequate ventilation.

In order to introduce or evacuate air to and from a building, the air supply or extraction is commuted via ducting, which may be connected to openings in the facade finished with protective grills. These form an important part of the building survey or TDD as they are areas which typically have a build-up of dust, pollution or leaves.

In older or unrenovated properties where no ventilation system is installed there may the option to open windows and allow this more 'basic' form of ventilation. However, this air is not filtered or treated, a factor which is of particular relevance to locations where there may be high levels of air pollution, such as city centres.

While the performance criteria for some facades allow for ventilation, this is something typically associated with low-rise or older properties. There an emphasis for more modern buildings to have 'sealed' facades in order to provide controlled internal climates and improve energy performance. There are often older commercial properties which have been fitted with full heating, ventilation and air conditioning services, but which also offer the possibility to open the windows. This can lead to the building's users opening windows while heating or cooling systems are in operation. This risks energy wastage and, with respect to cooling which is provided by cooled ceiling or chilled beams, condensation can form on these which may lead to water staining of the finishes. To avoid this happening it is 'normal' procedure in modern buildings to have fixed glazed facades or lockable windows where the key or locking mechanism is removed. In higher specification buildings where HVAC control is modular, it is possible to isolate this when windows are open. To effect this, it is necessary to have individual micro switches per opening window which are connected to the HVAC regulation system.

The surveyor should seek to establish the presence of micro switches to isolate the HVAC when windows are open through discussion with the building owner, property manager or tenant. This is an area where the building survey (architecture and structure) interfaces with that of the M&E consultant.

Resistance to the passage of sound

While the use of facades for the provision of ventilation may be considered passé, the resistance to the passage of sound is still considered a fundamental requirement of the facade. In most cases this is to prevent external noise pollution affecting the internal comfort levels of the end user or building occupier. Where there are openings in the facades, there will exist the potential for noise to pass from outside to inside and while this may appear obvious with open windows, it is less visible where there are defective windows or facade seals.

The initial observation of the survey will be to note the location of the building. This may be in a city centre or close to road and rail networks, which have the potential for noise disruption. The surveyor should also note during the external inspection the potential presence of noise from airports, particularly if the property is situated on a flight path. Where this is not an obvious issue, the surveyor should seek to establish any problems or noise complaints made by the occupiers or tenants to the property manager or building owner.

In some cases, noise is actually generated from within the boundaries of the building or site, due primarily to the operation of the property. This is not always due to manufacturing processes but is often attributable to the presence of chiller compounds with compressors creating high levels of noise. Almost all roof-mounted technical equipment, such as chillers, compressors or air handling units, will be placed in compounds or recesses which are clad for aesthetic and acoustic reasons. Externally, the facade cladding may have been designed to match that of the remainder of the facades but internally this should be lined with noise attenuating acoustic panels.

Two buildings with curtain glazed facades adjacent to a main line railway, which is a noisy location meaning that the facades require appropriate noise insulation (photo courtesy of Widnell Europe).

A roof-mounted chiller compound with external facades lined with acoustic panels (photo courtesy of Widnell Europe).

The surveyor should seek to identify the presence of acoustic insulation to the facades of technical compounds. Unless appropriately qualified, the surveyor will not usually undertake acoustic tests at the property but should seek to establish if there are any specific operating permits or restrictions pertaining to technical installations. Any evidence of acoustic tests and acceptance reports associated with these in the data room should be reviewed by the surveyor, often in collaboration with the M&E consultant. Finally, the surveyor should seek to establish if there are any reported noise complaints from the building occupants, neighbours or the general public. This should be done through discussion with the building owner, property manager or end users.

Resistance to the passage of heat

One of the largest influences on energy consumption and the EPC is the resistance of the building envelope to the passage of heat. The roof is one of the key building elements which is susceptible to heat gain and heat loss, and the facades can also contribute to this. The survey should establish the orientation of the property with respect to solar exposure. This does not have to be done solely when the sun is shining but can be established using a compass, map or internet-based satellite images.

The age of the property will invariably influence the ability of the facade to resist the passage of heat. Concerning low-rise commercial properties, those constructed pre-1920s are likely to have solid loadbearing walls which have high heat loss in the winter and moderate solar gain in the summer. It is important for the surveyor to establish the age category of low-rise 'residential' type properties as, despite the development of cavity walls, it was not until the post-1980s period that cavity wall insulation became the 'norm'. Therefore, the potential exists for most unrenovated, pre-1980s low-rise commercial properties to have facades with poor resistance to the passage of heat. Aside from the visual inspection, this should also be evident on the EPC. The transfer of heat through the facade is more exaggerated with window units, since both single glazing and the composition of the window frames have an influence on this. Despite the relatively 'old' technology, traditional timber-framed windows have better resistance to the transfer of heat when compared to steel or early aluminium frames, which exacerbated this.

Medium- and high-rise commercial properties are relatively modern compared to pre-1920s low-rise buildings but they still have issues regarding thermal inefficiency. These buildings are typically framed structures with the facades either infilling between the frame or clad over this. Some buildings constructed in the 1950s to the 1980s utilised solid facade panels, executed in reinforced concrete with no cavity or internal lining, which invariably has a negative impact on the passage of heat. Later cladding systems utilising curtain walling or glazing were also relatively poorly insulated. It was not until the early 1990s that the realisation of climate change and, in particular, the impact of the built environment shaped building regulations, influencing design criteria for more energy-efficient buildings. As a consequence, both glazing and facade cladding panels have undergone significant advances in achieving improved energy efficiency.

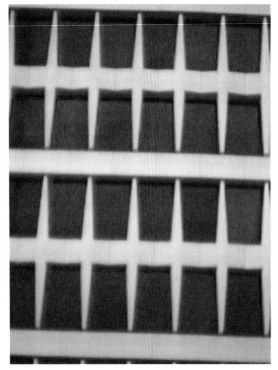
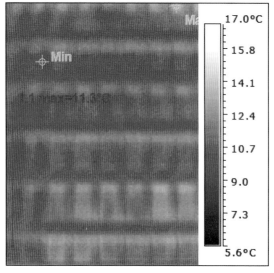

*A thermal image of a 1970s concrete facade and aluminium-framed single glazing units
taken in the winter identifies relatively high temperature levels to the facade with low
resistance to the passage of heat (photos courtesy of Geert Lybeer).*

When analysing facade cladding, panels or glazing it is important to acknowledge that
these equate to large surface areas of the building. Recommending replacement and upgrade
of facade glazing based on energy inefficiency alone is difficult to justify on a cost/benefit basis.
Typically, the cost to upgrade glazing compared to the energy saving means the pay back time
is often greater than the life cycle of the new cladding, or even the overall building. However,
where the facade cladding and, in particular, the glazing is defective, leaking or failing with
respect to any other of the functional requirements, it can be easier to justify replacement and
upgrade. One important consideration concerning the complete replacement of a facade is that
this often means that the building has to be vacated. The consequence of this is effectively a
loss of rental income throughout the duration of the works and this has to be factored into the
decision-making process of the building owner or investor.

Loadbearing and non-loadbearing

The majority of commercial buildings with loadbearing facades may be classified as low-rise
construction, with these rarely rising more than four storeys above ground level. In order to
support increased numbers of storeys above ground level it was common for the external walls
to be thickest at ground floor level, reducing in width with the placement of additional storeys.
The ability of thick walls to accommodate greater heights of construction is known as the
slenderness ratio. To be able to construct higher than four storeys without any structural frame
invariably means that the external walls at ground floor level would need to be extremely thick.
This is essentially the reason why the majority of pre-1920s commercial properties and those
built up to the mid-twentieth century are low-rise buildings with loadbearing external facades.

Structural frame technology was the principal factor in the development of non-loadbearing
facades and it has been established that these either infill the structural frame or over-clad

and hang off this. As a consequence, most non-loadbearing facades are likely to be evident to properties constructed from the mid-twentieth century and are medium- or high-rise buildings. There may be some exceptions to this with the advent of early iron and steel frames pioneered at the end of the nineteenth century and the surveyor should seek to establish what sort of frame is involved as part of the building survey of TDD process.

Common facade materials

There is a wide range of facade materials and, while some of these are intrinsically linked to building age and the relevant construction technology, others have been in used for commercial buildings throughout the building ages. These essentially include brick and natural stone, which have been adapted in their use with advances in construction technology.

Common facade materials include the following:

- brick
- stone
- copper
- zinc
- aluminium
- steel
- render
- concrete
- timber
- high density panels
- glass.

Brick

The use of brick for the facades of commercial properties is typically associated with low-rise buildings, where this is loadbearing and forms one of the principal structural components. Therefore, its use is often widespread in town and city centres, occurring often in the 'older' established districts. The origin of fired clay bricks stretches back many centuries for residential construction and it may be considered to be largely a vernacular material pre-Industrial Revolution. Bricks were largely handmade and the dimensions non-standardised. Their use was typically confined to geographic areas where there were shallow lying clay deposits.

The Industrial Revolution, typically associated with the mid to late nineteenth century, resulted in early factory production of bricks, which were kiln fired to a higher standard than the handmade version and with a degree of uniformity in size. Canals and railways typically associated with industrialisation meant that these could be transported to different geographic locations. Brickwork became the material of choice for creative and innovative architecture, with St Pancras railway station in London providing a notable example. In most towns and cities across the UK and Europe there are examples of intricate brickwork with ornate detailing to commercial properties constructed in the mid to late nineteenth century. As a construction material, brick continued to be popular for low-rise commercial buildings post 1920 to the current era due to its cost-effective durability, and the fact that it can also be utilised for loadbearing structures.

Fact File: BRICK

- ▶ Can be utilised as a facade material and structure when configured as loadbearing masonry.
- ▶ For non-loadbearing facades, early framed buildings utilised brickwork infill panels.
- ▶ Modern late twentieth and early twenty-first century buildings utilise brickwork cladding systems with non-loadbearing masonry on galvanised steel facade supports.
- ▶ Essentially, brickwork is a high-quality material which has a relative high construction cost due to the labour intensive construction process.
- ▶ Brick has a good life cycle with many examples of commercial and residential brickwork lasting several centuries.
- ▶ Factors affecting the life cycle are the brick type and quality, which relates primarily to its compressive strength and porosity.
- ▶ Most brickwork is laid and pointed in cement-based mortar and the strength of the mortar mix as well as the quality of the joints can have an effect on the overall life cycle of the brickwork.
- ▶ The position of the brickwork and exposure to wind-driven rain and freezing temperatures can cause deterioration of the material.
- ▶ Air quality and pollution can cause surface staining, which may require periodic cleaning and protection.
- ▶ Structural and non-structural movement of brickwork can cause damage, reducing the overall life cycle.

A pre-1920s (1901) commercial property with four storeys above ground level with the front facade constructed from loadbearing brickwork.

Reinforced brickwork cladding to the reinforced concrete structure of a modern (2016) office building.

Reinforced concrete frame with brickwork infill panels, typical of 1960s and 1970s construction.

While the majority of bricks used in construction are manufactured from clay, there is a small quantity of bricks which are manufactured from concrete or those known as sand lime bricks, which are made from calcium silicate. Clay bricks are manufactured using largely natural resources; however, concrete or sand lime bricks are mixtures of different materials which have very different visual appearance.

Concrete 'bricks', laid and pointed in cement-based mortar; note the staining caused by rainwater run-off from the cill above (photo courtesy of Widnell Europe).

Both types of brick became evident in the 1970s and concrete is more evident than sand lime as it can be cast or moulded into blocks which have the appearance of natural stone. Rushton (2006) considers sand lime bricks to be a deleterious material by virtue of the fact that they are prone to shrinkage and thermal movement; accordingly there is a requirement to ensure that sufficient movement joints are incorporated in the facade. Mortar strength is also important and, where this is stronger than the brick, any movement will result in fracture of the brick itself.

Low-rise loadbearing brickwork

Loadbearing brickwork facades originate pre-1920s and for commercial buildings typically from the mid-nineteenth century. Up to the early twentieth century, the facades were mostly solid wall with relatively shallow foundations. Some buildings from this era were constructed with brick basements, but this is often associated with better quality buildings and the basements were often limited to one storey below ground level. Up until the end of the nineteenth century, damp proof courses were uncommon, with the use of slate or blue bricks examples of early attempts to prevent rising damp. Traditionally, without cavity wall construction, these commercial properties have poor thermal performance and can be prone to penetrating damp, although commercially it is likely that many pre-1920s properties have undergone renovation and fit outs. This may include the retrofitting of damp proof courses as well as internal or external insulating of the facades to improve thermal efficiency.

The building survey or TDD should seek to establish the age category of the commercial property and, in the case of pre-1920s property, attention should be paid to the potential for dampness. On-site testing for dampness can be carried out with a moisture meter to the external walls, unless these have been dry lined. The visual inspection should note, wherever possible, any upgrade of the solid walls for thermal efficiency and this should be cross-checked with the EPC.

Late nineteenth and early twentieth century brickwork facades to high-quality buildings sometimes embodied faience. These decorative features are panels of terracotta which are typically non-loadbearing and used for aesthetic purposes. Faience is typically porous and where exposed to wind-driven rain and freezing temperatures this may be prone to erosion or decay. To buildings of historical importance, the faience may form part of the classification or listing and is therefore protected. When inspecting and reporting upon this, it may be prudent to seek additional advice from surveyors or experts working specifically with this material.

Ornate detailing to a public sector commercial building.

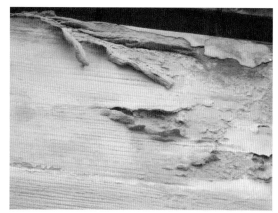

Faience detailing to the gable of a commercial property.   *Erosion and damage to faience, which is difficult to repair.*

Post-1920s brickwork facades were constructed from two leaves of bricks with a cavity in between, designed primarily to prevent water penetration, and the brickwork was tied together using steel wall ties. Foundation depths were greater, resulting in improved resistance to foundation movement and, importantly, damp proof courses were introduced to combat rising damp.

For low-rise commercial properties, this has been standard construction detailing for the past century, although foundation depths have become increasing lower and, while brickwork outer leaves have remained, the internal leaves are now usually constructed from lightweight concrete blocks. Problems associated with corroding steel wall ties have been remedied by substituting galvanised steel with stainless steel also being widely adopted.

Non-loadbearing brickwork

The presence of a concrete, steel or timber-framed structure often means that the associated brickwork facades are non-loadbearing. This is often the case with medium- and high-rise buildings, although there are examples of low-rise properties with non-loadbearing brickwork facades.

As these are non-loadbearing, the brickwork has important aesthetic qualities as well as providing the functional requirements for the building envelope. Typically, non-loadbearing brickwork was used from the 1960s or 1970s and is widely used in current construction. Embodying a cavity, the walls should offer sound weatherproofing to prevent water ingress and, overall, brickwork has good durability as well as few maintenance requirements.

Strength and stability is achieved with the placement of wall ties and modern brickwork often embodies steel reinforcement to improve its tensile strength. Infill panels rest on the structure and, where executed in small sections, are not usually susceptible to potential instability caused by wind load. They are also unlikely to suffer tensile force unless there is failure of the supporting structure. When undertaking a building survey or TDD inspection, infill panels may be easily identified where it is possible to see the frame surrounding the brickwork. This is usually evident externally and is also sometimes visible internally in technical rooms or shafts where there are no internal finishes. With all aspects of the survey it should be noted that the majority of the principal construction elements are concealed internally by the presence of finishes. This does impede or restrict the visual survey as, in most cases, this does not include provision for opening up parts of the building to allow detailed further inspection. Such works are not generally tolerated by the building owner unless there is evidence to suggest that defects exist in these concealed areas or there is reason to suspect that defects have been deliberately hidden or masked from view.

Brickwork infill panels to a 1970s office building which have been exposed during the strip-out for a renovation; note the holes in the facade where the HVAC system was formerly installed (photo courtesy of Widnell Europe).

An example of brickwork infill to a cast-in-situ reinforced concrete frame to a building also constructed in the 1970s. The presence of the formwork marks to the horizontal and vertical intersection with historic streaking of the wet concrete indicate that this was cast in situ.

The presence of infill brickwork panels is typically evident to medium- and high-rise buildings constructed in the 1960s, 1970s and sometimes later; however, post-1980s brickwork is more often executed as a cladding. One consequence of brickwork facades constructed in this period is that they are likely to have little or no cavity wall insulation. With most surveys it will not be possible to confirm this, as no destructive or intrusive tests are carried out. However, where holes have been made in the facade, to facilitate pipework, services or for other reasons, it is prudent to try to inspect these to see if they reveal any further information concerning the facades.

A hole created in the facade of a commercial office building in the technical room confirms that this part of the 1970s office building with brick infill panels has been executed utilising cavity wall construction with an approximate 50 mm uninsulated cavity.

Often, post-1980s medium- and high-rise properties with brickwork facades employ 'modern' cladding systems. These can utilise brickwork panels constructed on galvanised steel supports fixed to the building structure or internal walls, embodying a cavity and insulation. It is almost impossible to confirm if this is the case unless the surveyor has seen evidence of this during construction or can glean this information from documents located in the as-built file.

Two 'modern' post-2000 office buildings constructed with brickwork cladding to the facades (left-hand photo courtesy of Widnell Europe).

Level with the edge of a window opening and underneath the brickwork it is possible to see the steel section supporting the panel of brickwork.

Importantly, it is not the role of the surveyor to critique or assess the number, positioning or suitability of any brickwork supports as, generally, these are not usually visible during the survey. Unless the surveyor has been involved in the execution and acceptance of the facade, it would be wholly inappropriate to comment on the presence of these. Such a statement would invariably have a degree of inaccuracy and is therefore unhelpful to the client, leaving the surveyor exposed or liable for any contradictions in this information. In essence, the surveyor should describe or confirm the presence of these from any as-built documents available, making reference, where necessary, to the acceptance reports of the design team, control organisation and authorities.

Brick slips

As the inspection of a brickwork facade is only visual, it would appear straightforward by virtue of the fact that the surveyor should be seeking only to report visual evidence of defects. However, there are examples of facades which appear to be clad in brickwork but which are in fact finished with thin brickwork 'tiles' known as brick slips.

This technology is typically associated with properties constructed from reinforced concrete with the tiles applied as a decorative feature to appear as brickwork. Therefore, it can be suggested that the common use of brick slips occurred with the introduction of concrete buildings, typically associated with the 1960s and 1970s. However, brick slips are still in use with current modern commercial properties and it can be visually difficult to verify the presence of these during a building survey or TDD inspection.

ABOVE & RIGHT – *The use of brick slips to the facades of buildings which are almost impossible to tell apart from 'traditional' brickwork.*

On this medium-rise office building, the brick slips are only evident as this was observed during construction and it was noted that these are fixed to pre-cast reinforced concrete cladding panels with evidence that the concrete has been set and cured so that it 'keys' with the brick slips. This is evidenced by the sine wave key between the two materials, seen from above.

The brick slips appear glued to the existing facade with a cement-based adhesive. The presence of the brick slip is much more obvious than on the building on the left.

Brick slips are considered to be a deleterious material because, with early use of this technology, these were fixed to the concrete sub-base with polymer modified mortars (Rushton, 2006). Longer term differential movement between the concrete sub-base or frame and the slips can result in cracking of the mortar, allowing moisture ingress which can loosen the slips. Resultant dangers exist should the brick slips work loose and fall from the facade, putting both the occupants and the general public at risk. Therefore, it is very important to advise the client of both the presence of brick slips and also the potential for this danger to arise. However, it should be borne in mind that it can be extremely difficult to detect the presence of these through a visual inspection, particularly if this is limited to being carried out from ground level.

Other visible evidence that a facade may be finished with brick slips is where the 'bricks' do not appear to be 'typical' shape, size or colour. Most bricks used for construction have standard dimensions and the surveyor should be aware of these in the particular country in which they are working. In the UK, the standard brick dimension is 215 mm long, 102.5 mm wide and 65 mm high. Other countries have different dimensions but generally there appears to be uniformity within each respective country regarding the size of bricks used for construction. Therefore, if the surveyor observes or notes unusually sized bricks, which may be patterned, coloured or glazed, then it is possible that these are brick slips.

Unusually long and thin brickwork to a facade may indicate the presence of brick slips.

Ceramic tiles fixed to the facade of a 1970s office building which have come loose and fallen from the facade.

Brick defects

The purpose of the building survey or TDD is to establish the past, present and future in relation to all of the construction elements, including the facades. In essence, it is necessary for the surveyor to establish the historic construction technology used for each of these then establish the current condition and the evolution of any potential future defects or investment costs.

The very nature of brickwork used for facades means that this is exposed to the natural environment and prevailing weather. The location of the brickwork on a building is important regarding its exposure to wind-driven rain, solar gain and pollution. These factors are significant as they can have a detrimental effect upon the condition and life cycle of the material. Concerning the exposure of brickwork to wind-driven rain, this is usually determined by its positioning on the facade in combination with the location of the building. For low-rise properties, or medium-rise surrounded by taller buildings, there is likely to be some shelter from the external elements. This may be why, despite its excellent life cycle, there are very few high-rise buildings executed with brickwork cladding. Brickwork can be an attractive material but is rarely considered to have the same striking architectural appearance as curtain glazing. In essence, it is a solid, reliable but functional material.

Bricks used in the facades of older commercial properties may have been handmade and this is evident as they appear to differ in size from modern bricks. They are typically more porous, as well as having less tensile or compressive strength. This equates to bricks that are quite soft and prone to moisture absorption, which is one of the biggest factors affecting their life cycle.

As bricks were historically made from clay, which was readily available in certain geographic locations, their initial use was also localised, with style and construction techniques known as vernacular architecture. Therefore, when assessing historic commercial properties, it is often noticeable that the bricks or stones used are typical of the building's specific regional location. Industrialisation of the UK and many parts of Europe in the nineteenth century resulted in 'factory' production of bricks, with rail and canal networks transporting these to other regions. As a consequence, the quality of bricks used in the facades of buildings in commercial use today are of much better quality. This is one of the principal reasons why there are relatively large numbers of pre-1920s properties still in commercial use in the twenty-first century.

Marshall *et al*. (2014) give an excellent scientific analysis of brick defects, and the most notable problems are associated with the following:

- frost attack
- rainwater run-off
- efflorescence
- lime run-off
- lime blow
- sulphate attack
- wall tie failure.

Frost attack

The microscopic pores in bricks are prone to moisture absorption and, through capillary action, these can draw moisture into sections of brickwork, affecting significant quantities of the facade. During cold weather when temperatures drop below freezing there is the possibility for moisture within bricks to freeze. As the water turns to ice it undergoes an increase in volume, which reduces again once the bricks have warmed up and the ice turns back into water. The consequence of repetitive freeze/thaw actions is that the surface of the brick is weakened, causing this to work loose and become detached. This is known as spalling and is a relatively common sight where brickwork is exposed to high levels of moisture and freezing temperatures. This is typically to the upper areas of facades, such as parapet walls, or at ground floor level where rain or surface water can splash onto the facades.

Historic bricks, which may be handmade, in a 'Flemish bond' indicating solid loadbearing construction. It is possible to see pores in the bricks even on the surface.

Evidence of frost attack and spalling to the surface of brickwork. Note that some attempt has been made to execute surface repairs with a cement-based render, which is inappropriate.

The consequence of frost damage to brickwork can result in significant erosion of individual bricks or larger sections of brickwork panels. As this often has more significant effects on older, historic brickwork, repair of this may more complicated and costly if the building is listed or classified as having historical or cultural importance.

Brickwork used for modern low-rise, loadbearing facades or cladding to medium-rise properties is essentially constructed from bricks of superior quality to those used in pre-1920s properties. Manufactured under factory conditions, there is inevitably uniformity in terms of the dimensions and quality. There are typically varying degrees of classification concerning compressive strength and moisture absorption. Bricks which have high resistance to compression and low moisture absorption are often known as engineering bricks and are suited to exposed areas or for below ground construction. These bricks are typically identifiable by their smooth and almost 'shiny' appearance and they are rarely used for aesthetic reasons due to their relatively high cost compared to facing bricks.

When undertaking a building survey or TDD, it will be difficult to report upon the compressive strength or frost resistance of brickwork without having access to the relevant technical data sheets, which should be located in the as-built files.

Rainwater run-off

All brickwork facades, irrespective of whether they are loadbearing, infill or cladding, are prone to rainwater run-off. This occurs to exposed areas of the facade and is specifically linked to the design and execution of the facade detailing, typically at the intersection with the roof or with window cills and copings.

Modern architectural design of brickwork cladding or infill panels appears to favour 'flat' facades with clean, unbroken lines where protruding window cills or coping stones 'corrupt' the aesthetic. As a consequence of this, windows are often recessed into the facade with a flat cill incorporated into the facade's design. This is in direct conflict with traditional window cills or coping stones, which overhang or cantilever out from the facade, preventing water from collecting on the window openings and allowing this to run off and down the facade. Furthermore, it is imperative that copings and cills are manufactured or constructed to include a 'drip'. This is a recessed groove to the underside of the cill or coping stone which is marginally set back from the front of the stone and runs parallel to the cill and facade. Water collecting on the upper surface of the cill runs off this and is prevented from running down the facade by collecting on the front side of the drip.

A 'flat' facade with brickwork cladding: note the
soiling and staining from rainwater run-off initially
collecting on the recessed window cill. The dark
line to the mortar joint under the window recess is
moss growth which has loosened the mortar (photo
courtesy of Widnell Europe).

Similar staining to a brickwork facade originating from
rainwater collecting on the copings of the parapet wall
at roof level (photo courtesy of Widnell Europe).

The consequence of rainwater running down a facade in the worst case may result in water
absorption to the brickwork, which may make it prone to frost damage or erosion of the mortar
joints. The possibility of moss growth to the mortar joints has the potential to cause localised
damage and loosening of these, but this usually only affects the surface of pointing and is
unlikely to result in loosening of the brickwork. In most cases, rainwater run-off is likely to cause
staining to the facade with streaks of dirt or pollution occurring in vulnerable areas or localised
efflorescence.

The role of the surveyor is to identify the cause and effect of defects, with part of the report
dedicated to recommendation for remediation of these. With rainwater run-off to facades caused
by architectural detailing it is difficult to remove or treat the cause, therefore it is necessary to
treat the effect or symptom. This will mean cleaning the surface of the brickwork and, in some
cases, applying a protective liquid silicon-based coating to the surface. When costing this as part
of the immediate and any future investment, it is important to ensure that costs are inclusive
of access scaffolding as well as the provision of protection around window openings. It is not
unreasonable to foresee remedial cleaning of a brickwork facade being necessary twice within
a ten-year period; however, this will largely depend on the location, exposure to pollution and
cause of the rainwater run-off.

An angled brickwork facade which does not shed water as well as a vertical face. Note the localised erosion of the mortar joints with moss growth to these. There is no effective way to treat the cause of this as it is not possible or cost effective to remodel the facade. In context, this defect is present to brickwork which is over 40 years old, therefore the evolution of this defect has taken many years and it is unlikely that, having repaired this, it will be necessary to undertake significant works within the following ten years.

Rainwater run-off has caused moisture absorption to the facing brickwork of a perimeter boundary wall, which has resulted in frost damage and localised spalling of the brick face. With the only current remedy being to treat the symptom, it will be difficult to repair (replace) the defective brickwork without stopping a reoccurrence of the defect, therefore there appears to be justification in recommending the placing of a coping stone to the top of the wall. This could typically form an interesting point of discussion in an acquisition or TDD process.

Efflorescence

This 'defect' is largely aesthetic and is quite common to 'new' brickwork and is a result of the drying-out process which causes salts within the brickwork to be drawn out to the surface. The source of moisture and salts within the brickwork may come from within the bricks themselves or from the sand, cement or water used for the mortar. This could also be present where washing-up liquid has been used as a plasticiser to the mortar to aid its workability. Generally, efflorescence should be removed by brushing the surface of the bricks once these are dry and it is presumed that all of the moisture has been drawn out during the drying stage.

Where efflorescence appears to be established, existing brickwork this is usually caused by moisture being absorbed into the brickwork and subsequently drying out. This could be the result of wind-driven rain, typically associated with autumn and winter storms, affecting exposed facades which dry out in the warmer summer months. But it is more likely to occur due to another factor, such as defective guttering, rainwater pipes, overflow pipes or other causes of dampness. Therefore, it is important to report the presence of this and also to establish the possible cause.

White powdery salts, typical of efflorescence to a brickwork facade which appears damp, caused by defective rainwater evacuation. The salts are deposited at the outer edge of the dampness, where the brickwork is starting to dry out (photo courtesy of Widnell Europe).

An example of efflorescence where there appears to be no obvious cause; therefore, it is important not only to report on the efflorescence but also to establish the cause of this.

Lime run-off

Lime run-off is also an aesthetic defect and is not dissimilar to efflorescence. However, one noticeable difference is that this appears to be a more permanent staining to brickwork which cannot easily be brushed off the surface. In essence, this is due to water running through cement-based construction materials, such as mortar or concrete, which deposits dissolved calcium hydroxide onto the surface of the bricks. This reacts with carbon dioxide and hardens into calcium carbonate crystals. Typically, as this is caused by water running through cement-based products it is often present to parapet walls or retaining walls and it presents as visible streaks running down the face of the brickwork.

Lime run-off to a brickwork wall is characterised by thick, hard, white staining (photo courtesy of Widnell Europe).

Lime blow

The diagnosis of this defect is particularly difficult as the symptom affects the surface of the brick causing this to blow or spall, which could originate for a number of different reasons. This defect is caused by the presence of lime in the clay used for bricks and, when these are fired, this is converted into calcium oxide. As with many brick defects, the presence of moisture is an important factor and this can cause the brick surface to blow, with an appearance that is visually similar to spalling.

Sulphate attack

Sulphate attack is another defect propagated by the presence of moisture to brickwork walls, chimneys or even render. It is caused by sulphates, which can be present in ground water, pollution or the bricks themselves. Under excessive and sustained exposure to water, the sulphates can react with the chemical contained in Ordinary Portland (OP) cement. Typically, sulphate attack causes expansion, which presents as cracking to the joints in walls or 'bulging' render cracks along the hidden joint lines. This is also prevalent to chimneys where slow burning of fossil fuels can create sulphates, which can react with wet brickwork caused by condensation within the chimney stack or wind-driven rain to the exposed face of the chimney. Subsequent localised expansion, coupled with the force of the prevailing winds, can cause chimneys to lean and potentially collapse.

The construction of chimneys is typical to older, low-rise commercial properties and any assessment of these is usually done from ground level. Therefore, it is important to utilise the zoom function of optical equipment or cameras when seeking to identify any cracking and the possibility of sulphate attack.

Wall tie failure

Wall tie technology dates back to the late nineteenth century with evidence of early cavity wall construction developed in higher quality Victorian residential properties. This may also be the case for properties from this era which are being used for commercial purposes; however, this technology was not widely implemented until the early twentieth century. Initially made from iron and then steel, cavity wall ties were prone to corroding and, as a consequence, the stability of these loadbearing construction elements may be compromised. Later versions of cavity wall ties included galvanised steel and presently stainless steel is the material of choice. Therefore it should be recognised that the majority of modern low-rise commercial properties constructed post 1980 are unlikely to suffer from this defect. Symptoms of cavity wall tie failure are, typically, evenly spaced horizontal cracking to the mortar joints of brickwork and, in some severe cases, bulging or lateral movement to loadbearing walls.

In most cases, the surveyor will have no knowledge of the actual type of cavity wall construction and will have to rely largely on their knowledge of the relevant building technology which may have been applicable at the time. It is also critical for the surveyor to undertake a visual assessment of the property to seek any evidence of symptoms which may indicate this defect. In the event that there is sufficient visual evidence to suspect that this defect may be present, then it may be prudent to recommend further investigation.

A cavity wall constructed post 1960 which has been exposed during renovation works revealing large amounts of mortar 'snots' collected in the internal cavities during construction. These have the potential to bridge the cavity and may exacerbate corrosion of the steel. Removal of mortar 'snots' is a fundamental requirement during construction and is a good benchmark of the on-site quality control measures.

Horizontal cracking of mortar joints attributable to cavity wall tie failure (photo courtesy of Dennis Wilkinson).

Wall tie failure may be considered a defect of the past; however, this was brought sharply back into focus for modern post-1980s commercial properties in 2012. The sudden collapse of cavity wall infill panels to a school building in Scotland and the subsequent checks on other buildings revealed insufficient or incorrect wall tie use. This was widely reported on national television news, in press reports and also in professional journals, such as *Construction News*.

The sudden collapse was reported to have happened during high winds, probably with little or no warning. This raises important issues when undertaking a visual inspection as part of a building survey or TDD. Unless there is any visual evidence of imminent wall tie failure at the exact moment in time of the survey, it is likely that the cavity wall facade will be reported as functional and serviceable with no apparent defects. It will not be possible for the surveyor to analyse or report on the presence, numbers or spacing of cavity wall ties, as these are simply not visible. Therefore, the surveyor will have to rely and report upon any documents contained within the as-built files that relate to the design, execution or acceptance of this aspect of the build by the architect, engineers, control organisations and authorities. Unless the surveyor was present during the design, execution and acceptance of the works, they cannot be in an informed position to verify this.

Thermal movement

As detailed *Chapter 6 – Structure*, panels of brickwork can be prone to expansion and this typically occurs to south-facing facades where there is greater solar exposure. Therefore, when inspecting panels of masonry, either loadbearing, infill or cladding, it is prudent to note its orientation as well as any evidence of thermal movement. 'Modern' brickwork is likely to be executed in sections which have movement or expansion joints finished with flexible sealants, such as mastic. Depending on the location, UV exposure, thickness and quality of the mastic, this may be functional for 10 to 15 years, although it should be noted that some mastic may far exceed this. As it ages or degrades due to solar exposure, it tends to become harder and brittle with shrinkage cracks appearing. It may also become less flexible and is prone to tearing in the event of movement of the brickwork on either side. Cracking or opening up of the material leaves the exposed joint vulnerable to moisture ingress; therefore, it is not only important to report on the presence of flexible sealants but also to assess their condition, anticipated age and life cycle.

Two adjoining panels of brickwork without a movement joint. Subsequent movement of this has resulted in cracking and displacement of the bricks to the left-hand panel (photo courtesy of Widnell Europe).

Mortar

Often seen as the 'glue' used to bind courses of brickwork or stone, mortar is an essential requirement for all types of masonry, irrespective of whether this is loadbearing, infill panels or cladding. Mortar is used to hold bricks together and distribute load evenly; however, one fundamental characteristic of mortar is that it seals brickwork joints and prevents water penetration. Despite the evolution of mastic or flexible silicon-based sealants, mortar has historically been, and continues to be, seen as the most appropriate material for constructing brickwork and waterproofing joints. This is probably due to its simple sand, cement and water mix, which is relatively low cost as well as being easily mixed. Its workability can be improved with the use of plasticisers and the colour, texture or profile can be adjusted manually by altering the material content or through manual working of the mortar. Unlike mastics, it is not affected by exposure to extreme temperatures, which often result in cracking of these materials, therefore mortars have significantly longer life cycles. The advantage of mastic is its flexibility and this is why it is typically used to provide seals around openings in the facade or between panels of brickwork. In contrast, mortar is generally not flexible and any movement of the brickwork is likely to cause cracks to the joints. This is the case with mortars using OP cement; however, mortar use for historic buildings (typically pre-1920s) incorporated lime as the bonding agent and this resulted in mortars with a degree of flexibility which can accommodate movement. This is why it may be possible to observe distortion or movement to ancient brickwork facades which have little or no cracking to joints in the masonry or stone. Generally, cement-based mortar should be of a mix that is 'weaker' than the brickwork, as subsequent movement will result in cracking of the mortar joint rather than cracking of the brickwork. It is significantly easier to rake out and replace cracked or defective mortar than to chop out and replace bricks.

When undertaking a visual inspection of mortar joints as part of a building survey or TDD, the age of the building is important as many pre-1920s buildings will have lime-based mortar joints, which are softer than OP cement mortar and are susceptible to erosion. The surveyor should pay attention to any cracking to the joints and, utilising crack analysis, they should seek to establish any visual evidence suggesting structural movement. The in-situ strength of the mortar can be simply tested by rubbing the joints with a finger, pen or something hard to establish if this is crumbly and can be easily eroded. 'Soft' or friable mortar joints can be indicative of lime mortar or cement-based mortar which has insufficient cement content. This is likely to be susceptible to localised erosion and the surveyor should check to see if there is any evidence of erosion occurring to exposed upper areas of the facade or close to ground floor level where dampness may contribute to damage. The recommendation to repair mortar joints should include the stipulation that this is done to match the existing in colour, texture and profile.

Mortar joints which have been executed with OP cement where the cement to sand ratio is high will result in very hard but brittle joints. Any moisture collecting on the surface or edge of the joints will not readily be absorbed into these due to their hardness, but instead may be absorbed by the adjacent bricks. This can make the bricks susceptible to frost damage or spalling and is typically more evident in the case of historic properties. Where the brickwork has been repaired or repointed with cement-based mortar instead of lime-based, significant damage can occur to the brickwork.

Movement of this solid brickwork boundary wall, executed in English bond, has resulted in cracking through the brickwork as the mortar is stronger than the bricks.

A historic commercial building with a brickwork facade which has been repointed with cement-based mortar, which is significantly harder that the surrounding brickwork, resulting in moisture absorption to the brickwork and frost damage.

Mortar joints using OP cement can be formed or shaped to give different aesthetic appearance but it should be noted that sealing of the brickwork joints and the stability of the facade have greater importance. Most joints are 'struck' or 'bucket handle', which means that they are finished using a trowel or tool to gently angle or recess the joints between the bricks in order to aid water run-off from the joint. However, a popular type of pointing, which is used for brickwork but more commonly for natural stone, is 'ribbon'. This is pointing which is raised or protruded from the surface and needs to be formed by a bricklayer's trowel. Therefore, it typically utilises cement-based mortar, which is workable and has good strength when set. In contrast, lime-based mortar probably does not have sufficient strength to form 'ribbon' joints as it is softer and crumblier. It is a type of pointing that perhaps was 'fashionable' at a moment in time and is certainly not a practice that dates from pre-1920s, although it is typically found on this age category of building.

 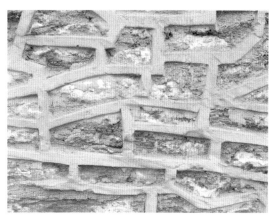

Typical 'ribbon' pointing to natural stone (LEFT), contrasted with the second image (RIGHT), where the raised mortar joints dominate the facade and are poorly executed with cement mortar splashes and smearing to the stone.

One of the consequences of raised and protruding mortar joints is that these do not readily shed water and rain can collect on their surface. This can potentially be absorbed into the junction between the stone or brick and the mortar, giving rise to avoidable frost damage. The damages may be exacerbated if the mortar is stronger or harder than the brick or stone, which is typically the case in older, pre-1920s properties.

The role of the surveyor should be to note the presence of 'ribbon' pointing and to establish any evidence of defects caused to the facade as a direct result of this. While it is necessary to comment upon the situation at present and link this to the historical construction technology, it is also appropriate to advise on the potential for future defects. The object of the survey is to give the building owner, property manager, tenant or investor sufficient information to make informed decisions on the built asset. This type of pointing has aesthetic properties which are subjective; however, objectively it is not normally in keeping with the historical building technology associated with pre-1920s properties and can introduce defects into the facade. In the case of buildings which are listed or classified, the presence of 'ribbon' pointing may contravene the listing. Therefore, clients should be made aware of this possibility and the potential consequences of having to reverse this alteration and put the building back in accordance with the listing.

Loadbearing stone facades

In the terms of materials used for commercial buildings, stone is probably the material most associated with pre-1920s low-rise commercial properties. It is, however, used for other commercial properties with medium- to high-rise office buildings utilising this hard-wearing and durable material.

Pre-1920s, the options for facades were typically limited to stone or brickwork and this was also required to perform a structural loadbearing function. Certainly, pre-industrialisation stone was sourced and used locally, as a vernacular material. With the advent of canals and railways the possibilities for transport opened up, allowing stone to be sourced as a material of choice, which subsequently led to it being seen as a material of wealth. The modern use of natural stone as a facade cladding is also seen as portraying an example of wealth or expense, with the additional benefits of an excellent life cycle. However, in line with financial constraints which are often placed upon investors or developers, this has led to the sourcing of natural stone from developing parts of the world. Here, the costs of quarrying and milling are typically lower than in developed countries. Combine this factor with large amounts of readily available natural resources, and lower cost stone facade cladding panels can be sourced. One unknown about this process is the quality and, in particular, the life cycle of the stone compared to that sourced from existing resources with evidence of durability spanning several centuries.

Certainly, with new properties it may be possible for the surveyor to access data sheets confirming the origins and quality of natural stone claddings. Furthermore, stone originating from developing countries, including the BRICS countries, was not readily available until the end of the twentieth century. Therefore, stone used before this period is likely to have originated relatively locally, from the 'native' country or continent.

The surveyor should be expected to identify the presence of natural stone used for a facade and be able to differentiate this from other facade materials. Although the visit may be brief and inspection is usually visual, there should be sufficient evidence for the surveyor to identify the broad type or class of stone.

Stone is the worked material and this is derived from rock, which is a natural material formed over many millions of years. There are, essentially, three types of rock formations and these are important with respect to the properties of the stone extracted from them. The rock formations are:

- igneous
- metamorphic
- sedimentary.

Igneous rock is formed from molten magma and the most common stone derived from this for the construction of facades is granite. This is a very dense and hard material which is difficult to work but has a very long life cycle. Typically, natural deposits of granite have occurred to mountainous areas, where it is used in blocks for facades as an example of vernacular architecture.

Metamorphic rock is formed through a combination of heat and pressure, resulting in a variety of different stone, including slate and marble. These two materials differ considerably in colour, texture and use, with slate often associated with roof coverings and marble considered to be a material for internal finishes. This is largely true for slate but there are several examples of buildings which incorporate facade cladding panels executed in marble, which is a high-quality, 'luxurious' material.

The most common type of stone used for facades is derived from sedimentary rock, formed by the collection of layers of sediment which are compressed over many millions of years to form stone. Examples of sedimentary rocks include limestone or sandstone and, when used as facade materials, these have varying degrees of durability.

In essence, stone used for facades can be classed as a high-quality material with a good anticipated life cycle. It requires a degree of skill in the manufacturing process as well as during on-site execution and, accordingly, is one of the highest cost facade materials.

As previously indicated at the start of this chapter, there are significant differences in the construction of facades which relate specifically to the height of commercial buildings. Low-rise properties with natural stone facades are likely to be historic buildings, with many constructed pre-1920s. To these properties, the stone facades are also loadbearing structural components and particularly concerning the design or visual appearance of these, the stone is either classed as rubble or ashlar.

Natural stone utilised for the facade of a pre-1920s commercial property. One part is a banking property and this is believed always to have been the case. The use of polished granite for the column detail and the ornate carving of sedimentary stone to the column heads are examples of exuberant materials 'imported' to this region in the nineteenth century. It is testimony to the quality of the materials that these are in such good condition some 150 to 200 years later.

Ashlar

Originating from the Georgian era, ashlar is the descriptive term for natural stone facades which encompass large, uniform blocks with fine and even joints. This type of construction is typical of the Georgian towns or cities, including Bath and Edinburgh, but is not only limited to concentrated areas of Georgian architecture as it is possible to find examples of ashlar facades in most UK towns and cities.

'Classical' pre-1920s architecture utilising ashlar to a building currently operated as a shopping arcade.

Ashlar detailing to the window and door reveals of a much later Victorian industrial property constructed with rubble walls and renovated for commercial use.

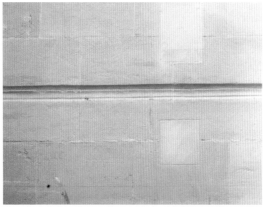

An example of smooth stonework with fine joints, typical of Georgian architecture.

Rubble

All other stone walls or facades are known as rubble. These are denoted by rougher stone pieces and are often associated with Victorian architecture. Local stone (vernacular architecture) associated with traditional, localised materials is typically pre-Victorian. As stone walls or facades to low-rise commercial properties are likely to be present on historic properties, these are often located in historic town or city centres. As a consequence of their location, they may fall within conservation areas or the buildings may be listed. This does have an influence on the repair of defects and it may be prudent, during the course of a survey, to contact a specialist surveyor or stone mason as well as the conservation officer in respect of remedial repairs. This can prolong the building survey or TDD, therefore the initial assessment of the property should seek to identify any significant defects in the natural stone facade with a view to raising this as a red flag issue at the earliest opportunity.

In many cases, stone defects are localised and typically caused by the presence of moisture, salts or pollution to the surface of the stone. Similar to bricks, stone is porous, therefore it can be susceptible to water absorption, which can cause frost damage or spalling as per masonry. There are a wide variety of different stone types used for low-rise, commercial properties, but typically these are limestones or sandstones and they have variable degrees of porosity. Some variation can even occur between individual stones on a facade, which might have been quarried and cut from different seams.

The significance of pore size is important to understand with respect to the defects which may occur. Stone with small pore size (microscopic) has high capillary action, meaning that it can draw water or moisture into the stone. In turn, this can lead to frost damage and spalling under the right conditions. Stone with larger pores is less susceptible to capillary action; however, water evaporating within the pores can lead to salt crystallisation, which in turn may present as spalling to the stone surface.

Certain stones derived from metamorphic rocks are highly resistant to moisture penetration and these include slate and marble. Both materials, in turn, have found their use in the built environment with slate almost exclusively used for roofing and marble as a high-quality internal finish. Most surveyors are inquisitive by nature and, by way of an open question, it is interesting to consider or form an argument as to why these materials have historically rarely been used for facade construction.

While undertaking an inspection for a building survey or TDD it is impossible for the surveyor to ascertain a material's pore size or its susceptibility to moisture ingress and the degree of related capillary action. The inspection is visual and the surveyor should be able to quickly assess the presence of evidence suggesting moisture ingress and spalling. Moisture entering a stone facade is likely to be present to exposed areas above ground level or from moisture in the ground. Having establish the symptom (effect), the surveyor should seek to establish the cause. With most building defects it is necessary to address or remove the cause before repairing the symptom.

Water collecting on an exposed stone cornice has run off and been absorbed into the vertical stone section, which has resulted in frost damage. Although relatively small, these stone fragments pose a serious hazard to health if they fall onto people below.

In contrast, an exposed stone cornice has been finished with lead detailing, including a 'drip' which prevents moisture being absorbed into the stone. Such detailing should be noted on the building survey/TDD report.

As the majority of low-rise, historic commercial stone facades are either limestone or sandstone, it is important to recognise the significance of these being sedimentary rocks. The natural fault lines in the rock can facilitate the splitting and milling of individual stones. However, when these are laid perpendicular to the bed (rotated through 90 degrees) this can expose the face of the bed, potentially allowing this to spall or delaminate from the surface of the facade. When observed during the survey, this may present as spalling to the stone surface and it may be difficult to establish the bed line to confirm this. Such a defect is likely to occur after a significant period of time, due mainly to these facades having been in existence for 150 to 200 years. The appraisal by the surveyor should seek to establish the past with evidence of any historic repairs, the present with those defects currently visible and the future with the likelihood of similar defects to occur. Such an assessment concerning the future evolution of defects can only be made on the basis of the evidence presented, anything else would be guesswork. In essence, the surveyor can only say what they see and contextualise these observations with the likely construction technology and effects of location, exposure and climate. Importantly, any evidence of cracked or spalling stonework to a facade should be reported immediately to the building owner or property manager. The surveyor should make an on-site assessment of the potential for this to fall from the facade, thus endangering members of the general public. If in doubt as to the potential for there to be a risk to health and safety, the surveyor should raise this immediately with the owner and recommend that immediate protective measures are put in place.

The photo illustrates the bed layer to sedimentary rock used in stone to a boundary wall. There is already evidence of some delamination or separation of the layers but, as this is in the horizontal plane, the surface of the stone is held in place by the stones above and below. However, turn the stone through 90 degrees and the surface of the stone would eventually fall away.

A visible example of a sandstone used to the facade of a bank. The building is listed and sits within a conservation area. Horizontal cracking in the stone has been caused by a soft bed; this is a layer of relatively softer stone formed by different organic materials during the formation of this sedimentary rock. The stone is raised approximately 1 m above the ground and not susceptible to moisture ingress, therefore cutting out and replacing the stone would cause more aesthetic harm than good. Currently, the stone poses no risk or threat of decay or coming loose, therefore it is prudent to recommend monitoring as part of routine maintenance inspection.

When undertaking a commercial building survey or TDD, it is important to note staining or soiling to the facade. In some cases this may be due to rainwater run-off from poor parapet or cill details and can be localised. However, with stone facades located in city or town centres these can be susceptible to soiling or staining due to pollution or the effects of acid rain. Twentieth century pollution can be a significant problem for historic low-rise commercial buildings, and there is an ongoing debate as to whether this should be cleaned or not. The risk is that cleaning the facade can cause more harm than good. If staining is present to the facade, the surveyor should note this in the report; however, with historic buildings it may be prudent to seek specialist advice from experts in this field. This is required to determine the appropriate repair or cleaning techniques and, in the case of listed buildings or those within a conservation area, the authorities will need to be consulted. Cleaning and restoration of a natural stone facade can be expensive and the role of the surveyor is to provide all of the necessary information to allow the building owner or investor to make an informed decision. Depending on the location and severity of staining, cleaning may need to be budgeted for twice within a ten-year cost plan. This is often something that is the subject of detailed discussion between buyer and seller in the acquisition process.

Air pollution caused by industrialisation can result in acid rain, which is a term used to describe acidic moisture within the air and includes rain, mist and fog. This can react with chemicals in the limestone, making it susceptible to erosion. This is described in significant detail by Marshall *et al.* (2014). In essence, acid rain can result in areas of exposed limestone detailing being eroded or even washed away. This may present during the site visit as missing or eroded stone details but can also be seen as blistering or peeling of the stone surface with erosion behind.

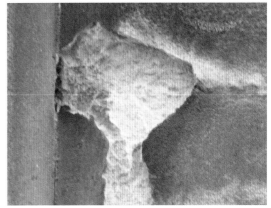

Complete erosion of a limestone window cill to a historic building (photo courtesy of Dennis Wilkinson).

TOP & BOTTOM – Acid rain causing blistering behind the surface of a stone facade.

The treatment and repair of stonework affected by acid rain can be complex and costly. As with most repairs to historic buildings, any listing or conservation area status may impose restrictions on this. It may therefore be prudent to seek a second opinion from someone suitably qualified in this field.

As a facade material, stone needs to embody the relevant functional requirements, including strength and stability, therefore one generalisation is that the thicker the stone, the stronger it is. Certain stone has greater strength than others, with granite being an obvious example. This is why this has traditionally been used for lintels or quoins to high-quality historic buildings.

Historically, the fixing of stone blocks for facades in pre-1920s commercial properties was done with iron or steel cramps and, when these start to corrode, this can cause damage to the facade. Similar to the process of corrosion associated with carbonation in concrete, corroding metal cramps expand, causing cracks to the surface of the stone and possible spalling of the material. This can lead to significant sized pieces of stone falling from a facade, resulting in a hazard to health. Any steel or iron fixings which are drilled or cut into stone can be susceptible to corrosion, which is likely to cause cracking of the stone. Therefore, the surveyor should seek to identify any visible evidence of metal components, advise on the presence of or potential for corrosion, noting also any actual cracks present to the facade.

A low-rise, pre-1920s commercial property with part natural stone cladding (TOP). Corroded steel fixings have caused cracking to the cladding (BOTTOM).

Corroded metallic cramps to a pre-1920s low-rise commercial property (TOP) and a corroded steel rainwater pipe passing through a stone facade (BOTTOM) have resulted in expansion and cracking to the stone (photo courtesy of Widnell Europe).

Painted stone

In some instances during the course of a building survey or TDD, the presence of painted external finishes to stone facades may become evident. This is typically undertaken to change the aesthetic appearance of a facade, mask or cover repairs to the stone and to prevent moisture penetration. This is often undertaken to historic pre-1920s properties and a consequence of applying a painted external finish to stone or any other facade is the requirement to repaint periodically. Therefore, when ascertaining the condition of external painted finishes, it is necessary to advise accordingly on the relevant costs associated with their routine maintenance. Typically included in the cost estimate for such works should be an allowance for access, which may include scaffolding as well as protection to glazing or other surfaces which are the subject of these works.

While painting an external stone facade has aesthetic consequences, it can also have adverse effects on the stone. Oil- or silicon-based paints are effective at preventing moisture ingress; however, where there is already moisture within the stone, these types of paint restrict its evaporation. This typically presents as spalling where the evaporating moisture pushes against the paint, which is impervious. Subsequently, spalled paint which is stuck to the stone causes the surface of the stone to peel or spall. If this defect is evident during the building survey or TDD inspection, the surveyor should note the cause and effect, advising the client accordingly via the report. Any proposal for remedial repairs or repainting should ensure that the specification of the paint is suited to the stone and fit for purpose. If the surveyor is in any doubt as to the appropriate paint, they should seek relevant specialist advice. In the case of listed buildings or those within a conservation area, this will need to be approved by the authorities with works undertaken according to the relevant permissions or permits.

Up close some spalling of the painted surface is apparent, due to moisture trapped behind the paint (photo courtesy of Widnell Europe).

Painted finish to a natural stone and brickwork facade has radically changed the appearance of this office building, originally constructed in the 1930s (photo courtesy of Widnell Europe).

Where the surveyor identifies evidence of 'recent' painted finishes to a stone facade, they should seek to review any relevant permits and acceptance reports, as well as reviewing any available data sheets for the paint held with the as-built file.

Stone cladding

Some medium- and high-rise properties have facades clad in natural stone and, in contrast to low-rise construction, these stone facades are normally non-loadbearing. Stone cladding panels may be considered to be high-quality, high-cost materials, therefore this type of material is usually reserved for commercial office buildings, hotels or civic institutions and other buildings where aesthetics are considered an important design factor. As a direct consequence of their cost, stone cladding panels are not usually specified to 'modern' buildings within the public sector, such as schools. The function and building life cycle also determine material choice, and this is the reason why low-cost, functional properties, such as factories and logistics buildings are highly unlikely to be clad with natural stone.

As discussed with brick claddings, there are several different ways to fix or secure this to the loadbearing structure but this is likely to be concealed during the building survey inspection. Therefore, the first thing to establish is the presence and type of natural stone; this is typically done through visual inspection. which can be difficult when the facade is at height and not easily accessible. However, where visible or accessible to touch, the surveyor can use these senses to establish the type of stone and its possible thickness or even the nature of the fixing.

Sandstone and limestone

As sedimentary rocks, sandstones and limestones are created from layers of sand or silt, which have been compressed over many millions of years. When extracted and cut into cladding panels for medium- and high-rise buildings, these are nearly always uniform in dimension, thickness and quality. Sandstone presents visually as blocks comprising many microscopic particles of sand, which are bonded through compression. It may be possible to see the formation lines of the layers of sediment in sandstone, but such 'irregularity' is not usually acceptably aesthetically when the requirement is generally for uniformity. Individual sandstone cladding panels are almost identical in appearance. Where facade cladding stones are fractured or chipped, it may be possible to see the fine particles of sand on the exposed edge.

Sandstone and limestones have generally excellent life cycles, as well as being examples of high-quality materials which are costly.

Shade differences to individual panels, which appear to contradict the aesthetic requirement for uniformity (photo courtesy of Widnell Europe).

Facade clad with a combination of sandstone and limestone. The upper panel is sandstone and this is chipped, revealing the individual sand particles. The lower panel is 'blue stone', which is a limestone. In this case, the blue stone is used to finish the lowest vertical metre from pavement level as it is more resistant to impact damage (photo courtesy of Widnell Europe).

The formation of limestone with organic matter and crustaceans often means that it is possible to see fossilised shells or parts of the historic organic matter within cladding panels. Normally this is not visible on upper facade areas with the naked eye from ground level and is only noticeable from close quarters, with the exception of 'Roach bed' limestone. This type of limestone is easily identifiable by the completely random and striking patterns of imprinted fossils and holes or voids created by water erosion through the material post formation. Architects have specified this material as both internal finishes and for facade cladding panels since the 1950s, with The Economist building in London (completed in 1964) being an example of this.

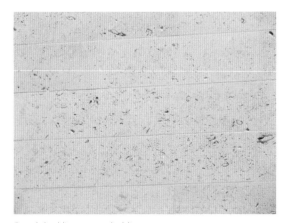

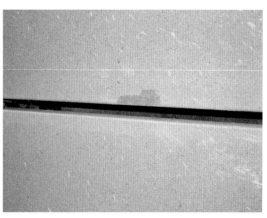

Roach bed limestone cladding.

Surface chipping to a sandstone cladding panel, which is only an aesthetic detraction. When advising a client, it should be noted that it will be very difficult to execute a repair to match the colour, texture and profile of the stone. The relatively soft stone has the potential for future accidental or deliberate chipping, and this is most likely to occur at ground floor level (photo courtesy of Widnell Europe).

Limestone is perhaps considered to be more resistant to impact damage or chipping but it also appears 'softer' than sandstone and can be scratched or marked more easily. It can also be prone to acidic erosion. However, it appears as widely used as sandstone for cladding panels to medium- and high-rise properties. Both materials are 'inferior' when compared to granite.

Granite

As a facade cladding stone, granite is an igneous rock and it is extremely hard wearing and durable. It also has low porosity, which makes it an ideal cladding for commercial buildings. Granite as a building material is used for both indoor finishes to floors, wall coverings or as vanity tops to sanitary rooms. It is also used for external paving and kerb stones. However, when used for facade cladding it can create a striking aesthetic with high performance and, as a consequence, it is a high-cost material. Unlike limestone or sandstone, which have relatively little variance in terms of colours or pattern, granite comes in a wide variety of different colours. When used as a facade material it often has a polished surface, which can make it glisten or shine in direct sunlight and, provided the surface is not the subject of rainwater collection or run off, the polished surface can remain smooth and shiny for many decades. Granite cladding can also be rough or textured, and this creates a totally different aesthetic. The recognition of granite as a high-quality facade material is not a recent concept as it has been used for the past 200 years for museums, town halls and other grand 'showcase' architecture. Typically, this was for classical columns and other detailing or features.

Polished pink granite cladding panels to a facade with a low-level plinth or band finished with limestone. Note the mechanical damage to the top left-hand corner of the left side limestone panel (photo courtesy of Widnell Europe).

Grey granite cladding panels with part mastic joints. The inquisitive surveyor should seek to establish the reason for the open joint (photo courtesy of Widnell Europe).

A striking modern office building with curtain glazing and beige granite cladding panels to the external reinforced concrete core (staircase) and columns (photo courtesy of Widnell Europe).

The gap between the base of the granite cladding and the paving (bottom left) prevents moisture or surface water from coming into contact with the cladding panel (photo courtesy of Widnell Europe).

Stone cladding and fixings

The stone cladding panels to most medium- and high-rise commercial properties are non-loadbearing. However, a facade needs to have intrinsic strength and stability, therefore it is necessary to consider the means of fixing this to the structure. Simple, small areas of stone panels to low-rise sections of a building can be fixed with cement-based adhesives or mortar, but there are usually also some stainless steel wall ties connecting the structure to the facade finish.

 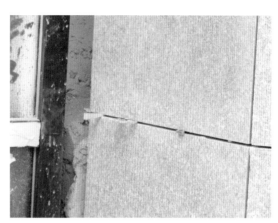

Renovation of a pre-1920s building (office use) with granite with granite cladding panels grouted to the existing facade and secured with mini stainless steel wall ties.

For medium- and high-rise buildings, the reinforced concrete and/or masonry walls may form the sub-base, with stainless steel fixing brackets used to secure the stone panels. These are screw fixed to the structure and have adjustable sub-components which can be altered to increase or decrease the cavity between the external cladding and the internal structure. This is important as the cavity can be filled with mineral wool insulation panels, which significantly improve the thermal efficiency of the facade. The external joints between the stone panels can be left open or sealed, and this is typically achieved with mastic. The choice of whether or not to seal the joints to stone cladding panels is usually made by the architect or designer and, while 'open' joints can be used to create clean lines with natural darkness attributable to shading and shadow, these can become soiled or filled with rubbish, dust or cobwebs, and allow moisture ingress. Closing the joints with mastic seals the facade and prevents moisture penetration from wind-driven rain as well as preventing rubbish from becoming wedged in these at low level. However, mastic joints have a typical life cycle of 10 to 15 years, with solar damage causing cracking.

Early medium- and high-rise buildings from the 1950s and 1960s were, in many ways, concept buildings and the fixing of facade cladding panels was achieved by cutting key joints into the stone and securing these to masonry walls. Timber spacers were placed to position and hold the weight of the stone while the fixing mortar set and the joints were finished with cement-based mortar.

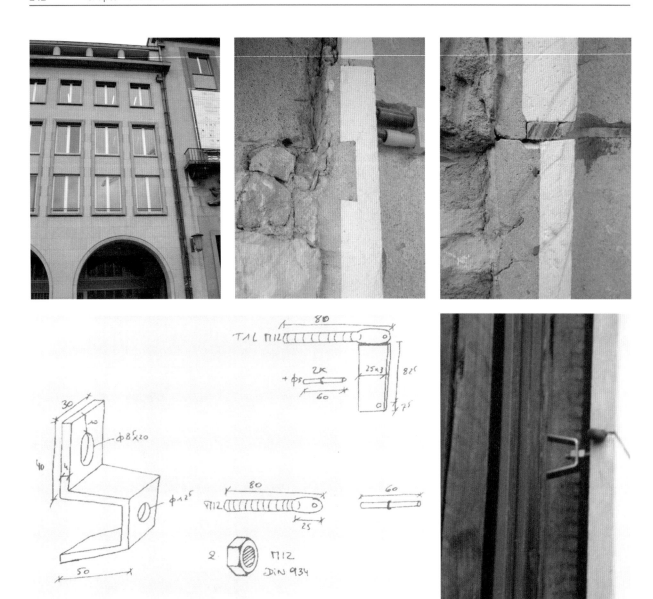

*A 1950s, stone clad office building (**TOP LEFT**) with fixings created from mortar 'key'
joints in the stone (**TOP MIDDLE**) and a wooden peg (**TOP RIGHT**) used as a support
prior to eternal pointing with cement-based mortar. In contrast, a detail of 'modern'
adjustable steel fixings (**BOTTOM LEFT**), which can be seen implemented (**TOP RIGHT**)
(photos courtesy of Widnell Europe).*

In most cases it will not be possible to inspect and ascertain the type, quality or positioning
of the fixings to stone cladding panels. This information may be held in the as-built files or
on technical data sheets for 'recently' executed buildings and the surveyor should refer to
this, if available. Unless they have been part of the design team or involved in the acceptance
procedures, the surveyor should rely on and refer to any acceptance reports signed off by the
architect, engineer and control organisations or authorities.

A similar approach should be taken regarding the placement of fire barriers in the cavity
behind the stone panels, as well as the type and thickness of any insulation. If there is any
doubt or reason to suspect omissions or defects with this, the surveyor may instruct further
investigation with a facade specialist, which could include intrusive inspection and opening up of
the facade. Such actions will have to be fully justified and evidence based as this is likely to slow
down the acquisition process.

For all its relative compressive strength, stone cladding panels are typically 30–50 mm thick and modern fixing with steel pins and anchors or fixings leaves the stone vulnerable to fracture if this is the subject of tensile force. Typical origins of such force are at pavement level, where the stone may be the subject of impact damage, either accidental or vandalism. Replacement of individual stone panels is not straightforward and often it is difficult to match the exact type and colour of stone. Other defects may include defective mastic joints, which can be easily replaced, although this may be costly and should be budgeted with any periodic cleaning of the facade. Stone panels may also become soiled and stained by pollution or rainwater run-off, and remedial cleaning of this can also be a significant cost. Locally, insulation in the cavity can work loose, although this is likely to be as a result of poor initial fixing during the execution of the works. Stone cladding can be the subject of moisture ingress and similar dampness that is attributable to capillary action in brickwork, therefore the execution of cladding panels should seek to avoid them being level with or below surrounding ground floor levels or adjacent surfaces.

Stone facade panels cracked at the fixing points, caused by the impact of the window cleaning cradle/gondola, which exemplifies the fact that stone panels are relatively weak when subjected to tensile force (photo courtesy of Widnell Europe).

Water absorption to a polished granite panel cladding to a facade, caused by the panel coming into direct contact with the adjacent external surface paving (photo courtesy of Widnell Europe).

A mineral wool insulation panel to the external facade cavity of a granite-clad building has come loose and is visible through the open joints; this will be susceptible to moisture absorption (photo courtesy of Widnell Europe).

Marble

Facades executed in marble cladding are relatively rare in the UK or central Europe. It is considered a high-cost, high-quality material which is perhaps reserved for 'lavish' internal wall or floor finishes. There are some instances where marble inserts of smaller panels are used to create decorative features but it is unusual to come across this to whole facades. Its use is often associated with buildings under the ownership of wealthy owners or investors, where these are landmark properties used as corporate head offices or within certain commercial sectors. High-quality facades in general, and marble in particular, are often evident on buildings which have historically been owner-occupied properties.

Marble should be treated like any other stone cladding by virtue of the fact that it is quarried and milled into panels which are then fixed to a loadbearing sub-base. A relatively common type of marble is Dolomitic, which has a bright white appearance and can be used with striking effect. It is durable and requires little maintenance, retaining its colour and smooth texture. However, one serious long-term defect associated with this material is a type of expansion or creep which is associated with the formation of the material many millions of years ago. Marble is derived from metamorphic stone, which is formed in a process involving high heat and high pressure. Having quarried the stone from below ground, this is then cut into thin slices and placed onto a facade exposed to the natural environment. The long-term consequence of this process is for the stone to heave or creep, resulting in expansion of the material. With the panels fixed or restricted by existing fixings, the marble deforms or buckles, with eventual fracturing of the material resulting in the potential for this to work loose or fall from the facade.

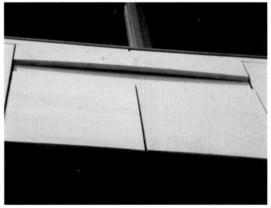

An owner-occupier building from the financial sector (TOP LEFT) constructed in the 1970s with a cast-in-situ reinforced concrete frame supporting dolomitic marble cladding panels. Creep or expansion of the panels has resulted in bowing or deflection (TOP RIGHT) and fracture of the marble with exposure of the steel fixings (BOTTOM LEFT) and a cracked and loose cladding panel which poses a life safety risk and should be removed immediately (BOTTOM RIGHT) (photos courtesy of Widnell Europe).

This defect is not typically linked to specific geographic regions countries or even whether the facade is north or south facing. The pathology of this defect as a deleterious material is examined in detail by Rushton (2006) and it is associated with stone panels that have a thickness of 30 mm or less.

Metal facade cladding

There is a widespread use of metal claddings across all of the commercial sectors for a variety of different buildings. The specifying of materials for all commercial properties is driven by cost, aesthetics and material life cycle. These factors, as well as the basic functional requirements of a facade, influence the choice of metal. In essence, the following are considered to be the most commonly used metals for facade claddings:

- steel
- aluminium
- zinc
- copper.

Steel

Perhaps one of the most popular of all metals used for cladding panels, steel comes in a variety of different guises but the fundamental problem of corrosion is the overwhelming factor affecting its life cycle. Furthermore, steel (as with all metals) is a good conductor of heat, which means that steel panels can influence solar gain within a building in the summer and heat loss in the winter. Therefore, all steel cladding panels are integrated into cladding systems to reduce the effect of temperature change.

Fact File: STEEL

▶ In most cases, cladding panels used for commercial facades comprise galvanised steel. They are often profiled.

▶ Steel cladding panels are typically used to industrial buildings or large sheds were there are large surface areas which require low-cost and rapid enclosing.

▶ Most steel cladding panels are composite or sandwich panels and between the inner and external leaves insulation is installed to improve thermal efficiency, eliminate cold bridging and prevent condensation.

▶ The insulation to sandwich panels could contain expanded and extruded polystyrene (EPS), which poses a high fire risk. Clients should be advised about this as it may render the property uninsurable. This type of insulation was replaced by polyurethane (PUR) and polyisocyanurate (PIR) panels, which are less combustible but produce large quantities of smoke. 'Modern' panels contain mineral wool which may be considered to have good fire resistance.

▶ Panels are typically screw fixed or rivetted to a supporting structure integrated into the facade.

▶ Single skin profiled steel cladding panels may be used as decorative cladding to the perimeter walls surrounding technical rooms or compounds to office buildings.

▶ Steel cladding panels have an average life cycle of, typically, around 35 years and the factors affecting this include the quality and thickness of the material as well as the integrity of any galvanised coatings.

Single skin, painted profiled steel cladding panels to the external cladding of a roof-mounted blockwork technical room create an attractive aesthetic (photo courtesy of Widnell Europe).

The effective use of painted, profiled insulated steel sandwich panels used to cover large surface areas of an industrial building. Note the lower facade section of the building, which has been constructed using pre-cast concrete cladding panels to reduce the potential for impact damage (photo courtesy of Widnell Europe).

Fact File: STEEL *continued*

▶ Inappropriate fixings can lead to wind damage and one of the biggest factors affecting the life cycle of steel cladding to industrial buildings is impact damage. This causes deflection and deformation to the panels, which can damage the galvanised protection leading to corrosion.

▶ Steel cladding panels, as per roof coverings, can be the subject of cut edge corrosion.

▶ Steel was historically considered a low-cost cladding material, although the price of steel has been noted to fluctuate significantly according to economic and market conditions.

▶ It may be considered a relatively low-skill material to install, which contributes to its low-cost image.

Impact damage to the profiled steel cladding panels on this industrial/storage building at low level has compromised the galvanised protection to the steel, causing corrosion (photo courtesy of Widnell Europe).

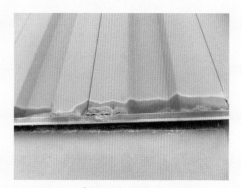

In-situ cutting of the steel panels above a door opening has resulted in damage to the galvanised coating to the steel resulting in cut edge corrosion. This is a defect that takes a long time to manifest and evolve; periodic or routine maintenance or treatment of the edge during the initial execution could have prevented this defect (photo courtesy of Widnell Europe).

When inspecting profiled steel cladding panels, it will normally be impossible to ascertain the age, type or presence of insulation. Historical building information, such as planning permissions or building permits, may assist in determining this and it has relevance as the potential for EPS insulation to be present can have a serious impact on the value of the asset. Should the building be uninsurable, it is unlikely to be lettable, thus rendering it unfit for purpose. If in doubt, the surveyor should recommend further specialist advice or opening up of a part of the cladding to enable sample analysis of the insulation. This course of action is likely to impact upon any acquisition process, prolonging the due diligence or building survey delivery.

Aluminium

Aluminium is a versatile and lightweight cladding material, with its use becoming increasingly widespread to a number of different buildings throughout the range of commercial property sectors. In addition to being a lightweight material, its good durability and relatively long life cycle make it an ideal alternative to steel. It is, however, costlier than steel.

Fact File: ALUMINIUM

▶ As facade material, aluminium has a similar visual appearance to steel cladding panels. It too is often profiled and executed as sandwich panels.

▶ Its use includes industrial or logistics buildings but it is increasingly popular as cladding to offices, schools and residential buildings.

▶ In some instances, when this type of cladding is not intended to provide thermal insulation, these may be single layered.

▶ As with steel panels, these are mostly held in position by screw fixing to a steel structure.

▶ Aluminium is likely to have a better life cycle than steel by virtue of the fact that it is not susceptible to cut edge corrosion. It is not unreasonable to believe that aluminium cladding panels can have a life cycle in excess of 45 or 50 years.

▶ Factors affecting the life cycle of aluminium cladding panels are the detailing, including joints and waterproofing.

▶ Systems designed and executed in the 1970s or 1980s are considered to be inferior to 'modern' cladding panels in terms of thermal efficiency.

▶ Aluminium panels can be the subject of impact damage as well as wind loading when used on high-rise properties or in exposed areas.

Aluminium cladding panels to a medium-rise office building. Note the minor misalignment of the vertical section adjacent to the window and the presence of the remnants of a protective film installed during manufacture which should be removed once installed on the building. Both are only aesthetic details but these do represent the levels of on-site quality control during the initial works (photo courtesy of Widnell Europe).

Corrugated, powder-coated aluminium cladding panels fixed in the horizontal plane (photo courtesy of Widnell Europe).

Fact File: ALUMINIUM *continued*

- ▶ Aluminium panels used for curtain wall cladding to medium- and high-rise buildings can typically become aesthetically passé and this may influence decisions to consider their replacement as part of a renovation.
- ▶ Ironically, used for many post-2000 renovations of 1960s and 1970s concrete-clad buildings, composite aluminium cassette panels have been found to be a fire risk, which is inherently associated with the insulation contained within these.
- ▶ Some environment-conscious investors are reluctant to specify aluminium for use in cladding systems due to the high levels of energy consumption used to produce and fabricate this material.
- ▶ Aluminium is a relatively high-cost facade material, which is high quality with good life cycle.

Evidence of water penetration above a window opening attributable to poor joints and sealing of the panels (photo courtesy of Widnell Europe).

Vertically fixed aluminium brise soleil panels in various colours have the long-term potential to be subjectively iconic or an eyesore. This aesthetic may influence decisions concerning long-term replacement of the facade. Technically, a missing horizontal profile is evidence of poor execution and on-site quality control (photo courtesy of Widnell Europe).

Presenting a similar aesthetic to steel, it is often difficult to ascertain if this is actually aluminium. This information may be available for new buildings on as-built plans or technical data sheets but these are not always available during an acquisition process. One way of establishing the presence of aluminium on site is to inspect the surface of the material in detail to look for scratches to the painted surface. These typically occur as accidental damage during installation or sometimes occupation. Scratches to steel panels often penetrate the fine layer of galvanised protection below the painted surface and, even with the newest of buildings, this will propagate the first signs or corrosion or rust spots. Likewise, any cutting of the facade panels around window openings or for the accommodation of services pipes, fixtures or fittings may reveal a cut edge. If the panels are steel then there is a probability that some corrosion to the cut edge may be evident, although this may have been prevented by sealing the edge, which is recognised as the correct process for in-situ cutting of panels. As a material, steel is denser than aluminium and the surveyor can use touch to seek to identify the difference by tapping or knocking on the surface of the facade panels; aluminium typically sounds lighter and more 'hollow' than steel. This is something the surveyor may learn to do and ascertain through experience.

Zinc

As a facade cladding material, zinc has a completely different aesthetic to aluminium and steel. It is rarely, if ever, given a painted finish and the reason for this is that is it more resistant to corrosion than steel and less susceptible to impact damage than aluminium. It is also a material of perceived beauty and, in its natural state, this 'beauty' can be enhanced by the patterns employed in fixing and jointing the panels. Without having to apply a painted finish, the regularity and uniformity of zinc cladding joints can create a striking aesthetic.

Fact File: ZINC

▶ Zinc can be installed to a facade as composite sandwich cladding panels, cassettes or fixed to a timber sub-base in a more 'traditional' manner. The cladding is typically non-loadbearing and insulation of the facade may be incorporated in the panels or provided by an insulated cavity between the inner loadbearing facade and the external layer.

▶ It is more costly than steel or aluminium and is therefore not usually used for high-area, low-cost buildings, such as those used for industrial or logistics functions.

▶ Because of its perceived high aesthetic value, it is often used with other materials to create a feature to a facade.

▶ When installed in a long strip arrangement, the joints between the zinc comprise a raised seam or rolled joints, which can accommodate movement as well as giving aesthetic value.

▶ The life cycle of zinc cladding is anticipated to be in excess of 50 years with acidic or alkaline attack being the most likely reason for reducing this.

Zinc cladding to a mixed-use development is effectively used as a central facade feature. This high-cost, high-quality material provides a striking contrast to the low-cost rendered facade on either side.

Relatively 'ornate' zinc cladding fixed to timber sub-base and forming an aesthetically pleasing facade detail to an office building (photo courtesy of Widnell Europe).

Fact File: ZINC *continued*

▶ Defective or damaged joints can also be a reason why replacement may be considered and, when in contact with lead or the rainwater run-off from lead, zinc bimetallic corrosion may occur.

▶ Zinc is a high-quality material with an excellent life cycle and, due to its cost, it may be considered a 'luxury' facade cladding material.

▶ Zinc has relatively poor resistance to fire. This is an important consideration since one of the functional requirements of a facade is to prevent the spread of fire.

TOP & RIGHT – Cladding comprising zinc cassettes, which are used to finish a reinforced concrete sub-base. The formation of white powder to the base of the cassettes was suspected to have been caused by corrosion attributable to a collection of cement dust during construction, this required further expert analysis. (photos courtesy of Widnell Europe)

Copper

The use of copper for cladding panels to commercial buildings is reserved for those of the highest specification or to create grand architectural features. It represents the very embodiment of a living material as the aging process causes a patina to build up on the surface, eventually turning the initial brown colour into shades of green. While copper is typically evident to roofs of higher specification buildings or those of social and economic importance, it is not widely used for facade cladding. Therefore, where this is placed on a facade to a commercial building it is usually done to create an architectural feature or a facade that is simply unique.

Fact File: COPPER

- ▶ Often installed to create a 'feature' to a facade, copper is probably the highest quality of all 'normal' metals used as a cladding. It is relatively rare to come across copper cladding in the scope of most building surveys or TDD inspections.
- ▶ It is typically used for commercial buildings, such as offices, although it is also evident to other buildings of historic importance, such as those in the banking, insurance or shipping sectors, where these owner-occupied properties were often designed and constructed as 'flagship' buildings with wealthy materials to represent their owners' success.
- ▶ It is the most expensive of the 'normal' metals used for claddings.
- ▶ As with zinc, copper can be installed in a long strip arrangement. The joints between the copper strips comprise a raised seam or rolled joints, which can accommodate movement as well as giving aesthetic value. Copper can also be welded or rivetted.
- ▶ Oxidisation causes the change of colour, with a layer of patina forming on the surface in various shades of green.
- ▶ The life cycle of copper cladding is anticipated to be in excess of 50 years with examples of this material currently still in use with an age closer to 100 years.
- ▶ Defective or damaged joints can be a reason why replacement may be considered or when moisture penetrates the timber sub-base, causing swelling and the potential for 'differential' movement which can buckle or ripple the copper.
- ▶ Copper is a high-quality material with an excellent life cycle and, due to its cost, it may be considered a 'luxury' facade cladding material.

Copper cladding with standing seam joints to the entrance of an office building; the high specification of this is in direct contrast to the low-spec render used to the remaining facades.

Copper finish to the door and window reveals of a city centre retail property. This appears to have been periodically cleaned or polished and has not therefore turned the green colour associated with oxidisation. The presence of a green tinge to the material is sufficient to verify the presence of copper.

Render

Developed for widespread use as an external finish to residential properties typically in the early twentieth century, render has been used primarily to seal facades and joints to prevent moisture penetration, mask poor or defective brickwork and create aesthetic value. Its use as a facade finish to commercial properties has been widespread throughout the twentieth century and it may be considered the material of choice for many property developers as it is highly cost effective.

Render can be divided into two different types, 'traditional' or 'modern', with the two presenting similar visual appearance but having significantly different properties. Traditional render is typically placed on brickwork or stone facades in two layers; these are made from sand and cement in varying degrees of strength. The base layer is typically 15 mm thick and is trowelled onto the facade with the surface being scratched or 'keyed' to enable adhesion of the top layer. In most instances, the surveyor will not be able to measure or verify the thickness of these layers unless the surface is damaged, exposing the layers of render. This type of render mainly performs an aesthetic function but for older, pre-1920s properties this may also serve to prevent moisture penetration as the facades were typically solid brickwork or stone with no cavity. One principal observation with render is that it is hard to achieve a uniform or even surface with manual trowelling and it is very difficult to blend repairs to the surface. Therefore repairs or evidence of movement to the sub-base or structure are more easily identifiable than with most other facade finishes. Traditional render has little or no flexibility; it therefore acts as a 'barometer' regarding movement of the building or facade.

The top 'finishing' layer to a facade finished with traditional render has eroded and is spalling from the surface. Note the evidence of efflorescence to the surface of this, suggesting the presence of moisture within the top layer.

Cracking over the top right-hand corner of an opening in a facade is exacerbated visually in the rendered surface. Prior to recommending a remedial repair and redecoration it is necessary to ascertain the cause and effect of any movement to the facade and only when this is deemed to have been stabilised can repairs be carried out. Anything else would simply be 'masking' the cracking (photo courtesy of Widnell Europe).

The surface of the render typically has a painted finish and this is done to create or enhance the aesthetic appearance. When assessing the condition of traditionally rendered facades to commercial buildings, the surveyor should note any visual evidence of defects, such as surface cracking, dampness or 'hollow' render, where a lack of adhesion has resulted in this coming away from the facade. The surveyor should seek to test the integrity of the render in situ by tapping on this with a folded knuckle or hard implement. If this results in a 'hollow' sound, this may indicate that the render has blown and has become detached. By tapping around the area of hollow render it should be possible to establish the extent and determine the outer edges of the defective area. This should enable the surveyor to estimate the approximate quantity of render that needs to be hacked off and replaced, to match the existing in colour, texture and profile.

Once an external rendered facade has been painted, it will be necessary to undertake periodic redecoration of this to ensure that it is maintained and fit for purpose. Often an area of discussion between vendor and purchaser in a commercial acquisition, painted render should be in an acceptable condition, irrespective whether this is 'only' an aesthetic detail. The costs of remedial painting of a rendered facade can be quite high as the access requirement inevitably means there is a need for scaffold. Concerning future periodic maintenance and any ten-year cost planning, it will be necessary to make provision for painting of a rendered facade typically every five years.

Modern render systems are still used to enhance the visual appearance of a facade but they are also used to improve thermal efficiency, which can inadvertently reduce construction costs. By placing insulation panels manufactured from mineral wool or other insulation panels onto concrete blockwork or a reinforced concrete facade sub-base, this type of finish can prove highly cost effective. The external mineral wool cladding is then dressed with a reinforcing gauze and a fine layer of render executed on top of this. This type of rendered facade is used for many different buildings within most of the commercial sectors and while it is highly functional, one criticism of this is that it can denote a low-cost facade and is often used to mask poor quality blockwork. This type of facade is also used to renovate existing external brickwork or concrete facades of 1960s or 1970s buildings, which may be considered an aesthetic as well as a thermal upgrade.

Above the renovation of the facade to a 1970s residential tower to improve the thermal efficiency and change the aesthetics. This is being executed with the placement of an external metallic supporting structure and the provision of mineral wool panels with a rendered finish.

A rendered facade to a modern office building which has been externally insulated with a painted finish.

Masking tape has been used to create a decorative yellow strip or feature to the facade and within this line it is possible to see the imprint or pattern of the reinforcing gauze. This low-cost facade serves a purpose but is the antithesis of many other cladding materials, which are superior in terms of durability and freedom from maintenance.

As well as the requirement for periodic painting of modern rendered facades, this type of material has very low resistance to impact or mechanical damage. Whether this is accidental or deliberate, damage to the facade is difficult to repair in situ while ensuring that this truly matches in colour, texture and profile. If the initial works are not finished correctly, with inadequate attention to detail, this type of facade finish can be a shoddy manifestation of its low-cost status.

An external render system to a city centre hotel. Where vandalism has exposed the cast-in-situ reinforced concrete sub-base, it is possible to see the type and thickness of insulation as well as the reinforcing gauze and painted thin layer of render (photo courtesy of Widnell Europe).

External insulating systems incorporating render are an excellent way to upgrade an existing commercial building in situ and it is possible to accomplish this while the building remains occupied. It changes forever the aesthetic appearance of a building and, once placed on the facade, there is a degree of permanence, meaning that it requires periodic decoration to maintain its visual appearance. In most normal circumstances, the surveyor undertaking a building survey or TDD inspection will have to rely on the construction drawings or data sheets in the as-built file to determine the type, thickness and fixing of the insulation. The visual inspection should seek to note any physical damage or evidence of building movement that may present as cracks in the render. Cost planning should make provision for periodic external decoration and within the budgets an allowance should be made for the provision of access scaffolding.

Concrete

The use of concrete as a facade material for commercial buildings and, in particular, offices is relatively common within the existing built environment. In terms of the timeline of construction, and specifically facade materials to commercial properties, concrete is a 'new' material. The use of concrete for building structures to medium- and, eventually, high-rise properties occurred typically post-Second World War but it was not until the 1960s and 1970s that it became prevalent as a facade cladding. 'Brutalist' architecture is viewed with admiration and distain in almost equal measure. The use of concrete facade cladding panels and exposed concrete structure may be considered a universal characteristic of brutalist architecture and, while the perception is that its name relates to this dominating, heavy and 'savage' style of construction, it does actually come from the French word *brute*, meaning raw. The term *'beton brute'* is still widely used in construction terms in France, Belgium and other French-speaking societies and is used to describe raw or exposed concrete. The brutalist movement was so much more than a series of unique and striking concrete buildings, which appeared globally between the 1960s and 1980. It was representative of a social movement, freedom and expression embodied in architecture. At the time, concrete was perceived as an ideal material by virtue of the fact that it could be moulded or formed into a multitude of different shapes. Importantly, it was also viewed as having no maintenance requirements.

History was not initially kind to brutalist architecture, with many high-profile critics voicing their opinion on its aesthetic as well as functional failings. Visual appreciation is subjective and it is only relatively 'recently' that society and individuals have come to admire these iconic buildings. In response to the actual and proposed redevelopment of these buildings, there has been a raft of high-profile listings or classifications to preserve their place in history.

Concrete facades or cladding panels are relatively simple to manufacture and, as concrete has relatively low tensile strength, they are always reinforced. While some concrete facades may be cast in situ, the majority are pre-cast and the principal difference between these two situations relates to the quality.

Discussed in *Chapter 6 – Structure* were the differences between pre-cast and cast-in-situ reinforced concrete and, applying the same principles to the functional requirements of a facade, it is important to assess whether this is fit for purpose. While reinforced concrete cladding panels generally have excellent strength and stability, good fire resistance and reasonable resistance to the passage of sound, it appears a flawed material with respect to its durability and freedom from maintenance as well as regarding the passage of heat. Initially constructed at a time when global warming, sustainability and energy consumption were irrelevant in the design criteria, many concrete facades or cladding systems utilised one outer skin of the material, which essentially formed a cold bridge between the external and internal facade. While the building survey or TDD report does not constitute an energy audit or EPC, the surveyor should be able to establish the potential presence of a cold bridge. By further analysing the EPC or any exiting energy audits held in the as-built file, the surveyor should have sufficient evidence to advise their client on the cause, effect and remedy of this.

Pre-cast reinforced concrete cladding panels and concrete frame to a 1970s commercial building

Internally it is possible to see cold bridging from both the structure and cladding with evidence of either water infiltration or condensation.

Concrete is susceptible to carbonation, which is a process that was not understood or considered in the 1960s, 1970s or even 1980s and while precast concrete is, on the whole, much more resistant than cast-in-situ concrete, it is still susceptible to decay. While carbonation is unlikely to prove 'fatal' for a concrete facade, if it is left unchecked it can have the potential to affect the strength and stability of cladding panels. Furthermore, the 'striking' and bold architecture characterised by the brutalist movement meant that large flat cills, window reveals and other detailing became widely acceptable aesthetically. However, the inability of these areas to shed rainwater effectively meant they were and currently still are at greater risk of carbonation.

Aside from architectural style or aesthetic requirements, the maintenance implications and the difficulty of undertaking sympathetic repairs to match the colour, texture or profile may be the reasons why concrete facades or cladding panels were not widely used at the end of the twentieth and beginning of the twenty-first centuries.

Pre-cast reinforced concrete cladding panels from a Grade II listed building completed in 1973. This example of brutalist architecture is far from maintenance-free, with soiling and staining of the facade and evidence of carbonation and cold bridging.

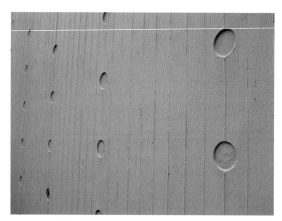

Cast-in-situ reinforced concrete complete with formwork imprint used as an architectural feature to a mid-1970s office building (photo courtesy of Widnell Europe).

Pre-cast reinforced concrete facade cladding, as seen by the clean, sharp edges to the detailing top right of the image. Note the staining to the surface of the lower cill and the horizontal lines appear to indicate the location of the steel reinforcement stirrups (photo courtesy of Widnell Europe).

Carbonation to the cast-in-situ reinforced concrete structure has exposed the corroded stirrups. However, there is little visible damage to the adjacent pre-cast reinforced concrete cladding panels (photo courtesy of Widnell Europe).

As with brickwork claddings or facades, concrete is also prone to staining caused by rainwater run-off and this is often exacerbated and more pronounced with concrete due to the large surface areas of cills, reveals or detailing. Angular pre-cast panels do not lend themselves to accommodate rainwater run-off and, combined with city centre pollution, this has the potential to cause significant staining and can propagate moss growth if conditions permit. The added and dangerous consequence of poor rainwater run-off provision combined with pollution is for this to be absorbed into the concrete, lowering the pH level and leading to carbonation. With most reinforced concrete facades which are susceptible to poor rainwater run-off, staining and carbonation it is almost impossible to prevent this. Therefore, the situation has to be managed and, if this is purely a cleaning issue, then the costs of this operation, including scaffold or access provision can be budgeted. However, where the damage includes carbonation and the presence of spalled or loose concrete, the intensity of the management required, including periodic inspection and repairs, is likely to increase, as is the cost.

A 'striking' modern twenty-first century pre-cast reinforced concrete-framed commercial building where the concrete structure is also used for the facade.

'Stress' fractures (widespread) which were deemed to have no impact on the stability of the building. However, staining caused by rainwater run-off has an aesthetic impact or detraction. A more sinister consequence is the potential for carbonation to occur within the concrete and any repairs to the facade will be difficult to mask and will significantly affect the visual appearance of the building.

When assessing concrete facades, it is important for the surveyor to establish an opinion as to whether these are pre-cast or cast in situ as this has an effect on the presence or likely occurrence of carbonation. There are several features or characteristics specific to reinforced concrete that can help in identifying if this is pre-cast or reinforced. By means of a checklist, the surveyor should assess seek to establish the following:

- evidence of formwork
- external surface or texture
- colour
- quality.

Although both pre-cast and cast-in-situ reinforced concrete utilise formwork, typically the formwork for cast-in-situ concrete is erected on site in a less controlled environment than pre-cast, which is factory made. As a consequence of on-site weather conditions, time pressures and possibly less stringent quality control, the formwork is mostly functional with less consideration given to aesthetics. When this is removed from the cured concrete, the imprint of the formwork edges is often left behind. With cast-in-situ concrete from the 1960s and 1970s it is often possible to see the imprint of the wood grain in the concrete as formwork from this era was 'basic', often involving the single use of timber planks. The residual imprint left behind by formwork may been seen as imperfections when this is exposed for a facade finish and this is probably one of the reasons why pre-cast concrete is more widely used for facade claddings. Ironically, some modern twenty-first century projects seek to replicate the timber-grained imprint for exposed fair faced concrete. This is achieved, but with the use of pre-cast concrete panels where sophisticated moulds are utilised to create these patterns. Sometimes when inspecting concrete panels it is also possible to see evidence of the anchorage points used to crane these into place; however, in most cases these are concealed by cement or grout once the panels are in place.

Spalled cement grout to a concrete panel of a car park has revealed the recessed steel fixings used to transport and place these pre-cast reinforced concrete panels (photo courtesy of Widnell Europe).

Pre-cast reinforced concrete panels with timber formwork pattern created during manufacturing; however, the sharp clean edges and almost perfect finish to the panels indicates the likelihood that these are pre-cast.

The appearance of the external surface of the concrete is key in establishing if this is pre-cast or cast in situ. It is almost impossible to create uniform, smooth surfaces to cladding panels cast in situ and, despite the most stringent on-site quality control, it is almost impossible to prevent voids or air holes to the concrete surface as part of the in-situ process. One key observation that can be drawn during a building inspection is that, where there are joints between concrete cladding panels or facade sections, this also indicates that these are pre-cast. In most cases, these joints are finished with mastic or flexible sealants.

Pre-cast concrete panels can also have surface finishes such as 'pebble dash' or washed concrete, where the outer surface has a level of exposure of the aggregate, achieved by washing out the fines during manufacture. Likewise, concrete with a coloured finish is likely almost always to be pre-cast as it may be difficult to achieve a uniform and even colour with on-site in-situ casting. In essence, the surveyor should be able to ascertain from visual inspection and close-up examination of concrete facades or panels the overall quality of the material. If this is free from blemishes with a smooth surface and sharp clean lines, then it is likely to be pre-cast.

The geographical location of the building and its exposure to pollution or moisture should be noted, as should the presence of any cracking or spalling of the material. If possible, the surveyor should examine cracking or spalling to advise on the severity or potential cause of this.

Although the inspection is visual and the surveyor is obliged to say what they see, there is also a requirement to look some way into the future. Particularly with concrete facades, and even if there is only evidence of relative minor defects, the overall long-term prognosis can be significant. There is the potential for defects to evolve and intensify as a direct consequence of the characteristics of concrete and its exposure to moisture and pollution. Concrete claddings to facades which directly overhang the public domain or access and entrance areas to buildings should be checked carefully for the presence of or potential for carbonation and spalling concrete. In the event that a piece of concrete, no matter how small, works loose and falls onto a person, this can cause serious bodily harm. There is a duty of care and corporate lability for building owners and their advisors to ensure the safety of occupants, users and the general public, and the surveyor should take this into account when performing an assessment of the facade.

There are many ways to address or repair carbonation, including the in-situ treatment of the corrosion and patch repair or cathodic protection or electrochemical re-alkalisation, but evidence suggests that this does not have a guaranteed success rate. Therefore, repairs to carbonation in reinforced concrete facade panels become an ongoing maintenance or management issue and it is important to try to predict the potential costs of this when advising a client.

Concrete and industrial buildings

While the use of reinforced concrete cladding panels is relatively rare for office buildings constructed post 1990, the use of concrete as a cladding material is widespread for industrial buildings. Buildings within the industrial sector are generally used for either manufacturing or storage and logistics. Some fabrication plants or factories can be quite sophisticated buildings, which are unique or bespoke in relation to their function. They may also be owner-occupied properties and accordingly these are likely to be better quality than typical buildings in this sector. The majority of storage or logistics buildings may be considered as low-cost, low-quality construction and this is where the use of concrete as a facade cladding system is a cost-effective way to achieve this.

Logistics and storage buildings are often referred to a 'sheds' and they usually have large surface areas but are restricted in height, typically to 11 m (internal eaves height), due to the limits of the storage racking which, in turn, is limited by the maximum reach of the fork lift trucks used on site. Situated on the highways of Europe, the buildings are often built in clusters or on industrial parks out of town or city centres. The development of industrial sheds almost appears opportunistic, and within industrial parks their presence is often evidence that these are built in phases with a new building constructed adjacent to the existing ones as and when demand requires. The builds are relatively simple, with either a steel or concrete structure supporting the roof; however, the external walls are often non-loadbearing. Although the walls need to be tied to the main structure, this is largely to achieve strength and stability against wind loads.

The concrete facade panels are always reinforced, with the majority of these being lightweight for ease of construction and because these are non-loadbearing. However, the principal disadvantage is that the panels have poor resistance to the impact damage associated with the manoeuvring of forklift trucks and heavy goods vehicles. Therefore, increasingly storage or logistics buildings have reinforced concrete panels at low level and lightweight versions of these higher up. Both types of panel are interlocking and slot into each other using the frame of the building to provide lateral support to form a single skin wall. If another building is added directly adjacent to the wall, then the concrete panels can be cut or removed to create an interconnecting door or left in place if separate occupation is required. The interlocking concrete panels have good fire resistance and can be used as a fire barrier or to seal fire compartments.

The drawback of this type of construction is that it has very poor thermal efficiency with the single skin concrete walls providing a cold bridge. However, energy efficiency is generally of lesser importance as the doors to these buildings are often open for large amounts of time in the day and the internal areas are rarely centrally heated. There is also a requirement for the external surface of the interlocking concrete facade panels to be sealed or finished with a waterproof barrier as the interlocking joints are rarely waterproof.

An external facade executed with interlocking reinforced concrete cladding panels; the lighter coloured panels at the upper level comprise lightweight concrete, with the lower level consisting of much stronger 'traditional' interlocking concrete panels. The red doors appear to be a suitable size for operating fork lift trucks, suggesting that this external wall could become an internal dividing wall if the shed is extended (photo courtesy of Widnell Europe).

A better quality industrial building with washed or pebble dashed reinforced concrete panels. These are uniform in colour, texture and profile as well as being consistent in terms of their strength (photo courtesy of Widnell Europe).

When inspecting concrete facades to industrial buildings or 'sheds' as a whole, the surveyor should seek to establish that these have sufficient lateral fixing or restraint. While it is not the responsibility of the surveyor to check and verify the wind loading capability or calculations, they should comment on the presence and type of restraint. Furthermore, the surveyor should seek to establish the presence of any deflection or movement to the facade, even if this is caused by mechanical or impact damage. Externally, the surveyor should seek to identify the presence and condition of joints between the concrete panels. It should be noted that flexible mastic joints have a typical life cycle of 10 to 15 years, depending on their thickness, quality and exposure to the sun. Replacement of these joints over large surface areas, including access requirements, can be quite costly. Finally, as these panels are made from reinforced concrete it is necessary to visually inspect these to check for the presence of carbonation.

Interlocking non-loadbearing lightweight pre-cast reinforced concrete panels used to clad an industrial shed (noted is the loose galvanised steel clip).

Multiple loose, screw fixed galvanised steel clips are compromising the wind load capacity and stability of the facade. Inadequate external waterproofing has resulted in water infiltration to the joints of the panels (photo courtesy of Widnell Europe).

Typical impact damage to the reinforced concrete facade of a loading bay to a logistics building despite the presence of a buffer plate. These damages are hard to avoid and almost impossible to repair with sufficient strength on the basis that the impact is likely to be reoccurring. The only solution is to periodically replace the buffer plates, which may be classed as 'disposable' (photo courtesy of Widnell Europe).

Solar damage to the mastic joints between pre-cast concrete cladding panels to an industrial shed. These appear adequate at present but consideration should be given to their replacement in the medium term. It may be suggested that this is not necessary as this is 'only' an industrial building but the same approach should be adopted as with buildings in the office and other sectors, thus allowing the client to make an informed decision on whether or not to replace these (photo courtesy of Widnell Europe).

Industrial buildings are considered to be low-cost, low-quality buildings and one of the difficulties during a TDD or building survey is that the costs or the recommendation for repairs tend to be 'dumbed' down. This is typically done by the seller or their agent, with the justification for this approach being that these are 'only' industrial buildings. While these may utilise lower quality materials or be the subject of inferior maintenance regimes in comparison to buildings in other sectors, the surveyor is obliged to act with honesty and integrity. It is therefore necessary to report evidence of any visual defect and this comprehensive process should seek to give their client all of the necessary information required for them to make informed decisions.

Timber

Timber cladding is typically used purely for aesthetic reasons and, along with stone, it is one of the few facade materials that can be classed as 'natural'. There appears to be a common perception among designers that timber is a 'living' material and this term is used to describe the process that the material undergoes as it evolves and ages with exposure to the natural elements. There are others who bemoan the use of timber cladding as a material that requires high levels of maintenance and is prone to decay. Despite these two different perspectives, one mutual consensus is that timber panelling incorporates individual planks or sections that are highly 'original'. This is in contrast to some manufactured high-density cladding panels which are patterned with timber effect.

Timber cladding executed in a variety of different plank widths with unique and individual grain patterns (LEFT) appears aesthetically soothing compared to the formal and 'sterile' timber effect panels (RIGHT) which have been manufactured from basalt (photo courtesy of Widnell Europe).

Timber is susceptible to bleaching by the sun and this discolouration essentially causes the material to fade to grey, which is typically something that is visually evident with cedar cladding. However, where this is then retrospectively treated with wood stain or timber preservative, the aging process is corrupted and replaced by a degree of uniformity.

When inspecting timber cladding as part of a commercial building survey or TDD, the surveyor should seek to establish the type of wood and the current age as this will help to determine the anticipated life cycle. It may be necessary to establish this information by cross-referencing this to any data sheets or certified plans held with the as-built file. However, if this information is not available, the surveyor should not speculate on the type of timber but focus on its current condition as well as any evidence of defects. Any maintenance recommendations made by the surveyor should make an allowance for the necessary access provision to the facade, as well as establishing whether this requires intervention in the short, medium or long term.

Fact File: TIMBER

▶ As a cladding material, timber is rarely used and this is probably due to its specific aesthetic qualities, which are not typically specified for commercial properties.

▶ External cladding systems are screw fixed or clipped to a supporting structure and can be used to encompass facade insulation.

▶ Panels or strips can be vertically, horizontally or diagonally placed for visual effect.

▶ It is a cladding system that is typically viewed as environmentally friendly and can come from sustainable sources.

▶ The average life cycle of timber cladding can vary significantly according to the type of wood, with a range estimated between 5 and 80 years.

▶ BS EN 350 (BSI, 2016) states 60 years for untreated Durability Class 1 Timber.

▶ Red Cedar is a Class 2 timber with 30 years' untreated life or 60 years when a coating is applied.

▶ Typical factors affecting life cycle are the exposure of the timber to the elements, pollution and air quality. As with many facade cladding systems, these are susceptible to wind damage as well as more specific material decay associated with rot. The life cycle can be extended by regular maintenance.

▶ Timber may be considered to have less fire resistance than other facade materials and has the potential to propagate vertical fire spread to a facade.

▶ The cost of timber can vary significantly between the different types of wood and has the potential to be a high-cost material.

Design genius or error? The upper section of this south-facing timber cladding is lighter and bleached by solar exposure. The shadow cast by a building in front means that the upper section is exposed to more sunlight than the lower. Is this a case of differential solar exposure or is there a maintenance issue meaning that only the lower section is treated with wood stain or preservative? These are questions the surveyor needs answer.

Horizontal timber cladding to a visitor centre and facilities at a country park, where timber appears a natural fit with the environment.

Timber cladding over 30 years old to a shopping centre; note that this is distorted, fatigued and 'end of life' (photo courtesy of Widnell Europe).

Glass

Used in some capacity for almost every type of commercial building in the different commercial sectors, glass is primarily used for windows and to introduce natural light into a building. The range of glazing options, characteristics, age, detailing and life cycle vary significantly according to the different applications. These can be divided into the following different sub-categories:

- curtain glazing
- windows and doors
    - softwood
    - hardwood
    - PVC
    - aluminium
    - steel.

Curtain glazing

One of the most striking forms of facade is curtain glazing and this has often been used to create the facades to some of the most iconic buildings in the world. As a facade system, curtain glazing has been widely used since the 1970s and, while there has been some evolution in the design, execution and detailing of this, the principle of this type of facade cladding has changed little since it was first used. Curtain glazing or walling systems have acquired this name because the external facade, like curtains, hangs off the building and is completely non-loadbearing. Fixings are typically provided off each floor level and, although the window profiles are often aluminium, the actual supporting structure is executed using steel as this has greater strength.

The glass panels in modern curtain glazing comprise sealed unit double glazing and, when assessing this against the performance criteria for facades, vertical fire separation is one of the key requirements along with resistance to the passage of heat.

As detailed early on in this chapter, it is not the responsibility of the surveyor to verify compliance with fire regulations, unless they were directly involved in the design, execution of acceptance of this. Concerning such compliance, it is necessary to review the acceptance reports of the design team, including the architect, engineers, control organisations and the authorities. The surveyor should note any visual evidence of abnormalities, which may appear as a breach of fire separation regulations, and draw this to the attention of the client as part of the red flag procedure during a building survey or TDD. Older curtain glazing design typically included a low-level opaque glass panel lined with a fibre cement (asbestos) sheet to provide a vertical fire break.

Another key requirement of full-length glazed facades is their resistance to impact damage, typically from people accidentally walking into the glass or impact damage from furniture or equipment being moved internally or externally close to the facade. Glazing typically complies with the relevant legislation at the time it was manufactured or installed and one general consensus is that rules or regulations have become more stringent over time. It is an appropriate assumption to suggest that there were fewer legal control or obligations governing older glazing compared to present-day regulations.

Therefore, when inspecting curtain glazing it is necessary to ascertain its age in order to comment upon potential health and safety issues. Sealed unit double glazing panels were relatively rare in the 1970s but nevertheless it is sometimes possible to find these. Ascertaining the exact age of sealed unit glazing can be possible by examining the internal gap between the glass where a date is sometimes imprinted. This is a good starting point when assessing the characteristics of the glass but it does little to prove if this is toughened and/or laminated to resist impact damage. Modern glazing often contains a date mark between the glass panels and sometimes there is also information detailing the manufacturer as well as specific glazing products. With this information it is possible to research the technical data sheets to determine whether this is 'safety' glazing. If there is no evidence to confirm the presence of safety glass, the surveyor should be risk averse in their recommendation to place security film on the internal surface to prevent fragmentation in the event of impact damage.

Laminated and toughened glass should be considered the norm for modern full-length windows or curtain glazing and these two characteristics fulfil different requirements. Laminated glass includes the presence of a safety film to the internal face of the glass on the side of the void between the two leaves of glass which make up the sealed unit double glazed panels. The purpose of this is to contain the glass in one piece if it is cracked or broken and laminated glazing which is damaged typically presents as cracks through the glass with little or no displacement of the pieces. Toughened glass is 'tempered', which means it is subjected to a process of heating and cooling multiple times during manufacture which makes it more resistant to impact damage. When broken, the glass fragments into hundreds of tiny pieces, which is designed to minimise the risk of cuts. The combination of toughened and laminated glass is the most appropriate solution for full-length or curtain glazing; when this is broken, it presents as hundreds of glass fragments contained in situ with a 'crazed' random pattern.

In properties where there appears to be no safety glazing to full-height glass panels, the presence of horizontal frames can be sufficient to prevent persons walking into the glazing. The height of the horizontal barrier is normally defined in the relevant building regulations and it is important to verify and check this on site.

A broken toughened and laminated glazing panel which is characterised by the multitude of glass fragments contained in situ by the laminating film.

A full-length aluminium-framed, single glazed window to a building constructed in the 1970s. While there is a low-level horizontal window frame, this is too low to prevent persons walking into the glass and another timber barrier has been fixed across the opening

Another important characteristic of curtain glazing, which is also part of the performance criteria for facades, is its resistance to the passage of heat. The admission of natural daylight is one of the key advantages of curtain glazing or windows in general, however, with this benefit comes the unwelcome by-product which is solar gain. This has significant potential to influence the internal comfort levels in a building, as well as impacting upon the running costs and energy consumption associated with increased cooling load. To reduce the effects of solar gain, modern sealed unit double glazing can incorporate solar reflective coatings internally in the glass. This information will be evident on data sheets contained in the as-built file for the building. If there are no data sheets, the surveyor can seek to establish the presence of solar reflection in the glazing system by seeking to identify the glass through any markings located in the gap between the panes of glass and cross-referencing these with data sheets of information made available by the glazing manufacturer.

In some instances it will be possible for the surveyor to observe the presence of solar reflective film, which can be added to the internal surface of the glass, and this is often evident in older, single glazed buildings or where there are early and basic types of double glazing present.

The presence of internal sunscreens or blinds should serve as an indication or warning to the surveyor that there may be issues regarding solar gain. Blinds are primarily installed to provide a remedy for solar glare, which can cause discomfort to or affect persons working on computers. When blinds are placed on the inside of the glazing, however, they do little to prevent solar gain as this has already been permitted to enter the internal space. Solutions to combat solar gain include the provision of external sun blinds, which can be manually operated or power assisted and even automatic. If these are present, then the surveyor will need to establish that they are fully functional by undertaking a random test/sample and also by verifying this through discussion with the property manager or tenants. External sunscreens can be costly to replace if they are not functional and it should be stated that, where there is an external element of the facade that incorporates moving parts, this may be liable to defects and may require a degree of maintenance.

Automatic external sunscreens are evident to the lowest shown floor of this office building, which effectively remove the internal solar gain. However, the upper two floors have deselected the automatic external blinds and there is widespread use of internal blinds, which reduce the effect of solar glare but are less effective in the removal of solar gain (photo courtesy of Widnell Europe).

Damage to an external automatic solar blind, which is a timely reminder that such mechanisms introduce an added maintenance requirement to the facade (photo courtesy of Widnell Europe).

An alternative to external sun blinds are brise soleil and part of the assessment of the facade should include a visual inspection to note if these show signs of defect. Traditional brise soleil are fixed in the horizontal plane above windows to break up direct sunlight or cast a shadow over the windows. Other forms of brise soleil can be fixed in the vertical plane and more sophisticated automatic systems can react to the position of the sun, moving into position automatically. Where there are working parts and the requirement for servo motors to move the fins, there will always be the necessity to maintain or undertake repairs.

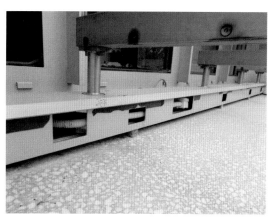

A stainless steel spindle is used to move the fins but detailed examination has identified corroded non-galvanised fixings, which may require treatment and protection across the whole installation if this defect is widespread. Such a relatively 'minor' defect can have significant short- and long-term cost implications.

Vertically hung, automatic brise soleil which react to the position of the sun.

More 'traditional' fixed brise soleil in the horizontal plane; the surveyor should seek to identify any evidence suggesting deflection, movement or failure of these (photo courtesy of Widnell Europe).

Windows and glazed facades require periodic cleaning and, while casement windows can be designed to open inwards to allow cleaning from the inside, curtain glazing is fixed, therefore cleaning of the glazing has to be performed from the outside and this is either done from the ground floor level with a raised lift (cherry picker) or from any external gantries provided and sometimes from above using roof-mounted window cleaning assemblies.

In the event that there are external gantries providing access to the facade, the surveyor should seek to access these areas as part of the survey, provided this can be done safely with the relevant life lines, safety fixings and personal protective equipment. The surveyor should seek to identify the presence of lifelines or safety fixings and advise accordingly in conjunction with the legal/technical findings of the report.

Facade-fixed lifelines to enable safe access for cleaning the facade. In this instance PPE is provided by the cleaning contractor.

A roof-mounted window cleaning assembly which supports a cradle or gondola; sometimes this is stored on site or is provided by the maintenance contractor.

The use of a raised lift is typically suitable for low- and medium-rise buildings (up to seven storeys); however, for high-rise buildings it is normal for there to be a roof-mounted window cleaning assembly.

Roof-mounted window cleaning assemblies are either static with parapet-fixed brackets used to support 'arms' which project from the facade and are used to support the cradle or gondola. Sometimes the cradle is located and stored either on the roof or in a storage facility within the building. However, increasingly common is the provision of the 'infrastructure' (cradle supports) within the building fabric with the window cleaning contractor providing the cradle. It is not the responsibility of the surveyor to comment upon the serviceability or appropriateness of the window cleaning provision but they should seek to identify any visual defects as well as reviewing any existing maintenance reports or safety certificates verifying compliance and fitness for purpose.

A cradle rail system to the perimeter of a building, allowing the roof-mounted window cleaning assembly to travel around each facade (photo courtesy of Widnell Europe).

A galvanised steel parapet-fixed mounting for a window cleaning assembly (photo courtesy of Widnell Europe).

Curtain glazing defects

The remediation of defects to curtain glazing can be extremely costly due to the access requirements for fixing these and also executing repairs to specialist glazing systems where matching glazing is required. The life cycle of curtain glazing is typically 30 to 40 years with deterioration of the joints between panels, internal waterproof seals and failure of individual double glazed panels being the main problems.

Advances in glazing technology have seen the U-value of double glazed panels drop from, typically, 3 kW/m²K in the 1970s to less than 1.4 kW/m²K for modern buildings. This has resulted in significant improvements in energy efficiency with lower heat loss in the winter and lower heat gain in the summer. Although these advances have typically reduced heating and cooling costs respectively, the cost benefit of replacing curtain glazing in order to improve energy efficiency does not appear to be a viable reason the force the change. Due to the high cost of replacing curtain glazing, and also the requirement to empty the building of tenants to accommodate the works, the pay back in terms of the costs saved through improved energy efficiency will take significantly longer than the anticipated life cycle of the new glazing. Recommendation for the replacement of curtain glazing may be justified if there are multiple broken panels and the likelihood that this defect will spread to the whole facade. Furthermore, it is also easier to recommend replacement of the glazing system if this is leaking, with water penetration evident, or where there are acoustic leaks or wind is causing internal draughts. In the event that there is a fire safety issue with the existing curtain glazing, and in particular compromise of the vertical fire compartments or separation, then this may influence the recommendation.

Evidence of leakage and water infiltration to the joints of sealed unit double glazing panels within a curtain glazing installation (photo courtesy of Widnell Europe).

Water staining or 'misting' created by failed or perished seals to sealed unit double glazing (photo courtesy of Widnell Europe).

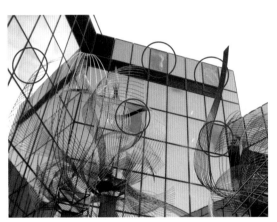

Defective and perished seals to five sealed unit double glazing panels to facade curtain glazing, characterised by dark patches when seen at distance or moisture and misting up close (photo courtesy of Widnell Europe).

It is a significant decision of the surveyor to recommend a full replacement of the facade curtain glazing and this will typically be viewed as 'extreme' measures by the vendor or purchaser. Likewise, this does have the potential to disrupt the acquisition process and there is likely to be pressure applied by the agent, buyer or seller to scale back the surveyor's recommendations. In this instance it is advisable to seek the second opinion of a facade specialist or contractor to establish alternative options.

Curtain glazing case study

A TDD was carried out on a high-rise office building which comprised 30 levels above ground. This was designed and constructed in the 1970s and had been the subject of internal refurbishment. The building was over 30 years old at the time of the TDD and the following observations were noted during the inspection and during discussion with the property manager and tenants:

- localised water infiltration from the glazing in office areas
- numerous cracked glazing panels
- evidence of rust staining to the external facade
- poor acoustic insulation provided by the glazing
- high solar gain internally in the office areas.

The curtain glazing was raised as a red flag issue with the purchaser, and subsequently with the vendor. An initial estimate for replacement of the curtain glazing was prepared with a figure of £11 million. However, this was based on the condition that the building would be vacated for the duration of the works and the rental income stream would be nil for the duration of the renovation.

Curtain glazing executed in the 1970s and, at face value, in relatively good condition (photo courtesy of Widnell Europe).

'Misting' between the glass to several panels indicates defective seals to the double glazing, making the window unfit for purpose as per the functional requirements of a facade (photo courtesy of Widnell Europe).

Localised cracked and defective panels requiring immediate replacement (photo courtesy of Widnell Europe).

Evidence of significant water infiltration and corrosion or rust stains indicating facade leakage and corrosion to steel components (photo courtesy of Widnell Europe).

In order to provide the purchaser (client) with sufficient information to decide on how to proceed, a specialist facade engineering consultant was appointed to undertake a detailed further investigation. An extension of time was agreed for the due diligence period to accommodate the further investigation and a specialist contractor was appointed to remove a glazing panel in situ to ascertain the construction detail.

Results from the investigation revealed the existing steel structure of the curtain glazing and aluminium frames to be in good condition. The external waterproofing to the lower panels, which essentially provided the vertical fire separation, comprised an EPDM waterproofing which was largely perished. Joints to the corners of the curtain glazing were also found to be 'end of life'.

The subsequent opinion and report of the specialist facade consultant was that replacing the waterproof membrane to the lower panels and replacing the sealed unit double glazed panels in situ but retaining the existing aluminium frames was a viable option. The works could be completed floor by floor with the tenants largely remaining in occupation but moving between floors to accommodate the works. Importantly, this compromise to renovate the facade as opposed to replacing it was essentially deemed to give an extra 15 years anticipated life cycle to the facade at a total estimated cost of £2.6 million. This was an acceptable solution for both vendor and purchaser with the work eventually being executed once the transaction was closed.

Windows and doors

While curtain glazing is typically associated with office buildings or feature glazing to retail or other 'landmark' buildings, more common to facades is the used of traditional glazing. This is either fixed or casement, with the latter being functional to provide fresh air or allow inward opening for cleaning purposes.

Windows can be fixed or casement and, where there is the facility to open these, there will always be the potential for defects to occur with their working parts. Modern office buildings are designed for maximum energy efficiency and to accommodate this there is a preference for glazing to be fixed with the internal climate provided by introducing filtered and treated air, which is cooled or heated accordingly. The presence of opening or casement windows to buildings which are fully air conditioned can cause problems. Human nature invariably means that occupants will open windows for individual comfort requirements, irrespective of the presence of a controlled internal climate. Untreated external warm air entering a cooled building in the height of summer can result in inevitable energy loss and inefficiency. However, where cooling is provided by convector units, cooled ceilings or chilled beams, the action of this relatively warm and moist air may be to form surface condensation to these elements. Therefore, where casement windows are installed to commercial buildings with controlled internal climate, these are often fitted with micro switches in the frame to switch off the HVAC secondary distribution locally. This is primarily done to preserve energy and, as the surfaces of cooling convectors, ceilings or beams take time to 'warm' up, the potential for condensation to occur remains.

Commercial buildings occupy a wide spectrum of use within the different sectors and there are also fundamental differences between low-rise 'traditional' construction and medium- or high-rise properties. Furthermore, there are a variety of different window types and styles associated with the different age categories, ranging from pre-1920s to post-1980s. In essence, the surveyor should seek to establish the types of window frames as well as the presence of single, double or secondary glazing. There is a relatively limited number of materials used for window frames to commercial buildings and these can typically be classified as the following:

- softwood
- hardwood
- UPVC
- aluminium
- steel.

Timber-framed glazing

Timber window frames may be considered to be the most classic or traditional types when used for commercial buildings and are widely used for low-rise properties in a variety of different town or city centre buildings. Certainly, for listed or classified historic buildings and those located in conservation areas, there is normally an obligation to ensure that these retain their character, detailing and finish. This can be problematic when the frames are in a state of disrepair or when considering upgrading these for improved thermal efficiency.

As a material, timber is either softwood or hardwood, with the latter having typically longer anticipated life cycles but with a higher material cost. Softwood timber for windows is currently widely used in commercial buildings due to its relatively low cost and it is derived from fast-growing evergreen trees, such as spruce or pine. This means that, when taken from sustainable forests, softwood timber may have a reduced impact on the environment. However, in line with the perceived low cost, it has also a relatively low anticipated average life cycle of 35 years. The application of softwood timber-framed glazing is typical to low-rise 'traditional' build commercial properties, such as high street shops or residential investment properties. The softwood windows may be fixed glazing, as is typically evident with shop frontage, or these can be casement windows. In order to maintain or enhance the life cycle it is imperative to carry out periodic and preventive maintenance, which essentially means painting of the frames to prevent decay.

The painted finishes to softwood window frames are likely to deteriorate most rapidly in areas that are exposed to sun and rain. In the UK, this applies to facades which are south or south-west facing. Cracked, blistered or peeling painted finishes are likely to result in localised rot to the timber frames and therefore provision should be made for periodic maintenance and repair, including repainting, in the ten-year cost plan.

Hardwood window frames suffer similar damage to softwood frames but, due primarily to the higher quality of wood, the average life cycle is increased from that of softwood to, typically, 50 years. There are, however, examples of painted hardwood windows in existence today which may date pre-1920; this is largely due to regular periodic maintenance and painting. On these older properties, it is important for the surveyor to be aware of the potential for lead-based paint to have been used. However, other than by taking a sample of the paint for analysis, there is little way of confirming this from a visual inspection.

Timber-framed glazing to low-rise commercial properties may be single glazed and this is often the case with listed properties or those situated within a conservation area. Such single glazing is unlikely to be toughened or safety glass, which raises the potential for impact damage and the associated risk of bodily harm in the event of breakage. There is also the issue of energy inefficiency which is related to single glazing. While both of these issues may be addressed with the retrospective provision of safety film or secondary glazing, with listed buildings this is likely to be in conflict with the legislation governing historic buildings. These are issues that need to be assessed by the surveyor and discussed with the client during a building survey or TDD instruction.

UPVC window frames

Developed for residential use typically post 1980, UPVC window frames are more usually employed in residential commercial buildings and are not widely evident to the facades of buildings in the office sector. Most UPVC windows have double glazing as standard and the presence of sealed unit, double glazed panels is largely evidence of post-1980 installation and certainly by the 1990s energy efficiency was seen as an important feature of the performance criteria for a facade. Window frames executed in UPVC have a thicker, 'chunkier' appearance compared to all other types of window frames. This aesthetic might be considered 'homely' for residential use but it perhaps lacks the slender, sophisticated profile and shape of other materials for use in the office sector.

UPVC window frames are not susceptible to rot or decay but can become brittle if subjected to extreme cold. However, it is rare to experience the severe and sustained below-freezing periods necessary to make this an issue in the UK or central Europe. It is a relatively 'new' construction material so has yet to be widespread and fully tested beyond half a century but it is assumed to have a life cycle of 30 to 50 years. Typically associated with varying degrees of quality control in both manufacture and installation, the inspection and assessment of UPVC window frames should largely be taken at face value as well as performing satisfactorily during in-situ testing of its operation and function. Also, when the subject of thermal movement, UPVC window casements can become distorted, misaligned or offset from their surrounding frames. This again is not necessarily a reason to justify wholesale replacement when perhaps some in-situ repairs might be able to resolve this issue. However, one general observation concerning the handles and locking mechanisms associated with UPVC windows is that these may often work loose or become defective in a relatively short period of time, when compared to window frames executed in other materials, and should be the subject of periodic maintenance and repair.

Steel window frames

Characterised by the advent of 'modern' architecture as early as the 1920s and typically in the 1950s, 1960s and 1970s, steel window frames are the antithesis of UPVC, with slender frames evident in the design of the prominent manufacturer Crittall. The obvious problem with steel frames is their susceptibility to corrosion and this is perhaps true with early versions of these. However by galvanising these, later versions were more resistant to this defect. The second significant defect with steel-framed glazing units is that the components are cast in whole pieces therefore when in situ they present a cold bridge which has poor energy efficiency and can lead to condensation. Another characteristic associated with steel-framed windows is the use of single glazing and, due to the slender nature of the window frames, it is not easy to retro fit these with sealed unit double glazing panels. The average anticipated life cycle of steel-framed glazing is 50 to 100 years, with replacement typically associated with corroded frames and energy inefficiency.

Steel-framed glazing to the facade of an internal staircase which is listed (photo courtesy of Widnell Europe).

Listed reinforced concrete and steel-framed single glazing which is approximately 60 years old (photo courtesy of Widnell Europe).

Example of modern twentieth-century architecture encompassing steel-framed single glazed windows.

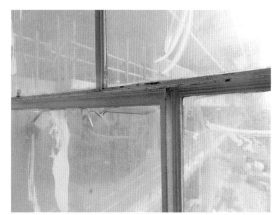

In detail, the slender framework of steel-framed glazing typical of Crittall style windows.

Therefore, the most obvious recommendation upon finding steel-framed glazing units during a building survey or TDD inspection may be to recommend immediate replacement. However, due to the very nature and age of the buildings that use this type of glazing, there is potential for these buildings and the glazing to be listed.

Aluminium-framed glazing

Typically associated with commercial buildings constructed from the 1970s to the present day, aluminium-framed glazing units have become one of the most widely used types of glazing installation. Early versions were used with single glazed panels and were also often devoid of a thermal break, meaning that cold bridging was widespread.

Aluminium-framed, sliding sash, single glazed windows to a building constructed in the early 1970s. High heat loss in the winter and solar gain in the summer are some of the principal issues associated with this glazing installation.

Evidence of water penetration to the corners of this aluminium-framed, single glazed casement window is an example of a facade element which is not fit for purpose and 'end of life' (photo courtesy of Widnell Europe).

As a design concept, aluminium frames may be considered to stand between steel and UPVC in term of aesthetics; they are thicker and typically more 'square' than steel without being as visually obvious as UPVC. Early aluminium frames were sometimes left in an unfinished state and the exposed metal used to produce an aesthetic affect; however, modern aluminium frames have painted or powder-coated finishes in a variety of different colours.

Aluminium is not susceptible to typical corrosion but oxidisation can cause a dulling or whitening of unfinished surfaces. Compared with steel, it is a softer material which is prone to impact damage or problems with misalignment between the casements and frames. It has an anticipated life cycle similar to that of steel frames, although this should be contextualised as the use of this material only became widespread in the 1970s. Therefore, in most cases this has not yet been widely evaluated in its performance between 50 and 100 years.

Window assessment – general principles

Cleaning

With curtain glazing it is recognised that cleaning is done externally from ground level or with the assistance of a roof-mounted cleaning system and cradle or gondola. Glazing units differ from curtain glazing by virtue of the fact that, if these are casement and can open in an 'inwards' direction, they can be cleaned from the inside. While this is an obvious advantage in terms of the maintenance or facilities management of the property, there are two important health and safety requirements associated with this function. First, the presence of opening windows to areas of occupation presents the possibility for occupants to fall through these, therefore the windows should be lockable with the access keys controlled by the building manager, or safety railings should be fitted across the window openings. Second, during the cleaning of these opening windows, there should also be safety provision for the operative or window cleaning contractor. This may take the form of permanent or temporary safety barriers and railings or anchorage points for the fixing of lifelines and other PPE. Generally, it is accepted that the landlord or building owner and occupier will provide the infrastructure for site safety, including barriers or anchorage points, and the contractor or operative provides and uses the appropriate life safety equipment. The exact detailing of this is normally contained within maintenance contracts and the associated health and safety risk assessments; this is not normally reviewed by the surveyor as part of the building survey or TDD process. In essence, the survey should note the presence of casement glazing, any potential risk of falling and subsequent provision of railings or anchorage points.

Functionality

The RICS guidance note (RICS, 2010) for commercial building surveys and technical due diligence recommends sample testing where there are multiple construction components or elements, such as glazing units. Part of the inspection process should be the interrogation of the property manager or tenants as to any reported defects with glazing units, and this may relate to breakage, leakage or disfunction. It is prudent to visit areas of reported defects to undertake testing and record evidence. However, where there are no reported defects then it will be necessary to undertake some in-situ testing of casement windows to establish if there are any issues regarding their function, as well as any defects to handles or locking mechanisms.

Top hung, aluminium-framed, sealed unit double glazed casement windows. During a random testing of these multiple loose window handles were observed and defective locking mechanisms (photo courtesy of Widnell Europe).

The overall recommendation was to undertake a full audit of all the windows and, subject to the findings, that an allowance should be made to repair or replace the window furniture immediately and periodically in the ten-year cost plan period (photo courtesy of Widnell Europe).

Single glazing

While the presence of single glazing or even old and early forms of double glazing does not comply with current opinion or standards for energy efficiency, it is not always appropriate to recommend wholescale replacement of the glazing. The cost benefit of changing the frames and glazing for energy savings alone has a very long pay-back period. Consideration should be given to alternative solutions, such as replacement of glazing panels while retaining the frames or installing secondary glazing. The following is a list of key points to consider before a recommendation for full replacement of the glazing units can be justified:

- material (frame) and anticipated life cycle
- frame defects (corrosion, rot or disfunction of casements)
- serviceability of opening mechanisms, handles and locks
- potential for cold bridging and the present of condensation

- leakage through the frames and seals
- the presence of single glazing and the effect of this on the EPC or MEES requirements
- the presence and quantity of defective sealed unit double glazing panels
- the absence of safety glazing to full-length windows or glazing panels.

A combination of the presence or poor condition of the above will need to be catalogued and assessed with respect to establishing a trail of evidence to justify replacement of the glazing units, which can be a substantial and significant cost.

Doors

The presence of doors within a facade is, essentially, to facilitate either entry to or exit from a building. Concerning the materials used for these, the same principles can be applied as for windows. There are some obvious differences with doors to industrial buildings, such as logistics, storage or manufacturing premises where these have larger apertures and typically utilise sectional or roller shutter doors to facilitate access.

Entry doors can be manually operated or automatic, with or without access control, and the survey should seek to establish that these are functional and fit for purpose. They may be fully or partially glazed, and in some cases solid, with or without vision panels. Concerning fully glazed doors, these will have to comply with relevant legislation regarding the provision of safety glass and consideration for health and safety. In most cases it is not a requirement to enforce building regulations retrospectively concerning the provision of toughened and laminated glazing. However, general observation and recommendation should be for the remedial placement of safety film to prevent personal injury or harm in the event of breakage to door glazing.

The principal entrance to a commercial building may also form the means of escape and this is something that should be assessed as a facade component as well as a legal/technical issue. The means of escape may also be provided by multiple other doors and it is generally accepted that these should open in the direction of 'travel', which is effectively outwards in the event of an emergency. These doors may be solid, glazed or partially glazed depending on the characteristics or function of the building and, while they may be locked for security reasons, these must have some override facility to ensure that the building can be evacuated. This is typically provided by push bars or 'panic' buttons and, as these are often alarmed, it is not always possible to test these during a building survey or TDD. Therefore, it is necessary to establish from the owner, property manager or tenant that the emergency exit doors are functioning and the subject of regular testing.

Draft lobbies are often created to the main entrances of buildings and these typically comprise two pairs of double doors providing an air lock which often contains floor-fixed entrance matting. The doors may be manual but are often power assisted and sliding; consideration should be given to their operation and function in both normal and emergency situations. With access-controlled doors where pass cards or badges are used to release electromagnetic door locks for entry, there are usually manual door release buttons on the inside to allow for emergency escape. With some systems there is a wired link between the fire alarm or detection system which automatically releases fire door locks for evacuation. Again, the serviceability of this will have to be verified by the owner, property manager or tenant as it is not normally possible to test this during a building survey or TDD inspection.

Revolving doors are often installed to hotels as well as some office buildings or residential properties. The purpose of these is to provide access and reduce drafts entering a building. These doors are typically unsuitable for disabled or wheelchair access and not conducive to evacuation in the event of an emergency. Where there are revolving doors as part of the principal access and the entrance also acts as a means of escape it is usual to have fire doors adjacent to the revolving door. Some revolving doors are designed so that the revolving 'fins' of the door can flatten during an emergency to create an opening in the door for a means of escape.

Roller shutter or sectional doors used for industrial buildings are often present for the 'goods in' or loading bays and the nature of the occupational use of these building types means that mechanical damage is often widespread. There may be a vast quantity of these doors present during a building survey or TDD inspection and in most cases it is not feasible to test these. However, observing these functioning during the survey will go some way towards establishing that they are serviceable and fit for purpose. This is another example of a facade component which should be discussed with the owner, property manager or tenant to establish serviceability. Periodic replacement of electrical motors used to power assist the operation of sectional or roller shutter doors is a relatively common occurrence. Therefore, part of the building survey or TDD process will be to establish the estimated cost and frequency of future replacement. Predicting future defects and replacement is a difficult task and the most appropriate method for doing this may be to assess the number of historical annual failures or replacements and to calculate an average number which can be applied to future predictions. It is typically difficult to find accurate cost data for the replacement of such a specific item as industrial roller shutter door motors, therefore historical price data should be obtained from the building owner, property manager or tenant.

 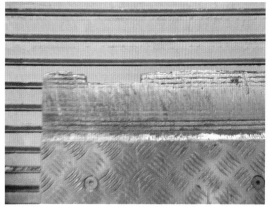

*Defective PVC 'baffle' to an industrial logistics building (**LEFT**) with an internally damaged hydraulic docking plate/leveller (**RIGHT**), which are high-cost items to replace.*

Facade cladding systems

Modern lightweight facade cladding systems are becoming increasingly common to commercial buildings. In most cases these are present on new build properties or as a retro clad and renovation to older non-listed buildings. These types of claddings have one thing in common and that is a supporting structure, as the facade panels are usually non-loadbearing. The retro fitting of these to an existing facade is primarily done to alter the aesthetics and improve the thermal efficiency and, despite this being the over-clad of an existing facade, it is important to assess this in the context of the functional requirements of a facade.

The materials used for the facade are important in the terms of its overall strength and stability, which is invariably linked to its durability and freedom from maintenance. Lightweight cladding panels are typically manufactured to include some of the following materials:

- terracotta tiles
- high pressure laminate (HPL)

- fibre cement
- UPVC.

They require a frame for support and this is usually aluminium, which is secured to the structure of the building. The facade cladding panels are then either clipped or screw fixed to the structure. There is often a cavity between the external cladding and the internal structure or facade, which is normally filled with insulation to provide good thermal efficiency. With the manual screw fixing of the panels, there is the potential for error in the positioning or number of fixing points, which cannot usually be seen during the building survey or TDD inspection. This is largely due to the fixings being covered up or concealed by the finishes of the panels themselves. The survey is visual only and therefore should seek to establish the presence of any movement or deflection of the panels, which could suggest defects. The surveyor will have to rely largely on the acceptance reports of the design team or control organisation, as well as any technical data sheets held in the as-built file. Manufacturers often give guarantees on their products, typically for ten years, and these generally provide for material failure or colour degradation but do not usually cover installation defects.

One very important consideration should be the provision of fire separation, which is one of the functional requirements of a facade. Therefore, it is important to establish the type of material and its rating for fire resistance. At face value, it is unlikely that the surveyor will be able to verify or confirm this from a visual inspection alone. Therefore, it will be necessary to obtain this information from any data sheets held within the as-built files. Any insulation used as part the facade cladding system will also have to be fire rated and the surveyor will have to obtain this information from as-built file. Specific attention should also be drawn to the acceptance procedure undertaken for the facade regarding compliance with relevant fire safety regulations. In all cases, and unless the surveyor is suitably qualified or has been part of the acceptance procedure, to confirm compliance with fire safety regulations would appear foolhardy and to be giving misinformation to the client.

The external placement of lightweight facade cladding systems also has to take into account the positioning and integrity of waterproof membranes to openings in the facade. The correct implementation of this should fall within the duties of the design team or contractor and, due to these largely being concealed during a survey, it is difficult for the surveyor to form an opinion on this matter. Evidence of internal dampness or water penetration may be an indication of the failure of or defects in waterproof membranes. Therefore, it is necessary for the surveyor to inspect internally around window or door openings as well as specifically requesting that the owner, property manager or tenant disclose any areas of suspected infiltration.

*Terracotta facade cladding systems which are visually attractive and highly energy efficient but the individual tiles may be susceptible to impact damage and cracking. (**RIGHT**) It is possible to see the supporting aluminium frame and insulation layer (mineral wool) placed between the inner and outer facade (photo courtesy of Widnell Europe).*

*These timber effect, modern high-density cladding panels are relatively 'untested' and time will be the judge of their suitability to meet the functional requirements of a facade. (**RIGHT**) 'Modern' fibre cement cladding, manufactured to look like timber weather boarding (photos courtesy of Widnell Europe).*

Modern architecture and the principal desire to create an architectural statement with the choice of facade cladding systems or materials. The materials discussed in this chapter may be considered those most commonly used and it is not unusual for there to be a combination of multiple different materials and techniques. One relatively rare example of a facade cladding material which appears to divide opinion with its striking aesthetic is weathering steel.

Weathering steel

The overwhelming aesthetic characteristic of weathering steel (often known by the trade name Cor Ten or Corten steel) is surface corrosion and this is what sets this material apart from any other. While it is considered 'normal' to prevent and treat corrosion, Corten steel utilises the unique pattern, colour and texture of surface corrosion as an aesthetic. The consequence of this corrosion is rainwater run-off staining the adjacent external surfaces, such as pavements or rainwater run-off from cills.

'Striking' aesthetic detailing using weathered or Corten steel.

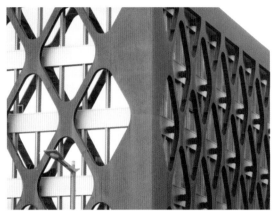

The steel provides only an aesthetic detail rather than fulfilling the facade's strength, stability and durability functional requirements.

The 'true' functional requirements of the facade are achieved with the inner curtain walling.

Rainwater run-off to Corten steel staining the external footpath.

As a material that is designed to corrode from the first moment it is placed on a facade, the obvious question should be raised as to the thickness of the steel and the potential for this to decay into a state of disrepair which will affect its life cycle. Corten steel utilises a mix of different alloys, which are engineered to corrode, and the very nature of this produces a protective layer to the steel. One example where the life cycle of Corten steel may be compromised or reduced is when this is placed in coastal locations (durability).

As a material and concerning the inspection of a facade clad in Corten steel, it should be treated like any other cladding material with the inspection seeking to identify any problems with the fixings, evidence of deflection (strength and stability) or failure of any waterproof membranes concealed by the facade finish. It is inevitably used as part of a facade system which should have been designed to provide both thermal and acoustic insulation, although this is likely not to be visible during the inspection. Steel is considered to have relative poor performance in the event of fire and, with respect to the fire resistance compliance aspects of this unusual or rare material, it will be necessary to form an opinion based on the evidence presented both visually and contained within documents. The surveyor will have to rely on technical data sheets and information held in the as-built file, including the acceptance reports or certificates from the design team and any control organisation.

One observation concerning Corten steel and the functional requirements of a facade is that this has relatively good freedom from maintenance if it has been correctly specified and is placed in a location that is suitable for its use.

Vegetation

Although not considered a facade cladding, vegetation can be used to mask, enhance or blight the external visual appearance of a commercial building. The type of vegetation associated with facades are typically climbing plants or those suspended from or fixed to the facade and this can create a sense of 'charm' in older properties or represent the contrast between nature and the built environment with modern properties.

Facade-fixed wisteria (green climbing plant) and Virginia creeper (red climbing plant) give a sense of 'charm' to this listed commercial (hotel) building.

A modern office building utilises Virginia creeper to the facades of the external areas to provide a natural contrast to this manmade structure.

As vegetation is a living thing, it is not bound by design or planning and will typically grow or propagate in areas where the conditions allow. However, most climbing plants are extremely hardy and, once they have taken hold on a facade, they can grow rapidly with aesthetic and performance consequences. The two most prominent climbing plants that appear on facades are derived from species of:
- Virginia creeper
- ivy.

The fundamental difference between these two types of climbing plants is that Virginia creeper attaches itself to a facade with small suckers, which do not have the potential to cause damage. However, ivy roots grow into the fine cracks or joints of a facade and this can damage the material or joints. Ivy is also an evergreen plant, which means that it is permanently in leaf but Virginia creeper is deciduous, shedding its leaves in the autumn.

Both plants, if left unchecked, have the potential for rapid growth and, while this may be used as a 'feature' to older buildings, it does introduce a maintenance requirement to trim or remove excessive growth. However, the process of removal inevitably leaves marks from the vegetation on the surface of a facade, which is an aesthetical distraction.

This eroded and damaged brickwork is significantly concealed by excessive ivy growth, with the fine cracks in the brick surface and mortar joints penetrated by this vigorous climbing plant.

Removal of an ivy plant has left unsightly dead vegetation clinging to the facade of this commercial building.

Despite management and removal of this Virginia creeper, rapid regrowth has returned to the facade. Note the presence of historic and dead suckers where this has been removed.

The presence of climbing vegetation to a facade should be noted by the surveyor, as well as any evidence of defects attributed to this. An allowance to maintain or manage the vegetation should be foreseen in any subsequent cost plan, which should consider yearly intervention.

Graffiti

The rise of 'urban art' in the late twentieth century has resulted in some quite magnificent and unique painting of images or slogans on the facades of many buildings. These are often used to express social or political opinion and are often profound yet simplistic in their artistic composition. One of the most famous 'artists' is 'Banksy' whose work can be found on the facades to many buildings in the UK and beyond. Such is the iconic and 'cult' status of the art and artist that in many cases the graffiti is not cleaned or removed from the facade but left as an individual 'masterpiece'.

*Examples of graffiti which may be considered as
contemporary urban art.*

The work of Banksy and similar urban artists is certainly viewed differently to that of other, more routine graffiti where spray paint is used to deface or mark the built environment in acts of territorialism, rebellion or vandalism. To most building owners or property managers, graffiti is an unwelcome blight or pest in the everyday management and occupation of buildings. It is unsightly and can be crass in the choice of words or slogans used and in most cases there is a requirement to remove the graffiti. This can be a labour-intensive process and costly, certainly in areas where repetitive graffiti occurs, therefore coatings can be applied to a facade to resist this and also to aid remedial cleaning.

Less 'inspirational' but maintenance-intensive graffiti which, if removed,
may return and require management (photos courtesy of Widnell Europe).

The surveyor should note the presence of any graffiti to the external surface of a building or the presence of this to adjacent buildings, which may indicate that this is a localised problem. The building owner or property manager should be consulted to ascertain if this is a historic problem to the building or area and to disclose any remedial works carried out to remove or protect against this.

Facades summary

The facades form a vital part of the external envelope to a building and this is one of the areas of a building survey or TDD inspection which potentially covers a wide range of different materials and techniques. There has been significant evolution in the different types of materials used for facades but the surveyor should seek to assess this in accordance with the functional requirements for a facade. By applying these standard principles to all facade types and components, it is envisaged that uniform and consistent thinking can be applied to specific individual situations, delivering evidence-based advice. As a construction element, the facade can equate to a large surface area which is often viewed from ground level as it may not always be easily accessible. Some important general considerations may be made concerning the facade:

- There are fundamental differences between the facades to low-rise commercial properties and medium- or high-rise, with low-rise facades generally forming a structural component.
- It is generally not possible to assess or verify the strength and stability of a facade as the fixings or components are often concealed by the facade finish. Therefore, the survey should seek to establish any evidence of structural movement, such as cracking or deflection of the

facade. Where available, the as-built file should be consulted to ascertain evidence of the design, execution and acceptance of the facade.

- For investors, property managers and also occupiers there is an obvious requirement for facades to have good durability and freedom from maintenance. Facades also have an important aesthetic function and sometimes this compromises the durability with the selection of 'inferior' or unsuitable materials. Furthermore, poor facade detailing can impact upon maintenance and the survey should pay particular attention to the joints as well as the principal facade components.

- The visual survey and document review should analyse the facade with respect to vertical fire separation. While it is not the responsibility of the surveyor to verify compliance with fire regulations, any visually obvious omissions or defects with the facade should be noted. In most instances there should be a trail of documentary evidence from the appropriate experts (designers and control organisations) which should verify compliance.

- The majority of facades will encompass openings for the admission of daylight and this is either in the form of windows or fixed glazing. In the case of windows, these may also be used for ventilation and the survey should be both internal and external to establish any evidence of leakage or defects. With multiple opening or casement windows, the survey should perform a series of in-situ tests to ascertain the functionality of these.

- While glazing is used to allow daylight into a building it also invariably allows for solar gain in the summer and heat loss in the winter. The document review should analyse the EPC to ascertain any impact that the glazing or facade may have on energy consumption. The presence of external solar blinds or brise soleil should be noted as a positive point although, if these are automatic and power assisted, this often results in an important maintenance requirement. The serviceability and function of such installations should be verified, in part visually and also with the relevant maintenance documentation.

- The acoustic performance or insulation of a facade is not always easy to establish during a visual survey of a property. It is therefore necessary to consult the building owner, property manager or tenant regarding any complaints or issues to do with the passage of sound through a facade.

- Remedial works to facades, including cleaning, repair or renovation, can be costly due to access requirements or temporary works, such as scaffolding. Major works to facades, such as cladding or glazing replacement, can be difficult to achieve with tenant occupation. Therefore, when advising clients of these potential works there will be a requirement to establish the feasibility as well as the impact of any loss of rental income which may occur should tenants have to vacate the building.

References

BSI. 2016. BS EN 350: 2016 *Durability of wood and wood-based products. Testing and classification of the durability to biological agents of wood and wood-based materials*, British Standards Institution, London.

RICS. 2010. *Building surveys and technical due diligence of commercial property* (4th edition, guidance note), London.

Rushton, T. 2006. *Investigating hazardous and deleterious building materials*, RICS Books, London.

Photo courtesy of Widnell Europe

Finishes   8

Introduction

Commercial buildings are usually leased on a 'shell and core' basis with the landlord ensuring the demise or leased area enclosed by the external walls and/or roof and structurally sound. In some instances, the tenant will undertake the internal fit out of the space with finishes to their desired specification. Alternatively, the landlord may fit out the space to the desired specification of the tenant as part of the lease terms or as an incentive to achieve a favourable lease duration. Ultimately, the responsibility and ownership of the internal finishes is determined by the lease contract and this can be a particularly confrontational issue when the tenant leaves or the lease expires. This situation can lead to claims and counter-claims between the building owner and tenant for the putting back of the finishes to the demise in accordance with its condition at the commencement of the lease.

The inspection, negotiation and settlement of claims concerning the putting back of the internal finishes is known as dilapidations, and this provides an important area of work for building surveyors. This specialist legal/technical income stream can prove to be highly lucrative as a sub-specialism. It has its own RICS guidance note, used to govern the principles and behaviour of surveyors undertaking this (RICS, 2010). By virtue of the fact that there are many publications supplementing the RICS guidance note on dilapidations, and specifically as this is an area of sub-specialism, this book will not seek to provide detailed comment on the processes involved. However, the principles of inspection, analysis and pathology associated with the materials used for internal finishes, as detailed in this book in relation to acquisition, can be applied to dilapidations.

Irrespective of who actually pays for the replacement or renovation of finishes, this is an area that can prove to be extremely costly. Therefore, it is critical during the building survey or TDD process to establish the ownership or liability for maintenance, upkeep or replacement of these items. This information is typically contained within the leases, which are held as part of the data room or are the subject of review by the legal conveyancing or legal due diligence teams. It is therefore often prudent to note the construction and condition of these during the site visit. Typically, site time is often restricted by the 'window' of due diligence and can be further complicated by tenants in occupation, as well as the restrictions that they may place on access.

Commercial properties can vary significantly in terms of their operating sector and also the age of construction. It has been established that a significant number of commercial properties are operated as offices and, pre 1960, most of these comprised low- or medium-rise properties. The building technology used to construct these may be classed as the same as that for low-rise 'residential' properties and, accordingly, the tenant demise may be all-inclusive encompassing entrances, lobbies, landings, sanitary rooms, kitchens and offices within one demise. 'Modern' (post 1960s) medium- and high-rise office construction tends to comprise separate office floors, which are linked or connected by common parts, including staircases, lift, lobbies, sanitary rooms and reception areas.

Other commercial building sectors, such as leisure, residential, industrial, retail or public sector, can be a mixture of operating space and tenant demise, with the provision of sanitary rooms, lobbies, lifts, staircases and staff welfare areas. In these instances, the ownership and responsibility for maintenance, upkeep and repair of these is typically charged to the tenant.

Therefore, it is important to establish the relevant ownership; however, the default position, if this is not known, is to inspect, report and provide cost estimates for this as an 'owner' or 'landlord' cost until proved otherwise. Taking a cost- and risk-averse approach, it is easier to include these in the initial building survey or TDD report with a view to taking the items out if ownership dictates this. However, it is always more difficult to introduce additional extra costs or issues post delivery of the initial report, particularly if negotiations have already started between a buyer and seller.

One important consideration in relation to any recommendations for replacement or renovation of internal finishes is the specification. Building survey or TDD reports should always seek to recommend works on a like-for-like basis with no 'betterment', unless there is a legal obligation that requires an upgrade for compliance. This principle should be applied to all aspects of the building survey; however, the specification of internal finishes is different, by virtue of the fact that this is often driven by market 'standards'. This typically relates to the selection and style of finishes which may be considered to be the 'norm', taking into account the type, location and sector of the building. Market standards are often established by real estate surveyors, who are engaged in the buying and selling of properties as well as letting floor space. Therefore, it is important to ensure that this is communicated to the buyer with respect to their intent for the building. Furthermore, any requirements or advice they may have received from their commercial agents regarding the desired standard of the internal finishes should be noted. In contrast, the seller is usually reluctant to reduce the asking price for a building when this will be the subject of an upgrade on the existing finishes. Consequently, the internal finishes are an area of the building survey or TDD which is the subject of detailed negotiation. While it is the surveyor who provides the possible options for repair, renovation and upgrade, it is the client who ultimately makes the decisions on how to proceed.

Common parts

While the finishes to the tenant areas or demise are usually the responsibility of the tenant, the common parts are often a landlord installation and this is particularly relevant in a multi-let building. Therefore, the responsibility for maintenance and refurbishment of the finishes is allocated to the landlord.

In the majority of instances, landlord ownership of the common parts occurs in office buildings which are let on a multi-tenant basis, where floors or half floors are leased by separate tenants. With 'modern' offices there are common areas outside the tenant demise, which are often arranged to form the core of the building. Lettable floor area is one of the most important factors to consider in property transactions. This ultimately determines the income stream for the landlord from the building, as well as dictating how much the tenants must pay for this. Measurement of the lettable floor area as well as that occupied by the common parts is governed by the International Property Measurement Standards (IPMS). These areas are critical in calculating the cost of both proactive and reactive maintenance of the internal finishes.

The building core is typically located at the centre of modern medium- or high-rise commercial properties. Contained within this enclosed space are often found the lift lobbies, which provide access to the office floors, as well as sanitary rooms, emergency staircases and technical shafts. Depending on the size of the floors plates, levels of occupation and distances required to travel to emergency staircases, common parts may also be situated to the perimeter of the building.

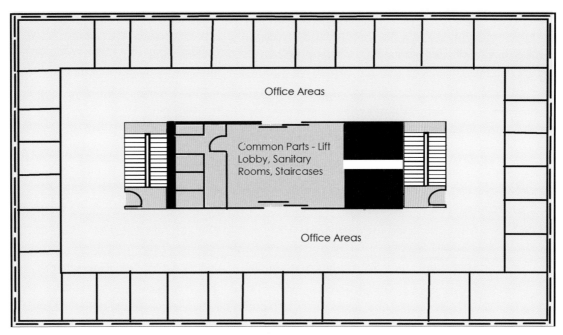

An example of a 'basic' office floor plate showing the common parts.

It is normal practice to situate the principal entrance on the ground floor of most commercial buildings, as this essentially provides the initial access to the building, lift lobby and staircases; these common parts are important areas in terms of the finishes. The design and specification of these areas are often high level and this space is typically used to showcase the building, tenants or sector in which the building is being operated. With the landlord often investing significant sums in the finishes of the principal entrance or reception of an office building, it is typical for the landlord to specify the same or similar materials for the upper floor lift lobbies. This provides an aesthetic 'connect' between the ground floor reception and the upper floors, which can be used to illustrate corporate branding or affirm the identity of the building.

Other areas that need to be considered regarding the internal finishes under ownership of the landlord are technical rooms and, while the finishes to these areas are likely to be largely pragmatic and practical, there is still a requirement to maintain or renovate these. The same principle can be applied to the internal finishes of basement car parking or internal emergency staircases, which may require upgrade or renovation for both functional and aesthetic reasons.

Surveying internal finishes

The RICS guidance note concerned with commercial building surveys and TDD (RICS, 2010) pays specific attention to the internal finishes by suggesting that the surveyor should note the general condition of the internal finishes and decorations. This recommendation appears vague on the depth of any investigation and, as the time allocated for site visits is often restricted, the surveyor should note as much information as possible regarding this issue.

As there is a need for evidence-based findings and an appropriate 'trail' of evidence, one important basic consideration regarding the survey is to apply the same methodical approach to the finishes, irrespective of the type of building and sector of operation. The surveyor should simply seek to note the construction detail, condition and defects to the:

- floors
- walls
- ceilings.

When collecting this information, it is also appropriate to establish the construction detail and condition of certain fixings. This may be a particularly 'grey' area of survey or TDD inspection as it is not always clear which fixtures or fittings belong to the landlord and which are the tenant's. For certain buildings, such as hotels, analysis of the fixtures, fittings and equipment (FF&E) is generally excluded from the survey; this is usually addressed in the contract instruction. FF&E is a particularly 'risky' area for surveyors to assess if they are not qualified to do so. Examples such as the fire resistance of furniture or the presence of safety glass to cabinets, shelves or table tops are impossible to assess within the remit and timescale of a normal building survey or TDD. There are many examples of problems within the hotel sector where guests or users have suffered injuries from defective or unsuitable FF&E. The assessment of this is a sub-specialism in its own right and surveyors working 'out of the box' regarding such assessment are likely to be liable for their findings. Therefore, unless they are qualified to deliver an opinion on this, it appears necessary for surveyors to resist any temptation to include this in the survey scope of works.

Regarding the presence of fixtures identified during a building survey or TDD report, the most appropriate approach may be to discuss this with the property manager or tenants when undertaking the inspection. In the absence of such authority, it appears necessary to note and record the location as well as the condition of fixtures, including these in the draft cost reports with a view to removing them if contrary evidence concerning ownership comes to light.

There has to be a degree of common sense regarding fixtures and fittings, with free-standing items such as tables, chairs or cabinets assumed to be under tenant ownership. Items which are fixed to the fabric or structure may be a landlord or tenant installation but, again, it will be necessary to apply some common sense to the matter, with things considered to be tenant installations usually being eliminated from the survey. Examples of such items can also include data cables, internal partitions, server rooms, comms cabinets and internal sun blinds. Fixtures may be assessed in terms of their 'permanence' and 'fixity' when trying to establish landlord or tenant ownership.

Considering that the majority of commercial space is let on a shell-and-core basis, the surveyor should assess what is the 'norm' and also what may be contained in the lease and any supporting schedules of condition. Each commercial building within the different sectors is likely to have different 'norms' regarding the base finishes, which are usually under the ownership of the landlord. Therefore, it may be necessary to establish what these specific norms are through discussion with commercial estate surveyors involved in the marketing and leasing of the space. For some building surveying consultancies, this may involve discussion with colleagues in multidisciplinary practices; for others it may be necessary to outsource this advice. In both cases, it is necessary to obtain such information in order to establish the relevant trail of evidence. It is perhaps appropriate to refer to the different commercial sectors discussed in *Chapter 4*:

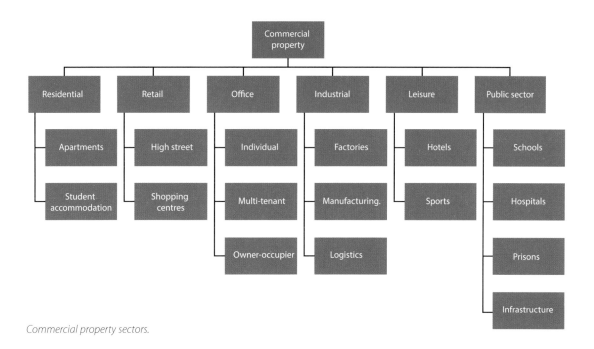

Commercial property sectors.

Included within these different property sectors and individual building types are a potentially large number of different fixtures. These can have a significant bearing on the short-, medium- and long-term cost forecast for the building owner. One further way to determine ownership and responsibility for maintenance and upkeep is to establish if the presence of the fixtures that facilitate the operation of the space or services are provided by the tenant. The following flow chart can be used to help ascertain the ownership of fixtures and fittings:

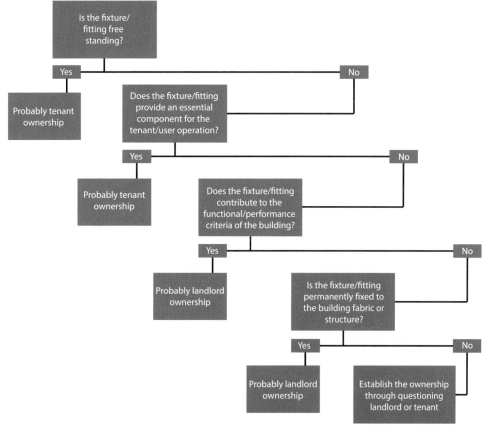

Ownership of fixtures and fittings decision tree.

Site inspection and survey

When undertaking a commercial building survey or TDD inspection, there is often a degree of repetition to the finishes and a methodical approach can be used to assess these. By considering each space, zone or office in terms of the finishes to the floors, walls and ceilings, a systematic check can be made. The nature of commercial building surveys or TDD instructions are characterised by highly restrictive 'windows' of due diligence, where typically insufficient time or access is permitted to visit all areas. Therefore, it is necessary to prioritise the common parts and central core, as these are typically under landlord ownership with potentially higher cost/quality finishes. Inspections of the finishes to internal tenant demises can be performed on a sample basis, although this may produce a general rather than a detailed analysis. Where access was not afforded to specific areas of the property, it is essential to note this in the report. It is also important to inform clients of any assumptions made concerning the finishes to areas which were not, or were only partially, accessible.

Floors

Evidence of deflection or movement to ground or upper floors should be noted in the part of the report dealing with *Structure*, although such evidence is often presented as damage or abnormalities to the finishes. It is necessary to have an understanding of the actual or likely construction technology beneath the floor finishes, as this may have an influence on material choice or the evolution of any defects that may be evident.

Low-rise commercial properties

Low-rise commercial properties constructed with 'residential' type materials are likely to have timber structure to the upper floors with timber floorboards or chipboard depending on the property age. The relevant anticipated building technology associated with low-rise commercial properties and, in particular, floor construction is discussed in *Chapter 4 – Commercial sectors* and *Chapter 6 – Structure*. Ground floors are either ground bearing concrete floor slabs or raised floors constructed from either timber structure or pre-cast reinforced concrete beam and block floors. Timber floors to the ground and upper storeys have a degree of flex and this will depend on the unsupported span of the floor, as well as the thickness and spacing of the floor joists. Due to the potential for flex or movement of the floors, the most appropriate types of floor finish should also be flexible, and these are typically carpet or carpet tiles and vinyl or linoleum (lino). Floor screed (similar to concrete) is relatively strong in compression but weak in tension; therefore, assuming it is not reinforced, it is not suitable for use with suspended timber floors. Suspended timber floors can be used to support ceramic floor tiles and this is usually done by incorporating an extra layer of sheet boarding to stiffen the floor structure, allowing for the placement of inflexible adhesives and grout, which should be resistant to cracking.

Ground bearing concrete floor slabs are generally reinforced with no flex and therefore any unevenness or localised deflection in the finishes should be treated with a degree of caution. Suspended reinforced concrete beam and block floors have a relatively small, almost negligible amount of flex and, although sounding or feeling 'hollow', along with ground bearing concrete slabs, they are suited to most types of floor finish. Most concrete floors are reinforced and the concrete component forms the structure; this is often finished with a sand and cement based floor screed, typically installed to provide a smooth finished layer which is then primed and prepared to take the floor finish.

Medium- and high-rise commercial properties

The floor construction to the majority of medium- and high-rise commercial buildings is likely to comprise pre-cast or cast-in-situ reinforced concrete, which is sometimes finished with floor screed or left to receive a raised floor. This is often the case with buildings constructed for office use, where a raised floor is often considered the 'norm' from an estate agency perspective, thus making the space more flexible for the sub-division of data and electricity supply via floor boxes recessed into the raised floor.

Raised floors

The presence of a raised floor should be an indication to the surveyor of the commercial flexibility of an office building and, while this may be considered the 'norm' by some agents, this will often depend on the location, sector or comparable buildings. A raised floor includes floor fixed and glued pedestals, which are secured to the floor slab with a degree of permanency or fixity typical of landlord ownership. While it is acknowledged that there may be different sizes of raised floor tiles, these are often standard dimensions, typically 600 mm x 600 mm. The floor tiles themselves are made from chipboard or MDF with perimeter edgings as well as a finish to the upper and lower face. The tiles are typically 38 mm thick and these require sufficient density to ensure appropriate acoustic properties. The 'hollow' sounding effect which can be present when walked upon is an acoustic characteristic of raised floors. High-quality raised floor tiles are encased in pressed galvanised steel boxes which have good acoustic properties as well as being very durable. These are inevitably expensive and low-cost alternatives may include glued vinyl edges with vinyl and/or glue fixed aluminium foil to the surfaces. Low-cost raised floor tiles are more susceptible to damage of the aluminium foil or vinyl surfaces, typically when carpet tiles are removed or through water damage. The characteristics of chipboard or MDF are such that, when they get wet, they can be prone to water absorption, swelling and possible disintegration. This inevitably means that water damage to low-cost raised floors often results in a need for localised replacement of raised floor tiles.

Exposed carpet tile edges have been caused by damaged pedestals to a raised floor which is loose when walked upon (photo courtesy of Widnell Europe).

A raised floor without finishes and it is possible to see the presence of floor boxes for the distribution of electricity and data (photo courtesy of Widnell Europe).

A damaged raised floor tile exposing the chipboard inner core (photo courtesy of Widnell Europe).

A raised floor tile has been partially removed for access (photo courtesy of Widnell Europe).

Typically, raised floors have a void of 80–120 mm to accommodate services trays and pre-formed floor boxes are countersunk to distribute the services on the basis of one floor box per module or according to the internal fit-out requirements of the tenant. The floor boxes themselves are used to house electrical supply, with sockets for mains electricity and, sometimes, an emergency or uninterrupted power socket. Floor boxes are also used for data distribution and this is an efficient way to provide services distribution avoiding wall-mounted sockets or trunking. Raised floors are also flexible by virtue of the fact that during an internal fit out it is possible to place floor boxes in practically any chosen location.

The presence of a raised floor inevitably means that it is not usually possible to inspect the concrete or screed below to establish any visible defects. Therefore, the survey is limited to establishing the type and quality of the raised floor, as well as any evidence of defects in this. Data sheets in the as-built file may be consulted to establish the floor specification, or this may be evident during the site inspection. Typically, a raised floor is covered with the actual finish and this may be carpet tiles, vinyl or a host of other materials; however, often there are no finishes to the raised floors in the shafts. This can be an ideal location to establish the floor type and characteristics as well as undertaking simple measurements to ascertain the dimensions of the floor void and thickness.

The inspection of the internal floor finishes often requires the surveyor to visit large amounts of the lettable floor area in a relatively short timescale. This is due to the inevitable short window for due diligence and time pressures placed on the inspection. Accordingly, a 'sample' assessment of the floor on each level should be inspected and the owner, property manager or tenant should be asked to disclose any areas of specific concern or defect. These areas should then be inspected to establish the cause and effect of the defects. Typically, raised floors can come loose when pedestals are defective or detached. This presents a symptom of loose floor tiles, which can eventually form a trip hazard. If this defect is discovered locally, the recommendations in the report may suggest a separate, more detailed audit of the entire floor area to establish if this is a widespread problem. Such a detailed investigation is typically not feasible within the timescales of an acquisition survey or TDD and could form part of the future property management or maintenance regime. The cost schedule in the report should seek to allocate a cost for the number of hours associated with the inspection and an estimated cost for an allowance to repair locally. The surveyor should seek to be risk adverse in estimating the cost of this; however, it should be accepted that there is a degree of guesswork in pricing the 'unknown'.

Raised floors may be considered a desirable characteristic from an agency perspective, although it should be acknowledged that the retrospective placement of these is not always simple. When combined with the dimensional requirements of a suspended ceiling, the overall 'lost' floor-to-ceiling height attributable to these elements may not be acceptable. The market 'norms' for clear floor-to-ceiling heights are typically 2.60–2.75 m for new CAT A office buildings and approximately 2.45 m for refurbished offices in the UK. Furthermore, 2.50 m is considered the minimum free height under health and safety at work regulations in Belgium, with 2.60 m being the typical requirement for the EU institutions. Therefore, individual countries, organisations or sectors may have different norms regarding required clear floor-to-ceiling heights and, in order to meet these requirements during renovation, it may be necessary to strip off existing floor screeds in an attempt to achieve conformity.

Channels and floor boxes

A predecessor to raised floors was the placement of floor boxes within floor screed, often to fit the building module or grid but connected to each other with floor channels cast or chased into the floor screed. This provides a solution or an alternative option to perimeter surface-mounted services, but the fixed nature of the floor boxes limits overall flexibility of the floor space. There is also an inevitable uncertainty or risk regarding the type and condition of the floor ducting, which is usually encased or concealed by a layer of floor screed with the possibility that this is executed in fibre cement and the further possibility of this containing asbestos. This risk tends to apply to medium- and high-rise office buildings constructed from the 1960s to 1999; however, the technology associated with floor channels does not appear to have been widely installed after the 1970s.

Industrial buildings

Buildings operated within the industrial sector are either used for manufacturing and production or for storage and logistics. These are typically large 'sheds', which often have a single ground floor level dedicated to the building's primary function, incorporating multi-storeyed offices located internally as mezzanine floors or constructed externally to the main building. While it may be necessary for some purpose-built or owner-occupied manufacturing buildings to have specific floor construction and finishes, the majority of sheds have cast-in-situ reinforced concrete slabs. These are often cast in bays to incorporate movement joints and one of the biggest problems with these type of floor is the presence of shrinkage cracks to the concrete or damaged slab edges. Often due to heavy industrial use and, in particular, the high point loads presented by fork lift operations, cracking to the floors slabs has the potential to worsen and, in turn, significant cracks can damage fork lift truck wheels.

In order to minimise potential damage to the floor slabs and eventual damage to fork lift trucks, 'sine wave' movement joints can be cut into the concrete bays. The surface of the bays is normally polished and in some cases epoxy paint is used as a surface finish or to provide circulation markings. One general observation with painted floor finishes is that these inevitably become worn, mainly due to the operation of the building, and, once painted, there is a requirement to budget for periodic remedial painting.

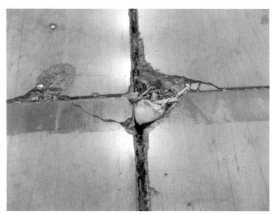

Polished industrial floor screeds to logistics sheds. The floors have been polished and, in part, painted with an epoxy lacquer. Defects often occur to the joints in the individual floor slabs, which have been cast in bays (photos courtesy of Widnell Europe).

Floor finishes

There are a wide variety of possible floor finishes for commercial properties and, while there are differences in the construction technology of the floors themselves between low-, medium- and high-rise properties, there also appear to be different techniques associated with the various age categories of buildings. However, the application of different floor finishes are widely transposable to all buildings. In essence, it is often the individual use requirements of buildings or individual commercial spaces which drives the selection of materials. Generally, commercial buildings have higher levels of occupation and 'traffic' when compared to residential properties and, as a consequence, the floor finishes are often prone to increased wear and potential damage. Some typical finishes include:

- carpet or carpet tiles
- vinyl and lino
- laminates
- natural wood
- natural stone
- ceramic tiles
- concrete/floor screed/granolithic finishes
- paint.

Carpet or carpet tiles

Probably the finish of choice to most office buildings as well as those in the leisure sector, such as hotels, is carpet or carpet tiles. This finish has aesthetic and functional properties which make it ideally suited to these sectors. For many occupiers or users of commercial offices and hotel rooms, carpet is the 'norm' and its use is widespread to central and northern Europe, where carpet has been long considered a popular choice for residential properties. It should be noted that 'carpet culture', while being widespread commercially, is less popular in some parts of Europe, with exhibit a preference for ceramic tiles, laminates or wood in residential dwellings.

Carpet is a very versatile and flexible material and its manufacture in rolls makes the installation relatively straightforward. However, this more traditional form of the material is usually installed with a layer of underlay, which often renders it more costly than carpet tiles. As it is supplied and laid in rolls, carpet is good at covering uneven or slightly undulating floors, which can also be used to mask defects as well as fulfilling an aesthetic function. The surveyor is not usually required to analyse in detail the type or specification of carpets but there are clear variations in price between the various qualities. The surveyor should be expected to identify the presence of underlay and this can be done by pulling back the carpet in a corner of the building or floor. Commercially, carpet has a life cycle of, perhaps, five to ten years and can be the subject of deep cleaning but, in most cases, carpets are replaced as opposed to maintained. Certainly, upon expiry of a lease or when tenants vacate a property, the carpet or carpet tiles are almost certainly going to be replaced for a new, incoming tenant.

Carpet tiles are easier to lay in comparison to carpet but these typically require a flat surface and are less accommodating of unevenness in the sub-base, as this can cause exposed edges or ridges to the actual tiles themselves. Carpet tiles are well suited to raised floors, where the surface should be completely flat and these are glue fixed in such a manner that the individual tiles are placed across the joints in the raised floor tiles to create the appearance of a smooth, seamless finish. One general observation is that high-quality carpet tiles appear to present less visual evidence of joints between individual tiles once laid. In order to ensure this result, self-levelling floor screeds are sometimes used to minimise or accommodate any unevenness. The use of self-levelling floor screeds is more common when carpet tiles are laid onto floor screeds where the base layer may be less even or uniform than a raised floor and certainly where the screed has been damaged through historic occupation or use.

When assessing carpet or carpet tiles as part of a commercial building survey or TDD process, this is often seen as a 'disposable' finish and is frequently the subject of contentious discussion as to the costs of replacement. It will be necessary to establish the market standard for the carpet or carpet tiles with regard to the type of building as well as the sector and specific requirements, such as anti-static provision, should be determined as these can have an impact on price. While the price per m^2 for carpets or carpet tiles can be discussed or negotiated, the quantities are typically fixed and this information should be verified and measured on site, where possible, or from as-built plans.

Vinyl or lino

Manufactured as impervious materials, both vinyl and lino have been specified and used for floor finishes to commercial buildings since the 1960s or 1970s until the present day. While lino is typically a sheet material used in a similar manner to carpet, vinyl tiles have been a popular alternative, used to create a different aesthetic. Both materials may be considered to be less aesthetically appealing when compared to carpet and this is evident in the hotel sector where it is almost unheard of to specify this for hotel rooms. They are materials that function particularly well in locations of high foot traffic or communal use, where the floor finishes are likely to become soiled or provide the interface between external and internal environments. This is why vinyl and lino are often specified for the entrance halls, lobbies, staircases and landings of schools or some office buildings. These are also well suited to the heath-care sector or certain manufacturing operations, as these floor finishes can be easily cleaned to ensure sterile conditions. They have relatively good life cycle (of ten to twenty years) and are comparatively durable as well as being easy to clean. The trade-off is that they may be considered quite dour, 'cold' and uninspiring materials, but this is offset by their value for money. This may be the reason why they are so prevalent within the public sector, where the budget available for the provision of renovation or replacement is typically lower.

Lino and vinyl tiles can be fixed to both floor screeds and raised floors by adhesives. Lino (like carpet) is more accommodating to uneven or undulating floors but it is important to ensure that sufficient adhesive is used or air bubbles and delamination of the finish may occur, which are almost impossible to repair in situ. While lino has relatively good durability, it can be prone to scratching or tearing, which again is difficult to repair. Vinyl tiles are not suited to uneven floors and are usually smaller in dimension when compared to carpet tiles. They are glue fixed to floor screeds and, while their appearance, design and specification of finishes is subjective, there does appear to be a preference for lino over vinyl for modern buildings. This is probably due to the installation costs, which are likely to be lower when rolling lino sheet floor finishes compared to placing individual vinyl tiles. As a consequence, vinyl tiles are a finish which is more widely evident in buildings executed in the 1970s and 1980s.

The vinyl tiles from the 1970s and 1980s, as well as the adhesive used to secure these, often contained asbestos. In most 'normal' circumstances this should be noted on the asbestos inventory and, provided this is not decayed or friable and depending on the type of asbestos, this could be deemed to be in a controlled or safe situation. It is important to ensure, as part of the building survey or TDD, that asbestos is assessed as per the contract instruction with reference to the relevant regulations governing this hazardous material. Unless the surveyor is also suitably qualified as an asbestos specialist, they should refrain from verifying the presence, condition and type of any suspected asbestos. However, if the surveyor has a suspicion that there may be asbestos present in vinyl floor tiles, they are at liberty to state this, subject to confirmation. Due primarily to the cost risk associated with asbestos removal and disposal, the surveyor should be risk adverse with respect to budgeting for replacement of such tiles. Therefore, it is normal procedure for any budget estimates to include the limitations that these are exclusive of costs associated with asbestos treatment.

Both vinyl and lino have relatively good life cycles but typically become brittle with age, which can cause them to crack or split, presenting potential trip hazards to the occupants of a building. Vinyl tiles are also susceptible to cracking or chipping due to impact damage with the replacement of these often difficult to complete with precisely matching tiles. Likewise, the patch repair of sheet lino is almost impossible to achieve without this being visually obvious, as patterns or designs often vary according to their age.

Damaged vinyl floor tile which contains asbestos. Any recommendation to replace the tiled floor finish should exclude specific costs for asbestos removal and disposal. Without specialist cost advice, this has cost risk.

Modern vinyl tiles are more flexible than their predecessors and one of the most effective uses of these is for common parts to buildings where visual image is important. The relatively sophisticated modern practice of laying vinyl in strips means that it can be used to create the effect of timber at a fraction of the cost. Timber flooring will be discussed in more detail later on in this chapter and, while this does have the advantage of being a natural material, it does need to be cut to shape and glue fixed together. It also requires there to be sufficient allowance for movement (shrinkage and swelling) which can in itself result in gaps in the material around joints. Natural wood is also susceptible to water damage and therefore not suited to sanitary rooms or kitchen areas. An alternative to natural wood is timber effect vinyl flooring, this is an extremely convincing substitute and is very hard to distinguish from the natural version at face value. It has the added advantage of being waterproof which improves the versatility of its application. Vinyl strip tiles are also very thin, at approximately 2–5 mm, and relatively easy to fit, provided the sub-base or floor screed is level. As these are glue fixed, they can be susceptible to lifting or blistering if insufficient glue is applied or if water manages to get under the floor finish.

Modern vinyl floor tiles have a variety of different patterns which can be used to replicate the visual effect of a number of different timber types as well as mimicking other materials, such as natural stone.

Laminates

Towards the end of the twentieth century, laminated floor finishes became a popular feature of residential properties and are still widely in use. The composition of laminate flooring typically comprises high density fibre board (HDF) with a laminated (glue fixed) plasticised top layer with, typically, a visual timber effect finish. The individual 'planks', which vary from 5 to 15 mm in thickness, include tongue and groove edging which is glue fixed or 'clip' fixings to create a 'floating' floor. This floor covering is most suited to existing floor screeds and, in order to enhance the acoustic properties, insulation can be placed under the floor with a vapour barrier for good measure on the warm side of the insulation to prevent condensation or moisture ingress from the floor slab.

The advantage of installing laminate flooring is its cost effectiveness but, as a product, it may be considered to be low quality. It is susceptible to surface scratching or chipping of the laminated top layer due to impact damage. When in contact with moisture or water, this type of flooring is noted to be susceptible to swelling and potential disintegration. Laminate flooring is rarely used in commercial buildings as, in areas of high foot traffic, it can be prone to the abovementioned damages.

Natural wood

As previously alluded to, natural wood as a floor finish has limited applications and this is due mainly to its inability to resist water damage or dampness. It is also significantly thicker than vinyl floor tiles or laminate, being typically 11–18 mm thick. The type of wood is also important, as hardwoods, such as oak, are more resistant to impact damage than softwoods, such as pine. Thickness is important as, with acoustic and thermal insulation up to 8 mm in thickness, the resultant floor thickness can amount to 20 mm plus. This will typically mean for the retrofitting of natural wood finishes to an existing commercial space that the internal door heights will have to be altered to accommodate the new floor thickness.

Natural wood floor finishes may be considered a high-cost material but the overall 'quality' can be subjective, as even the most durable of wooden planks are susceptible to scratching or denting with impact damage. There is also a requirement for there to be an expansion gap around the edge of the floor, which typically needs to be 15 mm although this does depend on the type of wood. The expansion gap is needed to accommodate the swelling and shrinkage of the material and, although the wood should be left to 'acclimatise' prior to installation to avoid unnecessary movement, this does often occur post installation. The visual consequence of this is for gaps to appear around joints or even the 'bowing' or lifting of planks locally where there is insufficient expansion space.

Natural wood is not a material in common use as floor finishes to commercial buildings although it does appear to have an attractive appeal for certain applications, such as hotels or some residential accommodation. It is probably most recognisable when used in the leisure sector or in school sports halls.

Parquet is a type of wood flooring which is sometimes found in older commercial properties and comprises small strips of hardwood laid in attractive patterns. It is a labour-intensive material to place and this may be why it is not widely used in modern commercial properties as a floor finish. As with all woods, it is susceptible to damage if it gets wet and therefore it is normally treated in situ with waterproofing agent or varnish.

'Engineered' wood floor finishes are a combination of laminate and natural wood, utilising a HDF sub-base with a thin layer of natural wood as a veneer. This has better durability than laminate at a lower cost than natural wood flooring.

Parquet flooring forming the finishes to a museum floor.

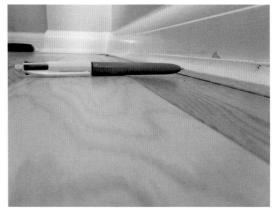

Oak strip tongue and groove floor finish with insufficient expansion gap causing the strips of glued timber to bow and deflect.

It has been established that low-rise, pre-1920s properties comprise 'residential' type construction with raised timber ground floors often evident. These can be finished with sheet carpet or lino as finishes which accommodate movement or unevenness; however, there is also the possibility for these floors to be left exposed with either wood stain, painted or varnished finish. Such finishes may be typical of town centre retail, pubs or restaurants, where there may be the perception of a certain 'charm' associated with historic natural floorboards. Georgian or early Victorian floorboards were typically executed in oak or elm and these are nail fixed to existing timber floor joists. In this instance, the floor finish also forms part of the floor structure and, assuming that this has not been the subject of rot or beetle attack, which is associated with this age of property, it can be an extremely effective floor finish.

Natural stone

As a floor finish to commercial properties, natural stone may be considered one of the highest specifications available to fulfil this. Historically, natural stone floor tiles or slabs have been evident in many of the great buildings, such as palaces, cathedrals and museums. This was largely due to there being few alternatives, with raised timber floors not being in evidence until the late eighteenth and early nineteenth centuries. Despite the development of raised timber flooring, as such floors were unable to realistically support the weight of natural stone, they were considered insufficiently 'grand' for civic or state buildings of architectural so natural stone continued to be the material of choice for such undertakings.

Commercially, natural stone is not a particularly versatile material. It has high density and is not suited as a material to finish the raised floors which are typically specified to modern office buildings. It is a statement material and is therefore used to finish the floors of building entrances, lift lobbies and reception areas. There are a variety of different natural stones which are used for this purpose but the more commonly used types are granites and marble. With polished surfaces and neat, grouted joints, natural stone can feel cold and clinical, which appears to suit the open space of reception areas that may also have multi-storeyed ceiling heights or atria.

Natural stone has excellent durability and, as an internal floor finish, it can last for many decades or even centuries. Its durability may also be seen as a negative by virtue of the fact that changes in interior design often mean that this feels obsolete or dated, depending on what alternative finishes are 'fashionable'. Invariably, the renovation of building reception areas with existing natural stone often results in inferior finishes being specified to accommodate current trends.

Aside from repolishing natural stone floor finishes and regular cleaning, they have very few maintenance requirements. The grouted joints between natural stone floor slabs can become soiled and stained with foot traffic, as well as possible localised cracking or wear to the joints themselves. Therefore, there may be a requirement to recommend periodic deep cleaning or long-term replacement of the joints. The obvious risk with highly polished floor finishes is that these become slippery when wet and part of the building management will be to risk assess cleaning and other occupational hazards. Although not considered to be a 'normal' part of the building survey or TDD, it may be prudent to inform the client of this particular cleaning characteristic in the interests of being risk adverse. Natural stone floor finishes are more likely to be replaced as result of becoming 'out-dated' as opposed to wearing out. Their presence should be noted as an example of a high-quality material.

Natural stone floor finishes to a shopping centre, which are cracked. These have been partially replaced with ceramic floor tiles, which represent a lower specification (photo courtesy of Widnell Europe).

High-quality marble floor finish to a hotel reception. This requires repolishing and the joints cleaning (photo courtesy of Widnell Europe).

Ceramic tiles

Commercially, ceramic floor tiles make sense for a wide variety of reasons and their application is typical to areas where there is likely to be the presence of water or a requirement for clean or sterile conditions. The obvious placement of ceramic tiles is to the floors of sanitary rooms or kitchens, but these are also commonly used for changing rooms in sports or leisure complexes. They are also installed to large retail areas, such as supermarkets or shopping centres (common parts).

Ceramic tiles generally have a long life cycle but can be susceptible to impact damage resulting in localised cracking or chips. It is quite difficult to undertake in-situ repairs or replacement of ceramic tiles without making the repair obvious; however it is often better to replace significantly damaged tiles rather than leaving them to attract dirt or suffer further damage. As with most internal finishes to commercial properties, they are materials which may be considered fashionable and while ceramic tiles have been in existence commercially since the Second World War, though their size, shape and dimensions have altered with time. Large tiles (typically 450 mm x 450 mm) are convenient in the terms of their speed of installation as well as their minimal amount of grouted joints; however, historic designs have included 50 mm x 50 mm and 100 mm x 200 mm, as well as mosaic tiles. Sometimes it is possible to determine the age of a building, in the terms of the decade of construction, based on the type of floor tiles present, assuming these are 'original'.

Tiles have similar characteristics to natural stone in terms of cleaning and the propensity for grouted joints to become soiled, although they can be manufactured with non-slip finishes and specified accordingly. Ceramic floor tiles are, however, inferior to stone in terms of their overall durability and, when the subject of point loads, they are more susceptible to cracking and fracture. One important consideration with ceramic floor tiles is that there is a requirement to ensure that these are laid in bays, with provision for movement joints for large surface areas dependent upon the manufacturer's recommendations. Without movement joints, large areas of ceramic floor tiles can be subject to movement resulting, in cracking.

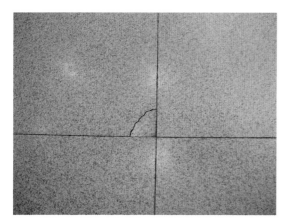

Typical damage and cracking to a ceramic floor tile caused by a point load, and possibly insufficient tile cement under the corner (photo courtesy of Widnell Europe).

Despite the presence of a movement joint in this ceramic floor, the 'parallel' cracking appears to indicate expansion or movement of the sub-base (photo courtesy of Widnell Europe).

Concrete/floor screed/granolithic/terrazzo

There are some instances when the floor finishes to commercial buildings are, effectively, the base concrete floors or floor screed. This may be as a consequence of design requirements, such as interior design to project an 'urban' look, as characterised by warehouse or factory floors. However, in most cases exposed concrete or screed is attributable to function, with industrial buildings being a prime example of this.

Exposed concrete or floor screeds are also evident in underground parking areas, as well as being typical in technical and storage rooms or archives of office buildings. For certain applications it may be necessary for the concrete or floor screed to have a polished surface; however, the most common problem with fair-faced concrete floors or floor screeds is surface cracks. These usually occur as a consequence of the curing process of the concrete, which generates significant amounts of heat, leading to evaporation of the water within the concrete and shrinkage. This problem can be exacerbated when concrete or floor screed is laid in hot conditions and it is good practice to lay concrete and floor screed in sections or bays with movement joints to help reduce cracking.

Movement joints are also typically evident to exposed concrete floors where these are in contact with the principal structural columns as there may potentially be some differential movement in these locations. Some polished concrete floors may have no evidence of movement joints because the floor screed is reinforced and forms part of the building's structural integrity. Therefore, cutting movement joints may compromise the structural capacity of the floors and it is therefore important to seek the advice of a structural engineer prior to undertaking these works. It will be difficult to determine from a visual inspection whether the floor screed is reinforced and this should be verified by reference to the as-built documents held in the data room.

Cracking to exposed polished floor finishes detracts from their appearance aesthetically but also, from a functional perspective, cracks are an obvious point of weakness. The occupation of the building and loading of the floors, particularly with logistics racking and fork lift operations can result in the exacerbation of crack widths or the breaking loose of localised sections of the concrete. This can cause operational problems with the wheels of fork lift trucks, which can be easily damaged by potholes or uneven surfaces that can occur in polished concrete floor slabs. In basement parking areas, shrinkage cracks offer a passage for water to penetrate the surface; this is typically brought into the parking areas on the wheels and chassis of cars during wet conditions. As discussed in *Chapter 6 – Structure*, surface water from the streets and roads contains pollutants and, particularly in the winter when salt is spread on the roads, this makes for moisture with high acidity. When this enters fine shrinkage cracks in basement car parking floors, the acidic moisture can be absorbed into the concrete floor structure, resulting in carbonation. Therefore, some basement car parks have epoxy painted finishes to the surface of the floor screeds or polished concrete, both of which provide excellent aesthetics or even the option of branding for certain car park operators, but which fulfil the primary function of protecting the concrete.

Many basement car parks do not have painted floor finishes and the building survey or TDD inspection should primarily seek to establish the presence of surface cracks as well as checking, where possible, the underside of the parking slabs to establish whether there is any evidence to suggest carbonation. Where cracking is evident, the surveyor should seek to establish the severity of this with a view to recommending the painting of a protective surface finish in epoxy paint (for very fine hairline cracks). More severe cracking or loose and damaged sections of floor screed or concrete should be the subject of localised repairs, and it may be prudent to seek the advice of a specialist concrete repair contractor to establish the options. However, as with any contractor who undertakes investigations as well as repairs, they normally have a vested interest in doing the works, so may seek to exaggerate the issue. Following repairs, it is normal to consider protective painting of the polished concrete floor finishes to maintain the protection.

Epoxy painted floor to a basement car park introduces the requirement to repaint this on a periodical basis (photo courtesy of Widnell Europe).

Widespread surface cracking of a polished concrete floor where no movement joints have been cut as the screed is reinforced and has structural value. The floor was laid in hot dry conditions and the cracks are attributed to shrinkage (photo courtesy of Widnell Europe).

A cast-in-situ finish of luxury is that of a granolithic floor finish. This type of finish was popular post Second World War and up to the 1960s and 1970s. It is a manufactured finish comprising cement, granite chippings and sand which is cast in typically small areas to reception areas or lift lobbies and stairwells. It appears to be a very durable material and, subject to routine cleaning, it does not appear to present any obvious wear. As with floor screed, any movement of the structure or sub-base may result in cracking to the granolithic finish but, on the whole, this material does not appear to suffer as much as cast-in-situ concrete or floor screed from shrinkage cracks. So widespread was its use in office buildings and residential properties in the 1960s and 1970s, that it was even developed into thick floor tiles as well as skirting edges. As with natural stone or ceramic tiles, it can be the subject of cracks or chips attributable to impact damage, although it appears more resistant than ceramic to impact.

Terrazzo is a high-class version of granolithic finish with marble chippings used instead of granite. The principal visual difference between terrazzo and granolithic finishes is that terrazzo has more sparkle, which is attributable to the marble chips as well as to the process which involves grinding and polishing the floor.

Both granolithic and terrazzo floor finishes are not widely used as modern floor finishes, despite their excellent durability, and it is supposed that their relatively high cost and skill required to produce them renders this an ineffective material compared to natural stone or ceramics.

Attractive granolithic floor finish to a museum which appears to have a shrinkage crack. This is believed to be in excess of 100 years old and is likely to be listed.

Cracking through a granolithic floor finish in line with a junction in the building (differential movement). The lino to the right of the granolithic finish is sufficiently flexible to accommodate movement.

Paint

The painting of floors for finishes is undertaken for aesthetic and functional reasons; however, introducing paint to the floors of a commercial building produces a consequent maintenance requirement. Depending on the location of the painted floor finish and the amount of wear that this is subjected to, a typical recommendation should be made to periodically repaint this every five to ten years. The paint itself is typically epoxy based, which is more expensive than 'traditional' paint used for internal finishes and, prior to application, the surface should be suitably prepared to ensure adhesion and to give the finish the best possible life cycle.

During a building survey or TDD there are often intense discussions concerning the painted finishes to the floors of technical or storage rooms and archives. These areas are often out of sight and, as a consequence, they usually suffer a degree of 'neglect'. The survey should seek to identify the condition of the painted floor finishes to these areas and, if they are soiled, worn or damaged, a recommendation should be made to repaint these. Depending on the condition, the surveyor will need to assess, with evidence-based rationale, whether repainting of the floor finish is required immediately or in the short, medium or long term. Having this cost entered in the cost schedule or recommendations will inevitably lead to the seller or buyer trying to diminish these findings on the basis that these areas are out of sight and 'insignificant' to the operation of the building. They may choose to come to a 'commercial' agreement to remove or ignore these costs and that is the prerogative of the buyer and seller. This is a good example of the surveyor giving the 'options' to the client without dumbing down or diminishing their recommendations and the client using this information as a potential tool of negotiation.

Painted ceramic floor tiles are peeling and these should be repainted as a periodic maintenance requirement. However, the impervious nature of the glazed finish to the tiles does not lend itself well to painting (photo courtesy of Widnell Europe).

Peeling epoxy paint to the floor of a technical room (photo courtesy of Widnell Europe).

Walls

The internal wall finishes to commercial buildings can vary considerably in accordance with the building type, sector or destination use. One general observation concerning internal wall finishes is that these usually become fatigued and worn quite quickly, typically requiring redecoration every five years. The type and specification of the finishes are also driven in some respects by functionality as well as fashions or trends.

Fair faced

When the internal walls are left in an undecorated or 'raw' state, this is known as fair faced and obvious examples of this are the internal faces of basement car park walls. In this example, the walls may have been constructed of out of reinforced concrete and simply left unfinished as 'fair-faced, reinforced concrete walls'. This may have been done as the walls contain a degree of dampness which is difficult to paint or simply left unfinished as painting these would have little or no aesthetic value. Brickwork can also be left fair faced and, although this perhaps may not have been considered the 'norm' post Second World War, more modern interior design appears to embrace this to create historic charm to modern or renovated properties. Stone can also appear fair faced and this is typically evident to historic commercial properties, such as pubs, restaurants or hotels, as post 1920 there are not many examples of solid stone walls or internal fair-faced stone cladding. Where fair-faced brick or stone is evident during a building survey it may be necessary to inspect this in closer detail as there is a range of facade cladding, such as brick slips or stone tiles, which can be applied to create this impression.

Fair-faced brickwork, concrete and stone can be painted to create 'painted, fair-faced' finishes; however, once this has been done, the results are largely irreversible as it is very difficult to undertake the cost-effective removal of paint.

When undertaking a building survey or TDD inspection and report, the surveyor is obliged to say what they see, with repairs or works largely recommended on a like-for-like basis. Therefore, it should be considered normal to recommend periodic redecoration of existing painted fair-faced finishes but generally not recommended or applicable to suggest the painting of fair-faced brickwork, concrete or natural stone.

Painted fair-faced concrete blockwork, laid and pointed in cement-based mortar.

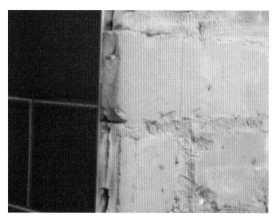

Fair-faced brickwork (TOP) which actually comprises brick tiles or 'slips' (BOTTOM).

Plaster and plasterboard

'Traditional' post-1920s commercial properties tend to have plastered finishes to the internal face of the external walls or to other structural components, such as concrete beams and columns. This often has a painted finish but can also be used as the sub-base for wallpaper or other fixed finishes, such as ceramic tiles or brick slips.

Plaster has a hard, smooth texture when finished but can be easily chipped due to impact damage and can also be eroded by water damage. It can be relatively labour intensive to install and requires a degree of manual skill to achieve a clean, blemish-free finish. As plaster is completely inflexible (like render), any building movement or movement of the sub-base is likely to result in cracking.

Similar also to render, plaster can become hollow or 'blown' when there is insufficient adhesion to the sub-base. It is also quite difficult to repair in situ without the repair being completely masked or hidden.

Gypsum is the main component of plaster and the methods of installation have remained largely unchanged for a number of decades. From the 1960s and 1970s there have been examples of asbestos evident in plastered wall finishes and, although this is something that will not be visible obvious to the surveyor, this information should be presented on the asbestos inventory.

Plasterboard is a relatively 'modern' material which is used as an internal finish to line external walls or create internal, non-loadbearing dividing walls. Manufactured from gypsum rolled and 'sandwiched' between paper to create stiff boards, this material has revolutionised residential and commercial construction. It is a material used in every sector of commercial building for both wall linings and to create ceilings. Susceptible to water damage, which tends to make it soft and weak, versions of this have been created which are water resistant for enclosing bathrooms and kitchens; this typically has a green colour to unfinished panels to indicate its nature. Specifically developed plasterboard can also be used to create or seal fire compartments and this has a pink or red colour to the surface. However, when installed in situ, most plasterboard will have a painted or papered finish which often conceals or masks its fire rating colouring. Therefore, when undertaking a building survey or TDD, it will be necessary to cross-check any fire safety compliance against the acceptance reports of the design team or control organisation as well as information contained in any as-built data sheets.

Plasterboard is also widely used as a 'dry lining', where this is fixed to a structure connected to the external walls and this can be used as an excellent sub-base to provide the internal finish where fatigued or uneven external walls are present. It can also be used to hide a multitude of different building defects or to close up openings, therefore the survey should seek to establish the 'where' and 'why' of such finishes.

By tapping or knocking on the internal wall finishes to a commercial property, it will quickly become evident where a dry lining or plasterboard has been used to mask openings or box pipework. Using touch and sound as sensory tools to reveal the presence of plasterboard results in deep, hollow sounds resonating from the material and the void behind. It is also less 'solid' and 'painful' when knocked hard but, as a consequence, it can be more susceptible to impact damage, meaning that it can be punctured with relative ease.

Plasterboard is a relative lightweight, low-cost material which has versatility in its application, making it one of the most widely used materials for internal fit out. It is not flexible and, in the event of building movement or movement to the sub-base, this can result in cracks to the joints in the panels. To mitigate the effects of minor movement, it necessary to reinforce the joints between panels with 'scrim' before plastering over the joint and plasterboard panels are specifically designed with a recessed or slightly bevelled edge to each panel to allow for this.

While it is the plaster or plasterboard that forms the base for the eventual finishes, the actual finishes can vary significantly in specification, with basic paint or wallpaper having the least resistance to damage and wear. These typically require redecoration every five years but finishes executed in ceramics or natural stone inevitably last much longer (as with floor finishes). In these cases, the replacement of such high-quality wall finishes usually occurs as the result of a requirement for cosmetic 'makeover' as opposed to them reaching their 'end of life' and one observation concerning this is that rarely are high-quality ceramics or natural stone replaced with materials of equal quality.

Internal partitions and doors

Investment surveyors and landlords have a preference for flexible space in commercial buildings. This effectively means that commercial space, and in particular office space, is delivered as shell and core with all available options for sub-dividing the space, as single offices, double offices, meeting rooms or open plan. Genuine open plan space is provided where the only fixed, loadbearing walls are the external facade or internal core areas. The subsequent subdivision is achieved with lightweight, non-loadbearing partitions. In most cases these are the responsibility of the tenant to install, maintain, make good and remove. These are not usually assessed as part of the building survey or TDD inspection but their principal construction detail should be noted in the event that this might become the subject of dilapidations or future renovation.

Partitions come in a variety of different materials and specification, which may include plasterboard, laminated panels or, more typically, glazing. In some cases, non-loadbearing partitions may be formed from lightweight concrete blockwork, although this practice typically dates from the 1960s and 1970s when these types of partitions were semi-permanent.

Internal doors are typically integrated into the internal partitions and made of similar materials. Some internal fit outs may utilise fire doors to individual partitioned spaces but these will only be effective if there are fire barriers above the suspended ceilings and below the raised floors. In many cases it is not necessary for there to be fire doors to individual offices or a requirement to sub-divide this space into such small fire compartments, this is usually dictated by the relevant fire regulations and approval of the authorities.

From a cost perspective, the majority of internal doors to partitioned offices are lightweight painted panels doors. The external surface of these doors is typically finished with a laminated surface with metal door furniture and fixings. Panels doors can be susceptible to chipping or cracking of the external laminated surface, typically from impact, and good practice would be to fit kicking plates to the base of the doors, particularly where wheelchair access is anticipated.

Ceilings

The placement of ceiling finishes serves two primary purposes and these are aesthetics as well as the concealment of services. The latter function is probably the most important, as the services to commercial properties comprise heating, ventilation and air conditioning (HVAC) as well as electricity, water and data supply as well as fire detection or alarm systems. Cables used to transfer data and distribute electricity can be concealed under raised floors or, where this facility is not present, they are often placed in the ceiling void. There is less flexibility with the placement of HVAC as the dimensions of distribution pipework, and certainly air ducts, are too great to place under the raised floor so these are nearly always installed in the ceiling void.

Industrial/urban design ceilings

Industrial buildings, such as warehouses and logistics centres, typically comprise large sheds and the nature of these buildings means that they are usually designed to be highly functional and practical. As such, there is little benefit to be gained from placing internal ceilings to these areas as storage racks are often placed up to the eaves level and there are relatively shallow pitched or flat roofs. Ceiling finishes to industrial buildings are also likely to become soiled and stained more easily than those of conventional offices so they are rarely applied, with the exposed pipework, ducting and light fittings appearing as standard. Similar ceiling finishes are also evident to retail properties, such as large supermarkets or shopping centres which resemble industrial sheds.

The industrial look has also been developed and used in other commercial spheres and is often described as 'urban'. In a bid to create an alternative 'edgy' image, internal designers appear in part to favour exposing the services as well as the raw structure of the building, which may comprise painted fair-faced reinforced concrete. However, instead of using a suspended ceiling to conceal the often 'messy' array of pipes, cables and ducts, these are now more uniformly arranged or tidied up and left exposed. Locally, sections of suspended ceiling panels may be installed to accommodate individual light fittings or ventilation diffusers. This is a specific design or image used to create an architectural style and should not be mistaken for situations where light fittings are placed below suspended ceilings. This may actually be driven by a lack of available depth to the ceiling void, which is attributable to low storey heights; where this is the case, it is also common to have exposed pipework and ducting.

Plasterboard ceilings

'Traditional' ceiling finishes, such as plaster or plasterboard, are typical of low-rise commercial properties which comprise 'residential' type construction. Older properties may utilise even more traditional techniques, such as lath and plaster, where fine timber strips (laths) are nailed to the ceiling joists and this is then finished with plaster. Typically, this labour-intensive, highly skilled operation has since been replaced by the use of plasterboard (as described above in the section on *walls*). Both plaster and plasterboard are then painted to create the overall desired aesthetic. On rare occasions plaster is applied to the underside of reinforced concrete medium- and high-rise properties; this is done to the underside of the reinforced concrete floor slabs in cases where no ceiling void is required and alternative means of distributing the services exist. The obvious and potential problem with plastering concrete is the perceived difficulty of achieving a uniform finish, which requires high levels of skill and is likely to be high cost. Furthermore (as with plastered or rendered walls), if the plaster has localised areas of poor adhesion, this can 'blow', causing it to fall. In such circumstances, plaster falling from a ceiling has the potential to cause personal injury that plaster falling from a wall.

Unusually, the complete ceilings to an office building have been finished with a plaster skim to the underside of the concrete floor above (soffit). Localised spalling of the plaster has a significant safety and cost implication.

There are several observations concerning the presence of plaster ceilings and the primary one is the inevitable requirement to paint these as they become soiled and stained. Particularly where extraction or supply grills are inserted in plasterboard ceilings, there is often a build-up of dirt or dust particles. In kitchens or washrooms, the presence of moisture and/or grease in the atmosphere may additionally lead to soiling or staining of the ceiling. Painted plasterboard ceiling finishes to commercial buildings, such as offices, are typically placed to reception areas and lift lobbies. This does appear to provide the most appealing aesthetic finish but it is likely to require the installation of inspection hatches if there are services, such as HVAC, located above the ceiling. Plasterboard ceilings are rarely placed in office areas, as these are often sub-divided with internal partitions which require fixing in part to the ceiling; a consequence of this is potentially unsightly damage to the ceiling when the partitions are removed. Plasterboard is also a comparatively weak material and this provides relatively little support for internal partitions. Therefore, to most office areas, where there is a requirement for flexibility in lettable floor space to provide open plan, single or double offices and meeting rooms, mineral fibre or metallic suspended ceiling systems are the most appropriate solution.

Mineral fibre suspended ceilings

One of the most common types of suspended ceiling systems utilises mineral fibre ceiling tiles supported by a lightweight grid, which is fixed by hangers to the underside of the soffit above. The ceiling tiles are uniform in dimensions and a common size is 600 mm x 600 mm, which can also be used to accommodate light diffusers, emergency lighting and HVAC extraction as well as supply grills.

The ceiling tiles rest on the supporting structure and, in most cases, the structure is exposed; this can be described in the building survey or TDD report as 'mineral fibre suspended ceiling tiles with exposed grid'. Mineral fibre is a low-cost material for suspended ceilings and can come in a variety of different textures or patterns but is predominantly white in colour. The lightweight nature of the ceiling tiles means that they can easily be removed to expose the services above, however they are not very strong and can be prone to breakage. They can also become soiled or stained relatively easily but are not easy to clean, therefore in situations where these form part of a claim for dilapidations, the recommendation is more often to replace rather than to clean. A low-cost alternative to replacement of mineral fibre suspended ceiling tiles is to paint these but this is rarely effective. When considering the localised replacement of mineral fibre suspended ceiling tiles, it may be necessary to take into account specific patterns or textures as these may not always be available, particularly with older ceilings. Therefore, in order to maintain visual uniformity, it may be necessary to remove whole areas of tiles to one room or separate areas, replacing these with 'new' tiles and thus providing sufficient 'spares' to replace other areas of the ceiling finish as required.

For kitchens or 'wet rooms', such as showers and bathrooms, mineral fibre suspended ceiling tiles can be sealed or coated with a waterproof impermeable surface to prevent moisture absorption. This is a matter for serious consideration as mineral fibre is a porous material which may be susceptible to moisture absorption.

Metallic suspended ceilings

More durable and higher quality than mineral fibre suspended ceiling tiles are metallic ceiling finishes. These can come as pressed cassettes with factory painted finishes or in strips/planks but both, irrespective of format, need a similar supporting grid to that used for mineral fibre suspended ceilings. In most instances, the grid to a metallic suspended ceiling is concealed, as the panels incorporate an edge strip or detailing to fix to the supporting structure.

Metallic suspended ceilings can incorporate perforations to the cassettes or planks which, combined with acoustic insulation laid above and integrated in the ceiling, can provide acoustic attenuation.

As with mineral fibre suspended ceilings, holes can be cut into the cassettes or strips to accommodate extraction or supply grills as well as standard of bespoke light fittings. One consideration concerning older metallic suspended ceiling panels which have integrated light fittings is the difficulties which may arise when replacing the lamp fittings. Typically, 'older' light fittings contain fluorescent lighting tubes and, as these are considered energy inefficient, there is a drive to replace them with more efficient alternatives, such as light emitting diode (LED) fixtures. Sometimes, the dimensions of these fittings make them difficult to insert into existing metallic ceiling designs, which may instigate the need for in-situ modification or wholescale replacement of the ceiling tiles.

Metallic suspended ceilings can come in a variety of specified colours and finishes for use in kitchen and bathrooms or shower areas. In essence, metallic suspended ceilings may be considered a higher specification of ceiling in comparison to mineral fibre. As a consequence they are likely to have better durability and life cycle but represent a costlier solution.

Specific use areas

Having established the various different materials used for the finishes to the floors, walls and ceilings of commercial buildings, their application to specific use areas should be considered. Concerning commercial properties these areas include:

- sanitary rooms
- kitchens
- staircases
- common areas (reception/lift lobbies)
- parking areas
- technical rooms.

Sanitary rooms

All commercial buildings will have the provision of sanitary rooms and these can vary from basic toilets with wash hand basins to extravagant facilities which may include showers. When undertaking an assessment of the finishes to these areas, it is important to note the construction detail and condition of the finishes to the floors, walls and ceilings. One important requirement of the floors to sanitary rooms is the need for a waterproof layer or damp proof membrane (DPM) under the floor finish or floor screed to prevent potential leakage passing through the floor to the floor below. In most cases, it will only be possible to ascertain this from as-built information, as a DPM will normally be concealed by the finishes. Higher specification sanitary rooms include a floor drain, which can assist with cleaning as well as providing a potential route for water leaks, which may mitigate their severity.

Basic and functional finishes may include lino or vinyl to the floors, with painted plastered or plasterboard walls. Individual toilet cubicles may be executed in masonry or blockwork but a more modern approach is to achieve this with the use of laminated panels. Ceiling finishes are either painted plaster or plasterboard, mineral fibre or metallic suspended ceilings. Walls can have painted finishes to plaster or plasterboard; however, the areas directly behind wash hand basins are typically finished with a course or two of ceramic tiles, and this is known as a 'splashback'.

Higher specification sanitary rooms have ceramic floor and wall finishes and, to an extent, this should be considered the 'norm' or standard requirement, certainly for office accommodation. As with all finishes, those to the sanitary rooms can vary significantly with high-spec or top-end finishes comprising natural stone, which normally has a polished finish.

Typical defects to the wall and floor finishes to sanitary rooms are soiling and staining and the condition of these is largely affected by cleaning regimes. However, despite having longer anticipated life cycles, ceramic and natural stone finishes can appear more soiled than vinyl or lino. This is largely due to the joints between the materials, which are slightly recessed and not particularly resistant to dirt and moisture. There is also a propensity for designers to specify grout which is a lighter colour, or even white; as a result, this often becomes soiled and stained. When undertaking a building survey or TDD inspection, the surveyor should assess the condition of the floor or wall finishes and, if these are soiled and stained, the surveyor should seek to include the short-, medium- or long-term deep cleaning of these, or of the grouted joints, as part of the cost plan.

All sanitary accommodation will have toilets and most male accommodation also includes urinals. During the inspection it is necessary to test the function of toilet flushes on a sample basis, although it may not be possible to inspect all toilets or toilet cubicles during the survey as these might be in occupation. Modern sanitary provision utilises suspended toilets, cantilevered from a rear fixing wall, often with concealed cisterns which differ from the more traditional 'pedestal' type toilets with exposed rear cistern. This is of significance when there is a recommendation or requirement to replace flush valves or defective washers, as it will first be necessary to access the cisterns by opening up or partially removing the finishes. One of the most common defects to toilet systems is leaking flush valves, attributable to defective washers. This causes limited but continuous water seepage into the toilet bowl. This is often a 'progressive' defect, with increased water seepage occurring if the washer is not replaced, and it can also occur or increase if there is a build-up of limescale to the underside of the washer. The symptom is the obvious trickle of water but this is often accompanied by water staining or streaking where the flow occurs. Although this may appear a small (but often repetitive) defect, the surveyor should seek to include this in the cost schedule.

A 'hanging' porcelain toilet with concealed cistern; note the staining to the toilet bowl caused by a defective flush washer, meaning that the cistern (concealed behind the ceramic finished wall) will have to be opened up to access this (photo courtesy of Widnell Europe).

A 'pedestal' toilet with exposed cistern dating from the 1970s: note the sticker on the cistern indicating that this contains asbestos.

All sanitary rooms should also have wash hand basins and these come in a variety of different materials and forms. More 'traditional' wash hand basins are porcelain 'pedestal' type but interior design often incorporates supporting structures known as vanity tops and these can be made from polished natural stone, laminated chipboard or other more contemporary materials, such as quartz, stainless steel, polished concrete or moulded resin. An alternative to porcelain insert basins is enamel, although this is less common than porcelain and more susceptible to impact damage or chipping. Wash hand basins or vanity tops are equipped with cold and/or hot water supply, with either individual taps or mixer valves, which are almost always manufactured from stainless steel.

The universal problem with all wash hand basins, vanity tops and taps is damage associated with water staining or limescale. This can build up on the surface of the basins or vanity tops as well as around the aperture of the taps. Typically, limescale is more pronounced to hot water taps and will be lessened if there is a water softener installed in the building, and the supply of the hot water may be from a central source or by localised hot water boilers. The type of hot water supply should be noted by the surveyor and the presence of individual hot water boilers should also be noted, as these typically have quite limited life cycles, depending on the hardness of the water. The potential replacement of these should be foreseen in the cost plan delivered as part of the building survey or TDD report. It is also very difficult to remove built-up and established limescale on taps, therefore consideration should also be given to the replacement of these.

Natural stone vanity tops are typically evidence of high-specification finishes and also have a relative long life cycle, which means that they are often replaced due to design or aesthetic reasons. One observation is that it is rare for natural stone vanity tops to be replaced by materials of equal quality. This perhaps reflects changes in design criteria, where more emphasis is placed upon aesthetics than life cycle and where wholescale replacement is seen as a preferred option, as opposed to renovation. Natural stone vanity tops can be removed from site for repolishing or they can be polished in situ as an act of restoration; however, timber vanity tops and laminated MDF are of much lesser quality. These may almost be viewed as 'disposable' and, while water can stain the finishes to timber, MDF is susceptible to potential swelling and disintegration if it gets wet. Depending on the quality and maintenance of the finishes to sanitary rooms, an allowance should be made to undertake full renovation of these up to every 25 years. However, this will vary significantly depending on the materials encountered.

High-quality polished black granite vanity top (photo courtesy of Widnell Europe).

Polished concrete wash hand basin with stainless steel taps (photo courtesy of Widnell Europe).

Low-cost laminated high density fibreboard basin to a budget hotel (photo courtesy of Widnell Europe).

Water damage to a natural wood vanity top with stainless steel tap (photo courtesy of Widnell Europe).

Kitchens

Most commercial properties in all sectors will have some kind of kitchen provision and this may vary from large facilities for total food preparation, known as 'warm' kitchens, to those used for reheating food, called 'cold kitchens'. However, most kitchen areas to commercial properties are for staff welfare and typically comprise a sink, worktop, fridge and microwave. The ownership of the kitchen, as with all finishes, needs to be established and, in many cases, this is often a tenant fit out but, in the event of this being under landlord ownership, it will be necessary to inspect and report upon the condition.

Modern kitchens utilise typically low-cost materials and the 'white goods' (electrical appliances, such as fridge, kettle, dishwashers and microwave) do not form part of the survey. Floor finishes are usually impervious and can be ceramic floor tiles or, more commonly, lino or vinyl and the choice of material should allow the floor to be cleaned with relative ease. Carpet or carpet tiles are not recommended for kitchen areas for both aesthetic and hygiene reasons. Walls may be solid or stud and there will typically be a splashback to the sink provided by an impervious material, with ceramic tiles being the obvious choice.

Worktops are typically chipboard or MDF with laminated melamine or some similar finish and sinks with hot and cold water supply are usually manufactured from stainless steel. As with the sanitary installation, there may be a centralised hot water supply or localised hot water boilers. Stainless steel sinks can be single or double and it is normal for these to be fitted with a drainer or draining board to one side, which drains into the sink.

The survey should seek to identify the materials used for the kitchen fit out, note the condition and any defects and establish a repair programme. The inspection should be largely visual with also a quick test of the taps to establish their functionality as well as the drainage provision. A check under the sink should be carried out to establish whether there is any evidence of leakage or water damage.

Kitchen installations can often be visually impressive when initially installed; however, these high-use, low-quality areas can become fatigued quickly, typically with soiled finishes, limescale build-up to the taps as well as the sink and damaged or loose cupboard doors and handles. The typical life cycle of kitchen installations is approximately 15 years and the relatively low-quality of the materials means these may be seen as 'disposable' fittings.

In direct contrast are the kitchen fit outs to hotels, restaurants and staff canteens, where food is prepared on site and there are also full storage facilities. Again, it is vital to establish who has ownership of these as there is a requirement for compliance with all relevant food preparation legislation and any operating permits. In some cases, the kitchen facilities and staff welfare canteens may be outsourced to specialist providers; in these instances, the responsibility for maintenance may be passed onto the operator. It is not the responsibility of the surveyor (unless suitably qualified) to verify compliance with legislation or regulations concerning food hygiene or operation of the kitchen itself.

Staircases

Most staircases to commercial properties provide an emergency means of escape, although some are used as internal features connecting different levels of occupation or use. The principal requirement of an emergency staircase is that this should be a fire compartment and therefore sealed from the connecting floors. This should be achieved with the appropriate rated fire door or the presence of an air lock (dependent on applicable regulations) created by two fire doors. The finishes are often simple and functional, often with exposed fair-faced or painted fair faced concrete or blockwork walls and fair-faced finishes to the floors. Most staircases comprise precast reinforced concrete and the survey should seek to identify the presence of textured or non-slip finishes and nosings to the edge of each tread. In some instances, the treads may be finished with ceramic floor tiles but it is important that these have non-slip provision.

Handrails should be installed according to the relevant regulations and these should extend beyond the end of the last step on each flight of stairs. Handrails may be executed in painted tubular steel or even timber, but should comply with the relevant regulations which were applicable at the time of construction or have been superseded by more recent health and safety or fire regulations.

Generally, the requirements for the maintenance of the finishes to internal emergency staircases are basic. This should typically include painting of the walls and possibly the floors (depending on whether these are already painted) as well as painting of the handrails. The decoration of these internal finishes should be undertaken perhaps once every ten years as these areas are out of sight and predominantly for emergency use only. However, this does not mean that the surveyor can discount these areas with respect to proposing maintenance or decoration. The costs involved with the recommendation for the painting of internal staircases are often debated between buyer and seller as these are seen as areas of least significance. However, it is the responsibility of the surveyor to provide evidence-based recommendations, thus allowing their client to make informed decisions.

Common areas (reception/lift lobbies)

Irrespective of the building type, the principal entrance to commercial properties usually has higher finishes or specification than the rest of the building. The main entrance or building reception is often seen as the interface between the public and private domains, it is a location where building owners and tenants alike can showcase their perceived success, status or sector importance. As these areas can utilise many different arrays of the individual finishes described in this chapter, it is important that the surveyor maintains an inspection in line with first principles. This should always include an assessment of the finishes to the floors, walls and ceilings. Attention may also be paid to fixtures, such as the entrance or reception desk, but individual items of furniture are normally excluded from the survey.

The highest specification of floor and wall finishes is probably natural stone and, as previously detailed, it is the floor joints to this that tend to become stained or soiled. It is also preferential that the stone is not slippery when wet, as this constitutes a risk to health and safety. It is normal for the main entrance floor finishes to incorporate matting, which allows visitors and occupants to wipe their feet on entry. Matting typically has a much lower life cycle than the principal floor finish and an allowance should be made to for its potential replacement.

Natural stone wall finishes should be considered low maintenance as the material is durable, impact resistant and easy to clean when soiled. Other wall linings, such as laminated boards, timber veneer or metals, also have good durability but may be less resistant to impact damage. Timber veneer and laminates are particularly susceptible to chipping, cracking or splitting if subjected to mechanical damage.

While suspended mineral fibre or metallic ceilings are considered the 'norm' to commercial office areas, they are rarely installed in building entrances or receptions. Therefore, in the majority of entrance areas to commercial buildings the ceilings are plasterboard and this can incorporate some quite sophisticated lighting installations to create a grand entrance.

Natural stone floor and wall coverings to the reception of an office building, which also utilises a 'green wall', complete with live vegetation to create an individual aesthetic (photo courtesy of Widnell Europe).

Attractive acoustic plasterboard cladding in the atrium of a multi-let office building (photo courtesy of Widnell Europe).

Corporate building identity is often achieved through the internal finishes to the building entrance and this can be extended to the upper floors by installing the same or similar finishes to the lift lobbies and separate entrances leading into individual tenant demises. In essence, the survey should seek to treat these areas the same as the main entrance and the evolution of wear or defects is likely to be the same in these two distinct areas.

Parking areas

The finishes to internal or basement car parking areas are almost always fair faced, with concrete or screed finished with circulation and parking bay markings. Better quality finishes to parking areas incorporate paint to the entire floors, which has aesthetic value but, more importantly, serves to protect the concrete from water penetration and potential carbonation. This type of finish uses epoxy paint as this is hard wearing as well as being more resistant to water and chemical attack. However, once painted, these internal floor areas will need to be the subject of continuous painting on a periodic basis, but typically every five years; this can add significant cost to the proposed maintenance budget.

Painted wall finishes are less common to parking areas but these are used to indicate parking levels or zones though the use of colour and this is typical where parking is leased to specialist parking operators. Painted wall finishes in parking areas can be susceptible to soiling or staining and these should be the subject of routine maintenance or redecoration; however, this is typically less frequent than for floor finishes. In the event of moisture penetration or dampness to the external walls of basement parking, then it is likely that the painted finishes may spall; therefore, prior to painting or redecoration, it will be necessary to treat the cause of the infiltration or dampness.

Technical rooms

The majority of finishes to technical rooms are fair faced with attention only really paid to the floor coverings. These are invariably functional rooms within a commercial building where there is often very little space between the vast array of different technical equipment. The rooms are normally situated either in basements or in roof-mounted rooms for functional reasons associated with the primary services supply and secondary distribution. There is often exposed structure to these rooms, as well as internal concrete blockwork walls or elements of the cast-in-situ reinforced concrete core. There are usually no internal ceiling finishes as these are largely exposed revealing the underside of the structure or roof slab, which is used to support and fix the equipment.

The floors are, however, usually painted as, with most technical rooms, water and other liquids often have to be transferred between equipment. The painted finishes help to provide an impervious barrier should a leak occur from the pipework; however, this measure is probably not enough on its own to prevent water seepage to the lower floors, if the technical room is on the roof. The survey should seek to establish the presence of a floor drain to all technical rooms where there is the potential for water leakage. The absence of this should be noted in the building survey report, as should the potential consequences of a water leak. Another important measure for mitigating the potential for water damage in a technical room is the placement of DPM between the floor slab and floor screed. It is not always possible to see this if it is concealed but often, where fitted, the DPM typically has a vertical upstand to the walls of the technical room. This can also be present to a low-level masonry or concrete wall which is used to create a bund.

The absence of a DPM, bund and floor drain poses a significant danger of water escaping from the technical room and infiltrating the floors below. On the premise that water typically takes the path of least resistance, this may be under the doors to the rooms, down the edges of the concrete floor screed or slab or even through the openings in the floor which are used as vertical shafts. It is almost impossible to place a DPM retrospectively into the floor slab as the difficulty is exacerbated by the presence of technical equipment which cannot be removed without causing serious disruption to the operation of the building.

The condition of the painted floors should also be noted, although commercially it may be suggested by the seller, buyer or commercial agents that the maintenance and redecoration of this is neither normal or required. From a technical perspective, the painting serves an important purpose and therefore should be the subject of periodic treatment. The painting of floors to technical rooms should typically be done using epoxy paint, which is more durable and resistant than standard paint. One observation concerning clean and well-maintained finishes to technical rooms is that this can also be representative of the overall maintenance regime within the building; it should therefore be noted in the building survey or TDD report.

Finishes summary

Finishes can be a high and significant expense for all commercial buildings, but this is typically less evident to properties in the industrial sector where, apart from some administrative areas, the finishes are largely fair faced. Concerning the survey of commercial properties and any subsequent advice to building owners, property managers and tenants, it is important to note and establish the following:

- Ownership and maintenance, repair or renovation obligations for the internal finishes should be a priority. Ideally, this should be established prior to the visit, although it is good practice to note the location, construction and condition of these as existing tenant-owned finishes may form part of a future dilapidations claim. Certainly, the finishes to vacant spaces should be assessed as there may be an immediate requirement to decorate these, hence investing cost. Where possible, the client should be advised of their potential cost risks and liability.

- While tenants may be responsible for the finishes to their individual demises, the common parts to multi-let properties are often the responsibility of the landlord in terms of maintenance and renovation.

- If in doubt, it is better to note and report on the condition of all internal finishes as this can be taken out or removed from the report upon further verification of responsibility. This is typically why there is a draft report, followed by a number of revisions up to the issue of a final report.

- The survey of the internal finishes is typically a methodical and repetitive process. To an extent (depending on the size and complexity of the building), the surveyor should adopt a 'global' approach to assessing the condition. However, there will be instances where specific individual or localised defects may be analysed to component level if they have a potential bearing on the overall condition of the finishes or investment strategy.

- One simple way to undertake the analysis of the internal finishes is to ensure that the floors, walls and ceilings are inspected in all areas, with individual differentiation noted between the different zones or occupation and uses.

- It is generally accepted that the recommendations for repair or replacement of finishes in the sphere of acquisitions or even schedules of dilapidations should seek to be 'like for like'. However, the surveyor should also take into account current market standards or changes in legislation, which may result in a need for improvement on the existing finishes when advising the client.

- The cost of finishes varies significantly according to the specification of materials used and, while the adjustment or manipulation of individual rates affect the overall budgets, the internal lettable dimensions are largely fixed. Therefore, the surveyor should seek to establish, as precisely as possible, the areas and quantities of materials included in budget estimates. These can come with a caveat on the basis that the inspection generally does not constitute a measured survey, with dimensions taken from existing plans, typically supplied by the building owner or property manager. However, it is prudent for the surveyor to undertake some checks of dimensions to assist in the process. This also serves as part of the trail of evidence which can be used to justify cost recommendations.

- Regarding individual materials and the specification of these, raised floors and suspended ceilings are considered the market 'norm' for commercial offices. This is an example of a requirement for 'betterment' where the presence of these is not evident during a survey. However, the surveyor should seek to take some check dimensions, including the slab-to-slab heights and finished floor level to the underside of the slab/structure. This is necessary to establish the feasibility of retrofitting raised floors and suspended ceilings.

- Natural stone floor and wall finishes are examples of high-specification materials with good durability and life cycle. These 'statement' materials are usually reserved for building entrances and lobbies as well as sanitary rooms. Lower specification materials include ceramic wall and floor tiles and, concerning sanitary rooms, this should be considered as the minimum requirement for most commercial buildings. Lino, carpet or carpet tiles are considered to be 'disposable' materials by virtue of the fact that these are often periodically stripped out and replaced with changes in tenancies. The cost of replacement is potentially very high, but this is usually due to the large surface areas involved in replacement. Comparably, per square metre, the renovation and replacement of finishes to reception areas, lobbies and sanitary rooms are amongst the highest for commercial properties.

Reference

RICS. 2010. *Building surveys and technical due diligence of commercial property* (4th edition, guidance note), London.

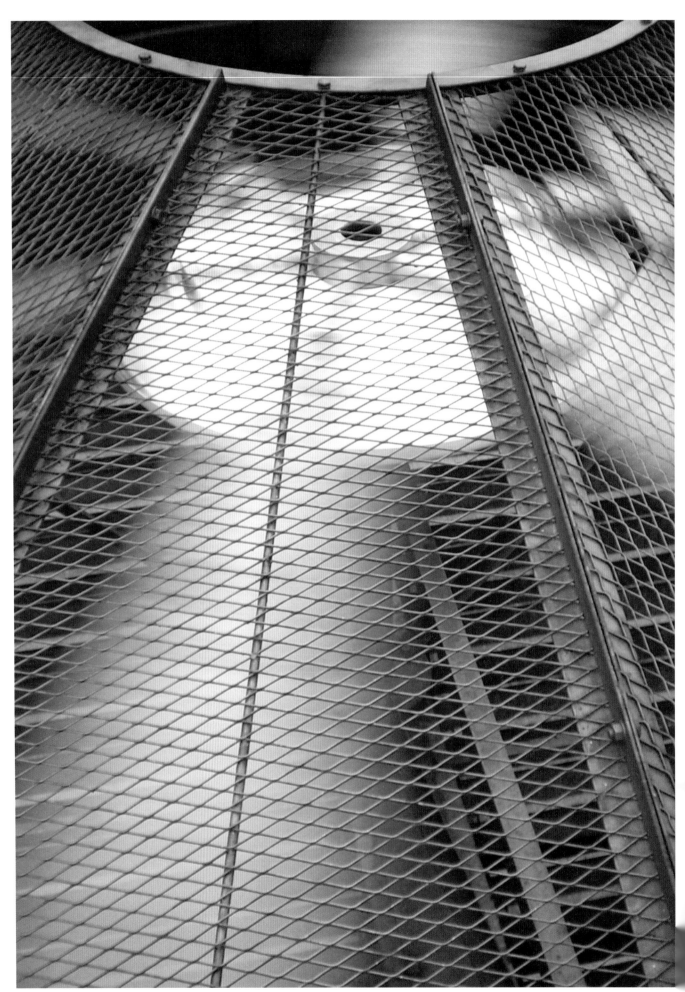

Photo courtesy of Widnell Europe

Building services
9

Introduction

The RICS guidance note, 4th edition (RICS, 2010) clearly states that, in normal circumstances, the surveyor is not expected to undertake a detailed assessment of the building services within the scope of a TDD or commercial building survey. In almost all situations this is performed by a mechanical and electrical (M&E) services engineer as this is a highly specialised area of operation. The appointment of the M&E consultant to perform the audit and report of the building services should be done in accordance with the relevant RICS guidance note. Accordingly, the contract instruction should make sufficient provision for professional indemnity insurance (PII) to cover the work of the M&E consultant and, where necessary, the providers of the surveyor's PII should be notified of this.

One important characteristic of the individual components and collective systems which make up the building services is their relatively limited life cycle and high investment costs. Building structures, facades and, to an extent, roof coverings have long life cycles dependent on the types of materials used. While it may be considered typical for medium- and high-rise office buildings to be the subject of full strip out and renovation every 30–50 years, the services installation will generally require renovation twice within this timescale. There is an degree of generalisation in this statement as there are varying levels of quality to the individual M&E components, with high-quality materials often lasting longer. However, irrespective of the underlying quality, failure to implement routine maintenance will inevitably significantly reduce the life cycle of the services installations. Therefore, an important aspect of the M&E survey will be to allow the building services consultant to establish evidence of the maintenance regime. This should be done through the checking of maintenance records and contracts and the collection of visual evidence on site to quantify this.

While the RICS guidance note stipulates that the assessment of the building services should be carried out by an M&E specialist, it does suggest that the surveyor should perform a visual appraisal of this to form an overall opinion on the visible condition and age and the need for further investigation. In order for the surveyor to do this, they should have a basic understanding of the principles of the following M&E or services installations:

- heating, ventilation and air conditioning (HVAC)
- electricity
- water/sanitary
- fire detection/alarm/fighting
- lifts.

The same approach should be adopted in assessing the services to all types and sectors of commercial buildings. However, it should be accepted that low-rise 'traditional' properties are likely to have significantly less complex installations when compared to medium- and high-rise properties. As with the construction elements associated with city or town centre low-rise commercial properties, the services installations are more likely to be similar to those used for low-rise residential properties.

Heating, ventilation and air conditioning (HVAC)

HVAC is the most complicated component of the services installations and it is often cited as being the biggest area of tenant complaint or dissatisfaction. Although these systems are designed to operate as a collective installation, it is important to understand that they consist of individual components and subcomponents. There is often a fine balance of performance between the individual components and it is not unusual for relatively minor component defects or failure to disturb the equilibrium of the combined system.

Internal climate within commercial properties is an extremely important consideration, as tenants who are paying to occupy and run their businesses in these buildings need to do this in productive environments. When designing and auditing the HVAC for a commercial property, it is important to acknowledge that comfort is often an individual subjective state. When asking the building owner, property manager or tenant to verify any complaints about internal comfort, it is typical that relatively high numbers of occupants have issues with this. Although this can vary between individual buildings or commercial sectors, it is not unusual for 20 per cent or one in five persons to experience discomfort with the internal building climate. There are several key factors which contribute to internal comfort and these include:

- temperature
- humidity
- air changes
- air quality.

The internal temperature for building occupation is not necessarily fixed but should be dependent upon the physical activity levels of the occupants and there should be a degree of regulation to allow individuals the possibility to control their environment. Humidity is a good thing, evidenced by the fact that when an artificial internal climate is provided in an occupied building, there is a requirement to introduce moisture into the air. When the level of humidity is insufficient, the air can feel 'dry' and occupants' complaints include having dry or sore throats. In confined spaces or sealed buildings, it is important to introduce fresh air, particularly where there are extraction systems removing used or 'dirty' air. For most 'traditional' low-rise commercial properties it is possible to achieve an air change by opening a window; however, this is not always possible in built-up town or city centres where this can introduce air pollution or noise into the building. Furthermore, it is recognised that, to achieve maximum energy efficiency, buildings should be as airtight as possible with controlled internal climates. Therefore, when introducing fresh air into a building this has to be of good quality, which means it has to be filtered prior to introducing humidity and is cooled or heated accordingly.

The consequence of building occupation and the effect of the built environment on the natural environment is the production of greenhouse gases (GHGs), which are attributable to the burning of fossil fuels, such as gas, coal or oil. It has been scientifically established that the production of GHGs is a significant contributor to global warming, which has had dramatic effects on the climate. Increases in global temperatures have resulted in rising sea levels and sea temperatures, droughts and crop failures in developing countries as well as changes in weather systems with apparent increases in the frequency of extreme weather events. Various initiatives to address global warming included the Kyoto Protocol in 1997, which set out to drastically reduce GHG emissions, but it was not until over 20 years later that the Climate Change Act 2008 in the

UK set a target to reduce the net carbon dioxide and GHGs by 80 per cent of the 1990s baseline. Pragmatic adoption of methods to achieve reduced emissions has resulted in the introduction of energy performance certificates (EPCs) to rapidly identify energy consumption as well as carbon dioxide emissions in both residential and commercial properties. This has enabled investors, building owners, property managers and tenants to benchmark different commercial buildings against each other in terms of energy efficiency and the potential running costs. Furthermore, in England and Wales, the Minimum Energy Efficiency Standards (MEES) have come into force in April 2018. This is derived from the Energy Efficiency (Privately Rented Property) (England and Wales) Regulations 2015, which stipulate that privately rented property must have a minimum EPC E rating before granting a new tenancy to new or existing tenants (BEIS, 2017). Therefore, there is a requirement for investors and property owners to consider energy efficiency as a priority in the design of HVAC. Building regulations or building codes have also been developed to enhance the energy efficiency of commercial buildings. When undertaking an assessment of energy efficiency regarding the provision of HVAC, it is necessary to evaluate the following:

- the building age (applicable regulations at the time of construction)
- location/orientation (external climates)
- building use (sector)
- building envelope (selection of materials and thermal efficiency)
- the type of primary production (heating/cooling/ventilation)
- fuel source for heating
- secondary distribution (pipework/ducting/presence of insulation)
- HVAC emitters (air cooling/radiators/cooled ceilings etc).

Despite the drive for increased energy efficiency of both building fabrics and HVAC systems, this goal often conflicts with market standards for commercial properties, dictated by the fact that tenants and occupiers have demands for certain minimum levels of internal comfort.

A series of design norms are typically applied to the provision of HVAC and these take into account the typical, average or maximum temperatures for summer and winter as well as the heat generated by building occupation and use. An 'example' of typical design criteria for HVAC is illustrated as below and it should be acknowledged that the design criteria will vary according to geographical location:

Performance criteria (thermal comfort)
- Base external climate (typically -8 °C to +30 °C):
  - internal temperature: winter 21 °C / summer 26 °C
  - humidity: winter 45% / summer 60%
  - ventilation:
    - 35 m^3/per person/per hour
    - 80 m^3/per hour (sanitary).
- Important considerations:
  - insulation of building envelope (solar gain)
  - heat gain from occupation:
    - people = 70 W/p (offices)
    - lighting = 10 W/m^2
    - office equipment = 30 W/m^2

In order to undertake a visual appraisal of building services to a commercial building, the surveyor should have a basic understanding of the individual components of the HVAC system.

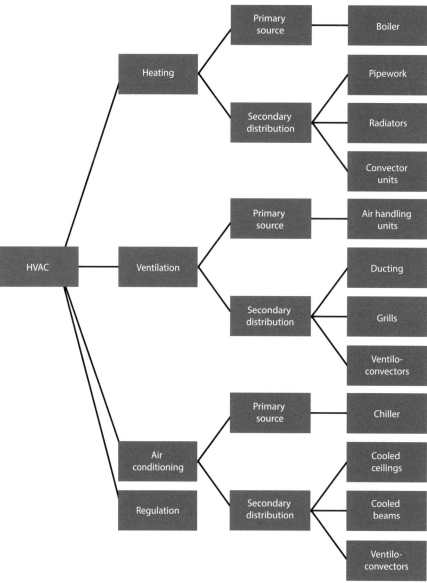

Principal HVAC components.

Heating – primary production

The generation of heat is mostly achieved with the burning of fuel to heat water and the common, applicable fuel types used include:

- gas
- oil
- solid.

The burning of solid fuels to provide heat for commercial buildings is almost never done, the exception being, in rare instances, the burning of wood pellets or coal, but the fuel requires large amounts of storage area. Solid fuel is also a 'dirty' fuel which produces relatively high levels of carbon dioxide and waste. Oil is another example of a fossil fuel that can be used but which, again, is relatively rare, with its typical use being in rural or isolated areas where there is no gas supply. It should be acknowledged that oil is not so readily used in the UK but is more common in other countries where there are limited gas networks or a different fuel culture.

Three chimney flues to an office building with oil-fired boilers (LEFT). Note the 'black' thick soot staining around the chimney and adjacent roof areas (RIGHT) which is symptomatic of this 'dirty' fuel (photos courtesy of Widnell Europe).

The use of oil as a fuel for the production of heat typically requires storage tanks with high capacity and the condition, defects and conformity of these should be established by the M&E engineer. The surveyor should be able to identify an oil-fired system and one sign is the presence of comparatively thin brown or black painted pipework supplying the burner as well as oil storage tanks.

Gas-fired boilers are the most common installation for the production of heat to commercial buildings. These are easily identifiable by the presence of yellow painted pipework and, when considering the production of heat, the surveyor should be able to verify and understand the principal components.

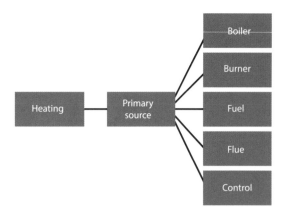

*The principal components which make up the primary source for a
heating installation (photos courtesy of Geert Lybeer/Widnell Europe).*

Concerning the operation of a boiler; fuel is introduced into the burner with the control unit initiating ignition and waste gas being emitted from the burner via the flue. Water is heated in the boiler and fed to collector circuits, which distribute this to the secondary distribution or heat emitters, and then returned to the boiler in a closed circuit in a typical 60 °C/40 °C (flow/return) design.

It is important to acknowledge that the M&E engineer will carry out a specific individual TDD or survey on the services installation. The surveyor is not expected to be able to assess in detail specific components or defects; however, there are some typical characteristics associated with primary heating production with which they should be familiar. Boilers (gas- or oil-fired) have life cycles varying from 10 to 30 years, but typical average lifecycles are between 20 and 25 years, though this does depend largely on the maintenance as well as the running time. It is not unusual in large commercial buildings for there to be more than one boiler and burner, with these being periodically switched to improve the collective life cycle as well as having at least one spare in the event of a malfunction. Burners have slightly lower anticipated life cycles with these being typically 18 years. In areas of 'hard water', where there are calcium and magnesium ions present, this has the potential to cause limescale which presents as a build-up of hard crystals to the inside of pipes, pipework joints and even boilers. This is one of the biggest factors in terms of reducing the life cycle of boilers. Therefore, in order to mitigate the effects of this, most commercial heating installations include the provision of a water softener. If a boiler lasts for approximately 25 years, the pipework forming the heating circuits is typically estimated to have twice this life cycle although, with the pipes themselves manufactured from steel and copper, they are also prone to accumulating a build-up of limescale or corrosion. The moving components joining the boiler pipework to the distribution pipes are the pumps and these are arranged to accommodate the relevant number and types of different circuits. These are known as collectors and comprise pumps with the number of circuits dependent upon the type or complexity of the building. One typical and basic provision may be for a building to have two circuits which could be identified as, say, 'North Facade' and 'South Facade'. As the pumps contain working or moving parts, these are prone to wearing or failure with life cycles typically of 15 years. They are often sealed components which are replaced rather than repaired when faulty.

Boiler with corroded joints and leaking is 'end of life'
(photo courtesy of Geert Lybeer/Widnell Europe).

Defective burner
(photo courtesy of Geert Lybeer/Widnell Europe).

A duplex water softener replacing calcium and magnesium cations in hard water to prevent limescale (photo courtesy of Geert Lybeer/ Widnell Europe).

*Two highly efficient condensing boilers (**LEFT**) used to replace two huge 'original' 40 year-old gas-fired boilers (**RIGHT**) (photos courtesy of Geert Lybeer/Widnell Europe).*

Heating – secondary distribution

The secondary distribution of heating is achieved using the following principal components:

- pipework
- radiators
- convectors
- regulation.

As established, pipework is usually manufactured from steel but this can also be copper and generally has a 50-year average life cycle. The connecting radiators or convectors have typically half this life cycle. 'Traditional' heating systems have static radiators but more modern HVAC systems utilise warm water circuits connected to air handling units to provide warm air for heating, or ventilo-convector units where air is circulated over warm water pipes to heat commercial buildings. Heat is regulated by controlling the flow rate through the heat emitter and typically this is done with radiators valves. Opening a flow valve increases flow rate and the higher the rate, the more heat is transferred into the emitter. This is a very crude form of regulation with the minimum requirement for basic heating systems being the provision of thermostatic radiator valves (TRVs), which open and close when a desired temperature is achieved. More sophisticated heating systems have motorised valves, electronically controlled by relay switches connected to a building management system (BMS) and localised thermostats.

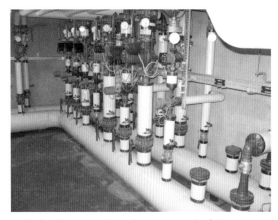

Multiple collector circuits and circulation vales for hot water distribution (photo courtesy of Geert Lybeer/ Widnell Europe).

'Traditional' static convector with a TRV (photo courtesy of Geert Lybeer/Widnell Europe).

Inside a wall-mounted ventilo-convector with a motorised flow valve top left in the photo (photo courtesy of Geert Lybeer/Widnell Europe).

A convector unit with no flow valves which has poor energy efficiency.

Inside a static radiator; note the fins to the left of the pipe which are used to generate the heat.

Ventilation – primary source

There is a requirement with all occupied commercial buildings to ensure that fresh air is introduced into a building and, in order to achieve this 'flow', there is a requirement for the 'old' air to be extracted: this is known as supply and extraction. In very basic terms, the introduction of fresh air can be achieved by opening windows. However, in order to improve energy efficiency, modern buildings should be sealed and as airtight as possible. Opening windows are not considered a viable option as this can introduce polluted air into a building as well as noise and uncontrolled draughts. There is also a more important issue, in that if a building is being heated and it is possible to open windows at the same time, this can waste large amounts of energy. Therefore, it is a requirement for most modern commercial properties to have managed and controlled ventilation.

As with most building services, there are primary source and secondary distribution facets for ventilation. The primary source can be illustrated in the following two images (overleaf):

Air intake grills to the facade of an office building (LEFT) and an air handling unit (ABOVE) to generate supply (photos courtesy of Geert Lybeer/Widnell Europe).

The principles of the generation of supply is that 'fresh' air is taken from an external source via a grill and ductwork to an air handling unit (AHU). This is then filtered to remove particles of dust and pollution before being heated to allow moisture to be introduced into the air with humidifiers; it is then cooled or reheated (depending on the temperature requirements of the end user) before being transferred via ducting to the secondary distribution.

Extraction works in a similar manner to supply, with used air taken into extract grills and distributed via ducting to extraction AHUs where this is then dispelled outside the property. Energy recuperation is possible with extraction AHUs and there can be heat exchangers (i.e. heat wheels) to recycle the heat energy. The extract and intake grills are often located on the roof of a building but these should be sufficiently spaced out to avoid extracted air being taken directly into the supply circuit. The extract and supply circuits are separate, each with independent AHUs and ductwork.

In some instances, and for heating a building, fresh air can be mixed with part extracted air and this can achieve cost savings for heating.

Ventilation – secondary distribution

The transfer of treated fresh air to occupied areas of a building can be achieved with galvanised steel ducting which is connected to either ceiling-fixed supply grills or wall-mounted ventilo-convector units where it can be the subject of additional heating or cooling by the hot or cooled water running though these emitters.

Another type of technology used to distribute ventilation is ejecto-convectors, which are supplied with temperature-controlled air forming combined heating, cooling and ventilation. Each ejector unit is connected to the treated air supply and, as a consequence, there are no individual fans (as with ventilo-convectors), making the system cost effective to install. However, there is less ability to control or regulate the temperature with this type of system.

Ducting can be a number of different shapes but square or round ducting are the most common and these are relatively bulky sections of galvanised, pressed steel. They take up a lot of space, which is one of the principal reasons why the clear floor-to-ceiling heights are so important when advising an investor or building owner. Ducting is usually concealed above the suspended ceiling and is not normally visible during site inspections. There are typically larger diameter sections of ducting in the vertical shafts which transfer air between floors and connected to this at each floor level is the principal horizontal ducting with smaller, and sometimes flexible, ducting connecting individual supply grills of convector units.

With limited or restrictive ceiling voids or clear heights, flatter ducting or ducting with reduced surface area can be used. However, this has a tendency to create noise, as a consequence of flattening or reducing the surface area of the ducting is to increase the air speed. The transfer of air requires the system to be 'balanced', and if end users cover up or block extract grills due to the perception of draughts, the air will transfer to other grills where the air speed will intensify. Many complaints by end users are of feeling cold, but this is often linked to air speed, which can reduce comfort levels if it is too high.

Chapter 11 – Legal/technical discusses the issue of fire and, in particular, the need for there to be fire compartments with fire separation, both vertical and horizontal. Pipework, cables and air ducting are examples of services which breach fire compartments and, in order to prevent fire from being transferred through ducting between compartments, it is necessary to ensure that these can be sealed or closed. This is achieved with fire dampers, which are fire resistant vanes that close automatically in the event of a fire. They are connected to the fire detection systems as well as the BMS.

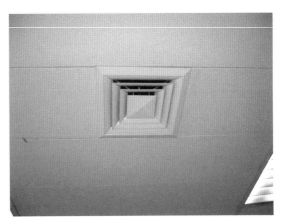

Ceiling-fixed supply grill
(photo courtesy of Geert Lybeer/Widnell Europe).

Typical galvanised steel ducting
(photo courtesy of Geert Lybeer/Widnell Europe).

Horizontal ducting leaving a vertical shaft noted is the
white mineral quilt and plaster used to fire seal the
compartment around the duct (photo courtesy of Geert
Lybeer/Widnell Europe).

A fire damper between a vertical and horizontal
section of ducting (photo courtesy of Geert Lybeer/
Widnell Europe).

A facade-mounted ventilo-convector unit
(photo courtesy of Geert Lybeer/Widnell Europe).

A basic extraction unit providing extract to a basement
car park (photo courtesy of Geert Lybeer/Widnell Europe).

Variable air volume (VAV) cooling

VAV is essentially cooling provided by pulsing cooled air into a zone of occupation from air
handing units which are supplied with cooled water. By effectively varying the volume and
altering the air speed, the system creates the impression of altering the temperature as this
creates a draught. However, this type of system is relatively low quality and is not particularly
energy efficient.

Concerning the individual components which collectively constitute the ventilation system,
these have varying anticipated life cycles Air handing units have an estimated average lifecycle
of 25–30 years but this is subject to frequent maintenance intervention as, typically, problems
occur with defective humidifiers. Galvanised steel ducting can last for more than 25 years as this
has no working parts, except for fire dampers. Ducting can be prone to an internal build-up of
dust and renovation or refurbishment of supply systems should allow for internal cleaning of
the ductwork. One of the biggest issues concerned with ducting is the presence of asbestos
in old fire dampers. This can prove to be highly costly and complex to remove when there is a
renovation or refurbishment of the system.

Air conditioning (cooling) – primary source

The cooling of most buildings is something which occurs throughout the seasons, although
during the winter months it is anticipated in the design process that there is a lesser demand
for this. It has become a perceived market standard for most commercial buildings and those
without cooling are likely to be harder to market, with lower rental incomes.

The production of cooling is achieved by compressing specific gases into liquids and
traditionally these were chlorofluorocarbons (CFCs) or hydrochlorofluorocarbons (HCFCs).
However, these gases are ozone depleting and were the subject of a worldwide ban as part of
the Montreal Protocol, which was signed in 1987. One of the most common HCFCs used for
cooling was R22 and, accordingly, this was supposed to be phased out by the end of 2014. Other
coolants have been developed and these include R134A or R407C, and the effect of compressing
these is, initially, to create heat. The heat is then extracted from the warm refrigerant by passing
this through a condenser, with a fan used to blow the warm air away from the chiller unit. The
refrigerant then passes through an expansion valve which causes it to cool significantly. This can

then be used to cool down a water circuit, which is connected to the secondary distribution circuits used to provide a cooled environment.

This process can be undertaken in combined chiller units, which are typically roof-mounted to allow the removal of the warm air around the condenser. Chillers are particularly noisy pieces of technical equipment (>95 dB) and are also very large, heavy items. Consequently, they are often craned onto a flat roof or placed within recessed or built-up technical compounds on the roof with acoustic cladding or panels to limit noise pollution.

Chiller units located in a roof mounted compound (photo courtesy of Geert Lybeer/Widnell Europe).

A galvanised steel roof-mounted chiller compound (photo courtesy of Geert Lybeer/Widnell Europe).

Compressor unit located in a basement (photo courtesy of Geert Lybeer/Widnell Europe).

Air conditioning (cooling) – secondary distribution

Cooled water can be used to supply air handling units or a series of secondary distribution devices such as:

- cooled ceilings
- chilled beams
- convectors
- air cooling.

Both cooled ceilings and chilled beams work on the principle that passing cooled water through these components results in their surfaces becoming relatively colder than the surrounding air. As the air closest to the ceiling panels or beams cools, it drops and is replaced by warm air below, creating an air flow or circulation. This is known as static beams and, irrespective of these two types of cooling distribution, there is a requirement for ventilation to also be present to ensure there are air changes in occupied areas. If ventilation is incorporated in the beams, these become known as dynamic beams.

Because the surface of the ceiling panels and beams is colder than the surrounding air, it is important that moist warm air is not introduced into the vicinity as this may condense on the surface of the emitters and water droplets may form. Moist air can also come from outside the building where there are opening windows, therefore it is essential that the facade is sealed or fixed and, where windows are operable, these should be fitted with micro switches to turn off cooling to their relative zone or module.

Convector units are one of the most common ways to distribute cooling and these can be ceiling or facade fixed. They are supplied with warm water and cooled water, which is selected according to their control or regulation devices (thermostatic valves). Fresh treated air passes over the pipes into the occupied zone or module to create the desired internal climate.

Ceiling-fixed cooled beams (LEFT) and wall-fixed ventilo-convector (RIGHT); both are methods of distributing cooling (photos courtesy of Geert Lybeer/Widnell Europe).

Split units

The HVAC systems described above are typical to large commercial properties where there is central production plant for the provision of hot and cooled water. However, there are situations where there are much smaller buildings or specific locations which require cooling and one way to achieve this is with split units.

Encompassing the compressor, fan and expansion valve in one relatively small unit, split units comprise two principal items. External to the building is the primary source and internally is the secondary distribution, which is often a ceiling-fixed 'cassette'. This is a relatively cost-effective way to achieve localised heating and cooling but this is often not connected to a fresh air source of evacuation, meaning that stale air is recycled.

In many instances, split units are under the ownership of the tenant and, if installed poorly, can cause damage to roof coverings, which can lead to water infiltration.

An external compressor unit placed on concrete paving slabs and protective matting to a PVC roof covering to mitigate damage to the PVC.

The identification plate reveals that the coolant is (banned) R22, which should have been replaced by the end of 2014.

The internal ceiling distribution cassette for this split unit.

Pipework

One important consideration for the flow and return of hot or cooled water is that the pipework should be insulated. There is the potential for heat loss in uninsulated hot water pipes and for condensation to form on exposed cooled water pipework. Typically, developers will seek to cut costs in execution by leaving pipes uninsulated and this is not always visible if pipes are installed above suspended ceilings without access. The only way for the M&E engineer to ascertain if insulation has been provided is to check the tender documents against the building acceptance reports and any technical data sheets held in the data room.

External pipework should be fitted with armoured external casings (usually in aluminium) and it is not good practice to finish external pipework insulation in silver foil as there is a risk of bird damage to the silver foil and any exposed insulation.

Regulation

Building tenants or occupants invariably complain about the internal climate in their zone of occupation and, to an extent, there is often no pleasing everybody. It is estimated that one in five persons consulted about the internal climate in their buildings have something to complain about. Personal comfort can vary considerably between individuals and the effects of layers or different types of clothing and activity all add to the perception of feeling too hot or too cold. Where an internal artificial climate is created, it is the regulation of this which appears to cause the most problems.

With respect to building services, the term 'regulation' effectively means the presence of manual or automatic devices to facilitate control of the internal climate. In simple terms, this could be the presence of a manual radiator valve used to turn the heat in a radiator up or down. However, if the radiator is emitting heat and a window is then opened to provide cooler air or ventilation, this creates the possibility of energy wastage. Some commercial buildings are much more complex than this with heating, cooling and ventilation all requiring regulation and offices, hotels and some residential properties have similarly complicated requirements.

Modern buildings or those which have undergone renovation may include energy consumption metering or measurement. This is typically displayed on a wall-mounted panel and can be used to calculate individual energy costs for tenants or occupiers. The M&E engineer is normally qualified to assess, interpret and analyse the information displayed on the metering equipment.

The HVAC to commercial buildings can be regulated:

- per floor
- per zone (sub-floor division)
- per module.

Regulation per floor typically occurs in buildings designed for open plan occupation; however, this is relatively basic with the HVAC regulated by a central thermostat. Sub-floor division may occur where there are open plan offices and meeting rooms or pods where high or dense levels of occupation require higher levels of cooling and ventilation. Regulation per module is the optimal requirement for most tenants or occupants but can lead to huge temperature differences on either side of internal partitions relative to personal and individual comfort levels. Regulation per module is also highly costly, requiring individual thermostats or control and valves.

In order to achieve regulation, this is normally controlled by a BMS which utilises computer-based software to electrically control mechanical valves to alter the variables in the HVAC installation.

Typical problems associated with regulation and BMS resulting in occupier complaints about internal climate occur when the regulation is designed for open plan occupation and tenants install partitions. As a result of this there may be one or only a handful of thermostats or control devices for a large number of individual end users. The cost to remedy this can be very high with, potentially, a need to install the relevant BMS infrastructure as well as multiple extra valves and controllers. The M&E survey is non-destructive, therefore it is not normal to inspect within ceiling voids to establish the presence of the relevant infrastructure and cabling required to partition the building, floor, zones or modules. The seller has an obligation to disclose this information to the purchaser and, if the investment memorandum (sales brochure) indicates the potential to sub-divide the internal space, it may be assumed that the BMS and HVAC system is equipped to do this. The inability to sub-divide the regulation from open plan to partitioned office space is a red flag issue, as the costs involved to rectify this can be high.

Other defects concerning the regulation of the HVAC can relate to defective software, which has not been possible to update. Of all the HVAC components, the BMS has one of the shortest life cycles; in some cases this may be less than 10–15 years. BMS installations become obsolete as advances in technology make IT upgrades difficult to achieve or physical spare parts become outdated or unobtainable. With relatively few manufacturers and installers of BMS, there appears to be little competition in this field. Accordingly, day rates for technical support are also high.

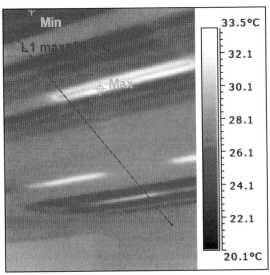

Faulty regulation has resulted in high temperature to this cooled suspended ceiling (photos courtesy of Geert Lybeer).

HVAC and commercial sectors

When analysing the different types of buildings within the commercial sectors, it is evident that not all are installed with the same HVAC systems. Different individual, sector-specific buildings typically have the following HVAC characteristics:

- low-rise 'traditional' offices and retail – central heating with radiators (landlord owned) and localised split units (tenant owned)
- medium- and high-rise offices – full landlord-owned HVAC system
- medium- and high-rise residential towers – full landlord-owned heating (district heating) and ventilation
- hotels – full landlord-owned HVAC system
- shopping centres – landlord-owned primary supply (hot and cooled water) with tenant distribution
- schools – landlord-owned heating installation and radiators with some localised split units and extraction
- logistics buildings – central heating system (landlord owned) to office areas and split units (tenant owned), suspended gas-fired hot air blowers or infra-red heaters to warehouses (tenant owned).

Electricity

The surveyor is required to have a general understanding of the electrical supply and distribution to a commercial property but it is the M&E engineer who will critically appraise this. All electrical installations have the following:
- primary source
- secondary distribution sockets
- secondary distribution lighting
- emergency power.

Primary source is effectively the connection to the public main or supply from the utility company. In traditional low-rise commercial properties this is often the electricity meter but in larger, more complex buildings this may be a transformer. There is always a main low-voltage electrical board, which is connected to individual 240V distribution boards hard wired to the secondary distribution sockets and lighting circuits. The M&E consultant will assess the electrical installations, noting their design and their conformity with any relevant regulations via the document review. One important consideration for large office floor plates are the number and locations of the individual electrical boards and the potential to sub-divide floors, as well as the electrical supply for multiple tenancies.

Modern or renovated buildings often have electrical energy meters per floor or half floor, which are connected to the BMS to enable tenants or users to be billed for their relevant electricity consumption.

Secondary distribution of electrical supply usually occurs through suspended ceilings with cable trays and vertical distribution poles or with cable trays and floor boxes to raised floors or via wall-fixed PVC or metal trunking. The surveyor should note where the distribution sockets are located and whether there is the potential to fully divide the space into individual partitioned modules while retaining secondary distribution.

A very important consideration in the assessment of the electrical distribution is the provision of lighting. This is a visual assessment undertaken in areas of occupation and it is not normal to take measurements regarding the lighting levels (lux) unless there are reasons or complaints about this. Lighting is an example of a secondary distribution of electricity which has a high investment cost and relatively short life cycle. It is also responsible for potentially large amounts of energy consumption, which has seen the 'traditional' florescent tube or strip lighting replaced with LED lighting. When recommending replacement of light fittings, it should be acknowledged that there might be a requirement to alter, amend or change suspended ceiling tiles too; however, modern LED replacement lamps can often be fitted into existing fixtures.

Many commercial properties have an emergency power supply (EPS) or uninterrupted power supply (UPS). These two facilities perform similar functions but through different arrangements.

An EPS traditionally is a diesel-powered generator that operates when there is a loss of electricity, it is a standalone unit, often with its own in-built fuel tank of approximately 100 litres. It requires an air intake as well as a flue to evacuate fumes. It is connected to an electrical distribution board, which then provides an emergency power supply to specific items, such as lifts, fire dampers in ventilation ducts, sprinkler pumps, fire detectors, fire alarm, access control, BMS, smoke or heat evacuation and sometimes sockets in zones of occupation. An EPS generator set usually has an operating or environmental/operating permit and should be the subject of yearly inspection and report. Typically, the electrical cables used to supply components from an EPS are fire resistant and these can be visually identified as being red or orange in colour. Halogen free cabling can also be used, and this can have a specific colour. The fire rating of all electrical cables supplying emergency or uninterrupted power should be verified in the as-built file.

A UPS is, typically, battery powered and, upon failure of the electrical supply, this switches on instantaneously, meaning that it is suited for use with IT/computer server rooms. The number of batteries can be vast and, in addition to requiring an operating licence or environmental permit, these need to be stored in a room with sufficient floor protection and bunding to retain the battery acid in the event of a spillage. The storage area should also be ventilated.

In the event of a loss of power, it should be noted that localised emergency lights should operate, although most of these are battery powered. The survey can perform a random test on these where there is a test button provided and this should form part of the annual fire safety test report with the relevant test certificates located in the data room.

Water/sanitary provision

Water supplied and distributed in commercial properties is either used as drinking water, or to supply toilets or showers and for the fighting of fire. Water used for firefighting appliances, such as hose reels or sprinkler installations, is normally separated from drinking or sanitary water with pipework painted red for firefighting water and (sometimes, but not always) blue for drinking water.

The M&E survey should seek to identify the entry point of the water, including any water meters, isolation valves and pressure pumps for high-rise properties or where the water pressure from the utility provider is too low. The presence of insultation to pipework is important as this will help to prevent condensation and also protect against frost damage if the pipes are exposed to sub-zero temperatures.

For large commercial properties there may be indirect water distribution, with storage tanks located on the highest level in technical rooms, but for smaller 'traditional' type low-rise properties this may be direct and without a storage tank.

The majority of distribution pipework is concealed until connection with toilet cisterns or wash hand basins and, in some cases, this is only cold water with hot water provided by localised electrical boilers above sinks or integrated into vanity units. Where there is the provision of hot water for wash hand basins, this may come from hot water storage tanks with either gas-fired or electrical immersion heating.

One important consideration for the supply of water is the potential for this to contain legionella bacteria and this typically occurs when cold water is left static for a period of time; this then warms up in the pipes allowing the bacteria to form. Legionella can also build up in a hot water supply where the temperature has dropped below 60 ºC. The M&E consultant will typically check any log books in the technical room which may contain data relating to legionella.

Water used to supply boilers and hot water production for sanitary supply or humidifiers for air handling units usually requires 'softening' in areas of hard water. This can help to prolong the life cycle of the heating equipment by preventing the build-up of limescale. Water softeners are technical equipment which removes calcium and magnesium cations out of water and replaces these with sodium cations. This requires salt (NaCl) to be added to the equipment and this is done inside a mixing vessel. As salt is particularly corrosive, it is important that this is stored in a PVC or plastic box to prevent damage being caused to the concrete or floor screed of the technical room. This should be inspected in greater detail by the M&E engineer.

Waste water evacuation is checked by the M&E engineer, although this connection is usually underground. For low-rise commercial properties it may be possible to lift the covers to manhole inspection chambers; however, for larger commercial properties there is often a siphon point in the evacuation pipework as this exits the building in the basement, although this may be concealed by a manhole cover.

Waste evacuation pipework can comprise vitrified clay in older properties and this is a particularly inflexible material which is prone to cracking and fracturing if there is lateral movement of the pipes. Another inflexible material is cast iron, and although this has an estimated 50-year life cycle, it can be prone to long-term corrosion and fracturing. More modern materials include PVC and high-density polyethylene (HDPE). PVC for underground use is typically orangey-brown in colour and HDPE is black. When considering the removal of all horizontal connecting pipework, as well as that located internally in shafts, this can prove to be a high-cost exercise and, in the event that the pipework's serviceability has not been verified, it may be prudent to instruct an internal camera survey.

Fire detection/alarm/firefighting

The M&E consultant will also assess the components which make up the technical fire engineering equipment. This may include the fire detection system, fire alarm and any electro-mechanical installations for firefighting.

Part of the survey will be to review the test reports and technical data sheets for the fire engineering installations as it will not be possible to test these during the inspection. Fire detection systems can include smoke detection, heat detection or a combination of the two, and the survey will seek to identify the type of detector as well as any visually obvious errors or omissions in the system. Older smoke detectors were ionic and contained a very small quantity of radioactive material. Where observed, these should be replaced and when costing this procedure, an allowance should be made for the cost of disposal of this hazardous material. In contrast, modern smoke detectors are optical or thermal.

Fire alarms can be connected to the fire detection system and, when either of these is triggered, an alarm will sound. This should be connected to the BMS, which will shut down various technical installations and close the fire dampers. The fire alarm will also be connected to the fire panel, which is often located on the ground floor at an entrance point. This panel offers the fire brigade the option to switch off the HVAC, open smoke vents and establish the zones in which the fire is active. Fire panels and fire alarms should have repeaters for other parts of the building or commercial complex to alert occupants to evacuate even when the fire has not reach their zone.

Firefighting equipment may be manual, such as extinguishers, or automatic where sprinklers are installed. All commercial properties will be equipped with fire extinguishers and, importantly, these should be checked annually, with a label typically attached to the device which is stamped or certified accordingly to confirm this. In larger commercial properties there may be hose reels installed, which are strategically located in shafts to be used by the occupants or fire brigade. These shafts may be 'dry risers', where there is no water in the pipes until the hose is opened, or 'wet risers', typically for high-rise buildings, where water remains under pressure and continually in the pipe and hose reel. Hose reels also require annual attestation to confirm their serviceability and the M&E survey will seek to check the labels on the hose reels or their cabinets, but on a sample basis where there is a significant number of these. This can then be cross-checked with the current fire safety report and reports of the relevant maintenance companies. Water used to supply hose reels, and in particular those to high-rise properties, may require additional pressure pumps and these are usually located in the basement of a commercial building close to the water meter with connection to the EPS.

Sprinklers are usually only installed as a specific stipulation under fire or building regulations and this is done where there is perceived to be excessively high fire loads, insufficient fire resistance to the building structure, difficulty in providing access for the fire brigade or where there is a requirement to provide a safe passage for evacuation. They are highly specialist installations and the M&E consultant will seek to establish the relevant evidence of testing through analysis of the data room documents. The on-site visit will assess the installation in situ and this usually comprises a storage tank, mains water connection, control equipment and pumps on the primary supply site. The secondary distribution of the sprinkler system includes ceiling-mounted pipework, which may be visible or concealed by suspended ceilings, and the sprinkler heads which are activated by heat. It is impossible for the survey to visit and check every sprinkler head, therefore, it is necessary to rely on the report of the maintenance contractor. There may be buildings with only localised sprinklers and this typically occurs in specific zones of particular concern, such as rubbish compounds or storage areas.

For computer server rooms, automatic firefighting is not normally done with sprinklers or water as this can cause huge amounts of damage to the operational equipment. Typically, this is achieved with the use of gases, such as argonite or CO_2, both of which are inert and either remove or displace the oxygen from the room, provided this is airtight. This type of automatic firefighting system is also highly specialist, with the storage and use of these gases requiring operating licences or environmental permits. The equipment and operation cannot easily be checked on site, therefore the M&E engineer will have to rely on the test reports located in the data room.

Lifts

Vertical circulation to most low-rise 'traditional' commercial properties is via internal staircases with some modern low-rise properties being equipped with lifts, depending on their sector of operation. Most, if not all, medium- and high-rise properties are equipped with lift installations and for modern buildings they are usually designed relative to occupancy rates as well as waiting and transportation times. It is the role of the M&E consultant to critically evaluate the capacity of the lift installations and any complaints of the occupants or tenants should be disclosed by the building owner or property manager.

Lift installations comprise the machine room, lift shaft, control or guidance equipment and the lift cars. Much of this can be visited during the survey, although the obvious inaccessible area is the shaft.

Lift shafts should be fire compartments in themselves as they pass vertically through all separate fire compartments of the individual floors. Typically, these are constructed in cast-in-situ concrete forming part of the building core; sometimes they are executed partially in concrete blockwork, but this is rare. Therefore, in most cases with fixed and reinforced concrete shafts it is difficult to alter or widen the shaft dimensions and this has an impact on the maximum size of the lift car.

The motors for lifts can be either hydraulic or electrical, with these located in machine rooms either placed at the bottom or adjacent to the bottom of the shaft at the lowest level or on the roof at the highest level. Some lift machinery is actually located in the shaft and, traditionally, for low-rise properties with lifts, there is a hydraulic motor under the lift 'pushing' the car vertically, while for medium- and high-rise buildings the electrical motor was situated above the shaft 'pulling' up the lift. In both cases, the switch gear or control and guidance boards are typically located in the lift motor rooms. These are a series of electrical boards used to process the lift command requests made by users pushing buttons in the lift lobbies on different floors and these instruct the lift to go to the correct floor in an efficient routine.

The lift cars or cabins are the metal 'boxes' used to transport users and are secured in the shaft by guidance rails. These can be refurbished relatively easily and the date in the cars does not always correspond to the date of the whole installation.

The installation, maintenance, repair and renovation of lifts are expensive, compounded by the fact that the limited number of lift contractors means potentially little competition and therefore high prices. The M&E engineer may review the lift maintenance contract but should seek to receive copies of the latest periodic test reports to verify serviceability and compliance with regulations. It is not the role of the M&E consultant to verify compliance, unless they are certified to do this, and their site survey will seek to check the lift motor for obvious signs of defect, correct threshold levels between the lift car and lobby as well as the age and condition of the control boards.

Historically, lifts have used asbestos in the braking systems as well as asbestos panels inside the lift doors. This should be identified or noted on the asbestos inventory and reported as part of the due diligence process.

Summary of building services assessment

When undertaking a TDD or commercial building survey, the building services are an important and highly relevant part of the process, concerning which it is necessary to note the following:

- According to the RICS guidance note (4th edition), the building surveyor is not expected to undertake a detailed assessment of the building services.
- Building services are the subject of specialist inspection by an M&E consultant.
- Building services have some of the highest investment costs with, relatively, the lowest anticipated life cycles for commercial buildings.
- Concerning internal climate and HVAC, approximately one in five persons is not content.
- Full HVAC is considered the 'norm' for most commercial offices, hotels and some residential properties but, when considering renovation, it is necessary to have sufficient clear floor-to-ceiling heights as well as areas of the roof with sufficient loading capacity for the equipment.

References

BEIS. 2017. *The non-domestic private rented property minimum standard: Guidance for landlords and enforcement authorities on the minimum level of energy efficiency required to let non-domestic property under the Energy Efficiency (Private Rented Property) (England and Wales) Regulations 2015*, Department for Business, Energy & Industrial Strategy, London, Crown copyright.

RICS. 2010. *Building surveys and technical due diligence of commercial property* (4th edition, guidance note), London.

Photo courtesy of Widnell Europe

External areas 10

Introduction

The external areas to commercial buildings can vary significantly according to the sector, type or use. There is always a physical point at which the building interfaces with its external surroundings and it is necessary to differentiate between what is the entrance to the building and the entrance to the site, plot or boundary.

With many town or city centre buildings, the entrance to the plot or site boundary is often also the entrance to the building, as these properties have direct street or pavement frontage. This is often the case irrespective of whether the building is in the retail, residential or leisure sectors. In contrast, out-of-town developments, such as retail centres, business parks or industrial zones, often have very large sites contained within their boundaries. With respect to both town or city centre properties and out-of-town sites, it is necessary to establish the site boundaries to be able to undertake a survey, ensuring this covers all relevant areas. This information is sometimes contained within investment memorandums. However, it will be necessary to fully verify this by obtaining as-built site plans or land registry documents. These are often held within the as-built file and should form an important part of the data room documents if the building is the subject of acquisition or disposal.

In the event that this information is not available prior to the visit, the surveyor should seek to establish this by questioning the property manager or tenant during the site inspection, although this information will still be subject to confirmation. Often, site time is limited with typically only one permitted visit, or the asset may be some distance from the offices of the surveyor. Therefore, it is important to be risk averse, using the visit to maximum effect by inspecting as much as possible, or preferably all, of the external areas within the permitted timescales.

The quantity of the external areas can be deceptive, as the site may have little or nothing to the front, but may extend significantly to the rear of the site, which may not be immediately obvious until inspected. In essence, the surveyor should be prepared for the 'worst case' regarding external areas, which may mean having to inspect potentially large surface areas. However, in many instances the presence of large surface areas often equates to repetitive external finishes, where it may be possible to offer quite generalised advice.

Another important consideration concerning external areas is the possibility of joint ownership or the leasing and sub-leasing of space for storage or parking. There may also be important issues regarding rights of way and easements. This is something that should be typically addressed in the legal due diligence or property searches. However, the majority of the legal advice concerned with property transactions is undertaken by means of a desk survey, with most legal advisors having little or no knowledge of the actual on-site conditions. Therefore, as part of the building survey or TDD visit and report it will be necessary to consider this legal/

technical issue. Initially, it will be necessary to establish from the legal advisors the presence of any joint ownerships or easements from a legal perspective. The technical application of this on site should be verified where possible, and any anomalies reported back to the legal advisors. It is not the responsibility of the surveyor to comment upon the consequences or conditions of legal rights of way, easement or infringements and it is reasonable to expect the legal due diligence or advisors to report upon these matters. Such instances are an example of the need for close collaboration between the legal and technical teams.

When inspecting the external areas to commercial property, the inspection should seek to assess the following:

- car parking
- access roads
- pavements
- patios and decking
- storage yards and hardstanding
- waste areas
- outbuildings
- lighting
- signage
- fencing and boundaries
- vegetation and landscaping.

Car parking

The allocation of car parking can be a contentious issue. Whether the site is located in town and city centres or out of town, there is invariably a lack of car parking provision to many commercial properties. Planning policy and the physical availability of land are the driving factors affecting the allocation of car parking to commercial properties and this does have relatively high commercial value from a marketing and leasing perspective. Most modern city centre buildings have basement parking, although this is typically restricted by planning policies aimed at reducing the pollution attributable to car use. However, despite the good intentions of this from an environmental perspective, there is often poor or insufficient public transport alternatives to make up for the reduction in car parking spaces. Furthermore, many owners of town or city centre properties with external parking seek to maximise the value of the land by selling or developing this for commercial use. Consequently, there is often very little presence of external parking to town or city centre properties. In contrast, out-of-town developments can have poor public transport provision and this is driven by, or a consequence of, the provision of significant areas of external parking provision.

External parking areas are usually executed in either tarmac, concrete or concrete block paving and these have to include a sub-base which is sufficient in thickness as well as compaction to provide the appropriate loading capacity. The top finishing layer is, in many instances, only as stable as the sub-base below. The sub-base can be formed with compacted layers of hardcore, but this needs to take into consideration the underlying subsoil. As previously discussed in *Chapter 6 – Structure*, subsoils are either cohesive or non-cohesive and these have different characteristics with respect to stability and the effects of moisture. While cohesive soils (clays) are more susceptible to swelling and shrinkage as a result of moisture changes, non-cohesive soils (sands or silts) are prone to erosion. Movement of the subsoil, or even the sub-base under external areas of parking, can have an effect on the parking surface itself; this typically results in localised sunken areas which are prone to 'ponding'. The effects of ponding, particularly where the car parking surface is cracked or broken, inevitably lead to water ingress into the material and sub-base. This has the potential to further destabilise the surface, with erosion of the sub-base or frost damage due to expansion and contraction in cold weather exacerbating the problem. Therefore, localised sunken parking areas should be the subject of immediate repair.

Concerning the materials used to form the top surface of parking areas, tarmac is probably the most widely used, due to it being a cost-effective material which is easily sourced and quick to construct. It has a good life cycle, which can typically be 25 years, although it is prone to cracking and crumbling as a result of frost action combined with the tensile and compressive force imposed by vehicles.

Concrete is a material which has excellent compressive strength, thus making it ideal for parking areas; however, it should be laid in bays with movement joints. Its relative weakness in tension means that it is prone to cracking unless reinforced with steel. Therefore, concrete is seen as a less cost-effective alternative to tarmac for external parking areas. It is also a material which has a degree of porosity, making it prone to potential moisture absorption and possible frost attack. As most concrete parking areas are poured in situ, the human element in the fabrication process introduces scope for inconsistency, such as the water to cement ratio and choice of coarse and fine aggregates. The curing process of concrete also generates relatively high levels of heat, which can result in the evaporation of water causing shrinkage cracks to occur; these cracks in the medium and long term offer a potential route for moisture penetration. Subsequent frost attack and spalling of the concrete inevitably leads to its degradation as well as weakening of the material, which can then be prone to further cracking.

Pre-cast concrete for structural frames and slabs is known to have better resistance to moisture penetration as this is manufactured in a factory environment where both the materials and the process are controlled, resulting in concrete will less porosity. It is rare that precast concrete slabs are used for external parking areas; however, precast concrete blocks (block paving) provide a viable alternative to tarmac and in situ concrete. Because concrete block paving comprises small, thick, brick-shaped slabs, these have excellent compressive strength and, since they are laid on a bed of sand or sand and cement, they appear to accommodate tensile forces and are thick enough to be resistive to cracking. The installation or laying of a parking area in concrete block paving is quite labour intensive with it also being difficult to achieve a completely level surface, which can eventually lead to localised ponding. One of the distinct advantages of concrete block paving is that this can be lifted or removed for the placement of underground services and reinstated with little or no visual alteration. Tarmac and concrete has to be cut or sawn to accommodate works below ground and, in most cases, the reinstated top surface is not only visible but often an area potentially prone to sinking.

The external edges or boundaries of all parking areas are finished with a physical edge and, in most cases, these are concrete kerb stones although granite is sometimes used for the parking area of more 'prestigious' buildings. These are often laid on a bed of cement-based mortar to help achieve the relevant levels but concrete kerbs are prone to fracture as they are weak in tension. All kerb stones are prone to sinking or displacement if they are the subject of impact damage or if there is a weakness in the sub-base.

Damaged, precast concrete kerb stones dislodged by lorries to a logistics site.

The existing tarmac hardstanding not fit for purpose.

When assessing external car parking, it is necessary to note the presence and condition of the lines marking out individual parking spaces or circulation routes. Although faded or worn lines may appear a relatively insignificant defect, the surveyor should establish their presence and if their condition makes them unfit for purpose. Any commercial negotiation to remove the cost of remedying worn lines to parking areas is the decision of the client and, unless otherwise instructed, the surveyor should seek to include this cost in their report.

Large surface areas of external parking can be particularly prone to flooding from surface water, therefore the survey should seek to establish any visible evidence to suggest this. During periods of wet weather it may be possible for the surveyor to note the formation of puddles or standing water; however, in dry conditions these may not be evident. Signs of water staining or the depositing of silt to uneven sections of paving may be indicative of historic ponding or flooding. The surveyor should also ask the property owner to disclose any knowledge of flooding to the site and, while they are not obliged to calculate the sufficiency of exiting drainage gullies, they should at least note the presence and condition of these. Underground drainage networks are also not surveyed as part of a standard building survey or TDD although manhole covers should be lifted where it is easy to do so. Historical drainage investigations or reports may be located in the as-built file or within technical documents placed in a data room; accordingly, these should be the subject of review.

Severe 'ponding' to an access road of a logistics site.

The damaged road is adjacent to a manhole cover and drainage network. Further investigation is required to establish the serviceability of the drain below ground.

Access roads

Access roads or routes around the site, including the separation of vehicles and pedestrians, should be assessed during the inspection. In essence, these are likely to be similar in construction to car parking areas, with tarmac, concrete and concrete block paving the typical materials used. On site access roads are usually only evident to business parks or industrial buildings, such as logistics properties. These sites can be particularly dangerous to survey as there can be a high quantity of heavy goods vehicle traffic, with minimal pedestrian separation or access. In some cases this may be limited to floor markings separating the two groups of road users. Drivers using these types of site are often not expecting there to be pedestrians present, which poses extra risk; therefore, it is essential to take the necessary precautions, including the wearing of high-visibility vests or jackets.

Similar separation of access roads can be made to accommodate cycle routes; this may be typical of large out-of-town developments. In these areas there may be lines or markings painted on the floor to indicate the separation but, if these are not executed in the appropriate paint, the surface can become slippery, potentially resulting in accidents. While the survey will not test or confirm the potential for slippage of cycles on such markings, the surveyor should seek to establish from the building owner or property manager any reported incidences and/or check the technical data sheets for the markings, where available.

There may also be access roads to large retail developments and often these offer better separation, with pavements or bollards as well as painted floor markings or zones. Typically, there may also be zebra crossings painted at strategic locations on the roads. No matter how trivial or minor it may seem, the surveyor should assess the condition of floor markings and recommend accordingly any periodic painting of these as part of the cost plan.

For some buildings there may be a requirement for a perimeter road around each facade as an obligation to provide access for the fire brigade. This is typically evident to industrial buildings but can also be the case for offices or buildings in other sectors. Any such obligation is usually documented with planning permissions, as well as being noted in acceptance reports. These roads can be constructed from compacted hardcore, or concrete, block paving or tarmac, and there is even evidence of access roads being executed with systems that utilise a structural sub-base seeded with grass for a landscaped finish. The obvious problem with this is that the grassed road can blend into the verge, which is not designed to take the load of a fire engine, thus resulting in the potential for a fire engine to get stuck if it drives off the road while attending an emergency situation. It is not the responsibility of the surveyor to verify the loading capacity or suitability of the perimeter roads for the emergency services. However, where visible, evidence of any aspect which may hinder or restrict the operation of the emergency services should be noted.

*A reinforced grass road with a 13 tonne loading capacity
provides perimeter access for the fire brigade. This was
a condition of the planning permission/building permit
(photo courtesy of Widnell Europe).*

Pavements, patios and decking

For commercial buildings located in town or city centres, there is an inevitable interface between
the public space and the plot or site boundary. This is often finished with a pavement, which is
typically the responsibility of the authorities in the UK. However, in other countries there may be
individual laws or byelaws which oblige building owners to have a degree of responsibility for
the pavement in front of their properties. This is particularly important where there are defects or
situations which pose a potential trip or slip hazard. This is one of the areas where it is essential
for a surveyor to know as well as understand the legal and technical environment in which they
are working.

   Pavements solely under private ownership are typical of large commercial properties, which
often have access roads or are situated within their own grounds. This is not exclusive to one
property type or sector; however, the basic construction of pavements remains largely the same,
irrespective of where they are installed. Pavements and paths are used for pedestrian access
and for maintenance. They can be constructed in isolation if they are used to cross a site or if
they border a road; however, many pavements or paths are used as the interface between the
buildings and landscaped areas.

   These are often constructed and finished with tarmac, concrete or block paving and similar
observations regarding their characteristics and defects should be noted as for car parking areas.
Higher specification pavements may be finished with pre-cast concrete paving slabs or even
natural stone. These are often laid on a bed of sand and cement, which is supported by a sub-
base constructed from compacted hardcore. The joints of the slabs can be finished with cement-
based mortar. In some instances, slabs may also be laid on dabs of mortar or plastic pedestals

and the joints left open. This is often the case where the external paving has been placed over a roof. The open joints are an indication of the possibility that there may be a roof below, as these are to facilitate rainwater evacuation.

Plastic pedestals supporting granite slabs to a patio over a flat roof. These have been removed to investigate a roof leak (photo courtesy of Widnell Europe).

Heavy duty' reinforced concrete paving slabs capable of supporting a car (photo courtesy of Widnell Europe).

Concrete pads and plastic pedestals used to support paving slabs (photo courtesy of Widnell Europe).

While cast-in-situ concrete and tarmac are relatively resistant to tensile loading, concrete paving and natural stone slabs crack easily when they are the subject of vehicle loading. They are also prone to localised sinking where there is consolidation, settlement or subsidence of the sub-base. This inevitably leads to the formation of trip hazards, as well as weed and vegetation growth where joints have opened up. Cracked, uneven or displaced paving slabs should be the subject of recommendation for immediate repair. Weed growth should be removed and an allowance made for periodic maintenance, including the removal of vegetation in the future.

Concrete paving can have a smooth or a riven finish and this is determined in the manufacturing process, with riven having a textured surface which does make this more susceptible to soiling or staining as it exacerbates the collection of surface water and dirt. Natural stone floor slabs can be polished, but this is particularly hazardous when it becomes wet, making the surface slippery. Therefore, most natural stone floor slabs do not have a polished surface and even where this is the case, natural abrasion from wind, rain and frost tends to dull the surface quite quickly. More bespoke finishes to external pavements and paths may include terracotta tiles, which may be plain in the case of 'quarry' tiles or have a glazed finish. These types of finish can sometimes appear visually similar to block paving, but they are much thinner with a typical thickness being 25 mm. As a consequence, they are much less resistant to impact damage as well as being more susceptible to cracking with vehicle loading. Another observation with glazed terracotta tiles is that these can become slippery when wet, posing a slippage risk to pedestrians.

Rainwater run-off and surface drainage is an important consideration regarding external pavement or patios as wet surfaces can pose a slip hazard and this risk may be exacerbated in sub-zero conditions. It should always be recommended to ensure that there is sufficient rainwater evacuation provision, although it should be noted that this is difficult to rectify retrospectively as it is not often feasible to re-lay large areas of paving or patios. It may therefore become a management issue by which building users are pre-warned of areas prone to become slippery and, in freezing conditions, salt or grit is used to minimise this risk.

External ramps and steps are areas of particular concern if the materials are prone to becoming slippery when wet. The treads of external steps are particularly hazardous and it is typical for older buildings or developments that these were compliant with the original regulations at the time of their construction, but do not comply under current regulations. While the enforcement of building codes or regulations retrospectively may not always be appropriate, external steps or staircases are typically also covered by health and safety laws. Therefore, the building owner is obliged to ensure that these are fit for purpose and this is a legal/technical area in which the surveyor should be risk averse. Where there is sufficient evidence, the surveyor should recommend immediate repair, renovation or replacement of external steps and staircases, if they appear to pose a risk to health and safety. Concerning the presence of hand rails to staircases and external ramps, the surveyor should also feel obliged to recommend the installation of these to mitigate potential health and safety risks. In essence, the building owner and/or tenant is typically responsible for safe access via staircases and ramps. This duty of care and any subsequent claim for negligence may be passed down to the technical advisor if insufficient advice was given on this matter in the scope of a building survey or TDD report.

External staircase to a shopping centre constructed from pre-cast reinforced concrete.
The crumbling and decayed nosings to the treads make this unfit for purpose
(photos courtesy of Widnell Europe).

Timber decking is another form of external surface that is used for commercial buildings and this is often installed to external terraces or roof gardens. The advantages of using timber is that it is easy to work, meaning it can be cut to size thus allowing for the quick construction of external finishes. The main disadvantage is its relative short life cycle and the requirement for maintenance with periodic application of wood preservatives necessary. Timber used for decking can be either hardwood or softwood, with the former typically being more expensive but having a better life cycle. Treatment with timber preservatives is essential to enhance the life cycle as well as maintaining aesthetic value. If left untreated, the timber is prone to water absorption and, with built-up dirt or algae on the surface, it can become dangerously slippery. Therefore, untreated or poorly maintained timber decking should be approached with care during the survey, as well as being the subject of a recommendation for immediate and periodic treatment. Modern composite materials with a wood-effect finish are readily available as decking and these provide a credible alternative to timber with a greater anticipated life cycle and are relatively maintenance free.

Storage yards and hardstanding areas

Usually associated with industrial buildings or some large retail properties, there may be areas for delivery, storage or the manoeuvring of goods and materials. These are typically constructed from in-situ concrete or tarmac as these materials are the most cost-effective options for such large surface areas. They are, in essence, susceptible to the same wear, tear and damages associated with tarmac and concrete road surfaces. As these are not utilised as access roads, sometimes insufficient consideration is given to the provision of rainwater drainage. The relevant gradient or slope of the external surfaces is critical in providing for the evacuation of surface water and it is surprisingly common to note such surface draining sloping towards the buildings, particularly when these are distribution or logistics sites where vehicles reverse into docking areas. In such instances, it is often common to find grill or slot drains creating a protective channel in front of the facades and it is essential that these are kept in good condition. However, the nature of industrial buildings, and even the 'behind the scenes' areas of retail parks, is that these areas are typically the subject of poor maintenance or even neglect.

Cracked and defective drainage channel to a logistics site.   *Defective tarmac access roads.*

Logistics and industrial buildings also utilise docking bays and it has been discussed in *Chapter 7 – Facades* that these are usually the subject of significant occupational impact damage. The external surfaces are also the subject of significant impact damage, with raised concrete kerbs, concrete bollards or dividing walls routinely being struck by vehicles. Although there is an inevitability in this which is associated with the specific occupational use, it should not be discounted during the survey and any defects or damages should be noted in the cost schedule. Likewise, the 'apron' in front of the docking stations or the hardstanding areas used for trailer storage should be inspected for damage. Where these have been executed in tarmac, there is the potential for localised sinking or potholes to form where the retractable legs of trailers point load the surface. Therefore, the most effective material for these areas of hardstanding is concrete and, where damages exist to tarmac surfaces, it should be recommended that these are replaced or upgraded using concrete.

Damaged tarmac hardstandings caused by the point loads of retractable legs of lorry trailers. The apron should be chopped out and replaced with concrete.

Waste areas

Typically, the collection and storage of waste may be considered a by-product of the building use or operation. Accordingly, there may be lease obligations placed upon the tenant or building user to ensure that this is done in accordance with all relevant legislation or regulations. The purpose of the building survey or TDD inspection should be to establish the condition of the property and part of this is concerned with the storage of waste materials, which is typically located in the external areas.

Areas of concern are the use of the site and the potential for users, occupants, visitors or the general public to be exposed to hazardous materials. Typically, the user of the site is bound to conform with the relevant health and safety regulations and, depending on the location of the property, specific national or European regulations may relate specifically to individual procedures. In the UK, specific emphasis is placed on the responsibility of the site or building operator for compliance with the Control of Substances Hazardous to Health Regulations 2002 (COSHH) and this should include site waste.

Waste is often collected in skips, bins or containers, which are located in external areas or within secure compounds. Although exposed to the elements, some waste areas may also be enclosed under shelter to prevent their deterioration. Waster can be categorised into common recyclable products, such as paper, cardboard, glass and metals, and more hazardous products as well as biodegradable food waste or general rubbish. The survey should not seek to identify the types and quantities of hazardous materials as this information is often contained on operating licences or permits. However, the survey should seek to assess the potential for fire to combustible waste or rodent infestation to food and general waste. For example, it should be common practice for smoking areas to be far removed from waste areas. Discarded cigarettes, as well as the potential for arson, can pose significant risk to waste storage areas and, if these are placed too close to the building facade, there is a real danger that the fire may spread. Typically, this should be noted during the building survey or TDD survey and discussed in the report.

Of real concern with regard to external waste is the potential for this to get into the drainage system and eventually into water courses. This would inevitably lead to environmental issues and potential prosecution by the authorities. This issue also applies to the storage of petrol, oils and lubricants; therefore, irrespective of whether these are waste products or not, such storage containers or areas should be noted during the survey. The storage of hazardous liquid chemicals or polluted water is more likely to be evident to industrial buildings, and in some cases remote sites may encompass the storage of fuel. Here strict operating conditions will be in force, with one typical requirement being the placement of a ring of protective drainage or bund around the facility with the possibility of collecting any fuel spillage into an interceptor. This situation may also be present to retail sites where there is a fuel station and in both instances it will not be possible to verify the presence and serviceability of any fuel interceptors. Therefore, the survey should note and report the presence of these installations, seeking further confirmation of the presence of any interceptors as well as conformity with all relevant operating regulations. This is typically a red flag issue, which should be communicated early on in the reporting process for the building survey or TDD.

Waste bins adjacent to a facade pose a fire risk and, despite the wall-mounted facility for putting out cigarettes, there are numerous cigarettes ends on the floor. To mitigate the potential for arson, the bins should be locked, as it is not possible to remove them further from the facade.

Outbuildings

During the site inspection, it will be necessary to establish the presence of outbuildings, which are typically located on sites that have sufficient external surface area, so these are often present to industrial buildings, retail parks or buildings situated within their own grounds or plot. With city or town centre commercial buildings, outbuildings are unlikely to be placed to the front or on the street facades, as these is usually under ownership of the local authorities. Therefore, such buildings may be present to the rear facade within the plot boundaries.

Outbuildings can be present for a wide variety of different uses and therefore exist in a number of potential forms. The building survey or TDD inspection will seek to establish the presence of outbuildings on the site and it is not unusual to discover that these are not located on site plans or within planning permissions. Where unrecorded outbuildings are noted during the site visit, this should be cross-refenced with the legal due diligence or advisors to verify the significance of such potential non-compliance. Any subsequent discussion with the legal advisors will typically seek to establish the size and degree of permanence or fixity of the outbuildings. Therefore, the survey should note the construction technology, size and use of the outbuildings. Clearly, timber sheds or storage buildings placed on an existing hardstanding have less size and permanence, as well as not being physically connected to or in/on the ground, therefore it may be argued that these are temporary structures. In contrast, a brick-built storage building, with strip or raft/slab foundation may be seen as a much more permanent structure, although its size is clearly important. It is not within the remit of the surveyor to comment on the compliance of outbuildings with any existing planning requirements imposed upon the property, and therefore this should always be referred to the legal advisors.

All outbuildings should be treated as mini buildings in the context that every one of their relevant construction elements should be assessed with respect to their condition and defects. Accordingly, it will be necessary to assess the construction detail and condition of the following:

- roof
- structure
- facades.

In many cases, access to the outbuildings may not be permitted as these may be used for storage and they may be locked during the survey. If no access is permitted, the surveyor should note this in the report. Generally, the materials used for outbuildings may be somewhat inferior to those used for the main building on a site; however, the more permanent the materials, the greater the anticipated life cycle. The timber shed alluded to in the previous example will typically have a shorter anticipated life cycle than the brickwork outbuilding and may also be susceptible to more onerous maintenance regimes.

An 'outbuilding' to an industrial building for the storage of cardboard and packaging. The size, fixity and degree of permanence mean that this is a 'real' building.

Lighting

Unless the building survey or TDD inspection is undertaken at night or in particularly gloomy conditions it will be difficult for the surveyor to establish the suitability and serviceability of external lighting. Therefore, the inspection and report should seek to establish the presence of external lighting as well as differentiating, where possible, between 'normal' and emergency lighting. For commercial buildings, external lighting serves the purpose of illuminating both the entry and exit points as well as other external areas, such as car parking, access routes or hardstandings. Lighting may be fixed to the building facade or free standing. It may illuminate from below with uplighters, from the side with bollard lighting or from above. External lighting may also be used for security with passive infrared or movement sensors used to initiate this.

As it is not normally possible to confirm the serviceability of external lighting, the surveyor should establish this through dialogue with the building owner, property manager or tenant. The type of lighting and any typical maintenance expenditure allocated to the periodic replacement of bulbs should also be established. For example, LED lighting has both a significantly longer anticipated life cycle than traditional lighting filaments used in sodium or halogen bulbs and much better energy efficiency. With respect to the findings or recommendations of the survey, it may be appropriate to recommend the replacement of sodium or halogen external lighting with LED fixtures as an 'upgrade'. While this may appear to go against the typical 'like-for-like' recommendations of the survey, this can be justified as an improvement to reduce energy consumption and potentially improve the EPC. Such works would ultimately be beneficial to both the landlord and tenant as well as having a more positive impact on the environment.

Typical problems associated with external lighting are defective bulbs, but these are usually replaced as part of routine maintenance works. More severe problems relate to cracked and broken diffusers, which can be as a result of accidental impact or vandalism. Broken diffusers typically result in moisture ingress to the fitting, which can prove terminal and, where even the slightest of cracks are evident in diffusers, the recommendation should be to replace these. A common problem associated with uplighters is rainwater ingress or even condensation if there are problems with the seals around the light fittings. Where present, water ingress to uplighters is sufficient to justify replacement of the fittings. With street light fittings, the most common problem is damage from vehicle impact or corrosion of these when the lampposts themselves are manufactured from steel or the galvanised protective coating is compromised.

Emergency lighting is either battery powered or fed from an emergency generator or uninterrupted power supply. In many cases it is not possible to verify and test the functioning of this, therefore it will be necessary for the surveyor to get confirmation from the building owner, property manager or tenant regarding serviceability.

Emergency lighting to an external facade over an emergency escape route; test reports from the owner are required to verify the serviceability of this.

A water-filled, damaged external light diffuser which requires immediate replacement.

Signage

Primarily, signage within the realms of external areas relates to that around the site and is associated with vehicle circulation. However, there may also be some instances where signage may be present for advertising or connected to the site or name of the occupants. In all cases, the survey should undertake a visual assessment of the condition of this with typical risks associated with defective or loose signage being that it could suffer wind damage and fall onto vehicles or pedestrians using the site. Most signage is manufactured from steel, with the actual signs typically screw fixed to a supporting structure or to the building.

Signage plaques, fixing bolts or posts can be the subject of corrosion and therefore, upon discovery of any evidence of this, the recommendation should be to treat the corrosion and apply a protective painted finish. Signage used to indicate circulation routes around a site is typical to industrial zones or retail parks and is likely to be the responsibility of the site owner or operator. However, signage to individual commercial properties indicating the name and type of business operation are normally the responsibility of the tenant. While ownership of the signage equates to the responsibility for maintenance and repair, surveyors have a general duty of care to inform the owner, property manager or tenant if there is evidence that defective signage may present a health and safety risk.

External signage to a retail park which requires painting of the external supporting structure. The height of this will require a 'cherry picker' for undertaking maintenance, which significantly increases the cost (photo courtesy of Widnell Europe).

Fencing and boundaries

The building survey or TDD inspection does not normally include the measurement and verification of the site boundaries; this is normally defined or indicated by the site plan or title deeds and is something that is usually covered in the legal due diligence. However, it is unusual for the legal advisor to visit site to verify this and, in the event that a request is made by the client or legal team to confirm this, the surveyor should seek further specialist advice. Land surveyors appear to be the most suitably qualified professionals to undertake this check as it requires the use of highly specialist equipment and techniques to ensure accuracy. It is not unusual for land surveyors to identify significant anomalies between information contained on planning permission plans and as-built plans. Typically, developers simply change the title box on plans from 'Execution' to 'As-built' without instructing a measured survey. Presumably this is done to save costs; however, it can prove costlier in the long term to rectify this, certainly if there are errors in the documents.

The role of the surveyor is to assess the presence and condition of the boundary and this may include access gates, fences or walls. Concerning commercial properties located in town or city centres, the boundaries to any external areas may include walls and fences which are effectively party walls. This means that there is a joint responsibility to notify the appropriate adjoining owner when works are being undertaken along the boundary. This is specifically covered by relevant party wall legislation and, although perhaps this is associated with laws in the UK, many other countries have similar legislation to cover such issues. Therefore, it is the responsibility of the surveyor to have an understanding of the relevant legislation in their area of operation. If in doubt, any damage or potential infringements to the boundary or party wall should be recorded and raised as part of the building survey or TDD inspection during the red flag procedure.

Commercial buildings which are located within their own grounds or plot may have surrounding boundary walls or fencing and, to an extent, the same knowledge concerning construction technology and material science should be applied to these as for the other construction elements.

Boundary walls are constructed from natural stone, brickwork or blockwork and, as with the same materials used for facades, they will be susceptible to movement, cracking, efflorescence, frost damage and a range of similar defects. One general observation concerning boundary walls is that for large sites, where there is a requirement for security as well as enclosure, masonry walls are not a cost-effective method to secure the perimeter. These walls would typically have to reach a height of 1.80 m and utilise 'traditional' construction methods, including concrete strip foundations and laying the masonry in cement-based mortar. Such works will be labour intensive as well as requiring large quantities of materials; therefore, fencing is probably the most cost-effective alternative. Irrespective of this, where constructed, boundary walls are still required to include a DPC to prevent rising damp as well as coping stones with sufficient detailing to prevent rainwater run-off and penetrating damp, which in turn may lead to frost damage.

Fencing is the system of choice for the containment of large surface areas and is evident to logistic sites, industrial zones, retail parks and other more bespoke commercial sites, such as schools and health-care buildings. Most of these fences comprise steel mesh supported by plasticised or painted steel posts, although pre-cast concrete posts are also common. Timber may be a relatively common material used for fencing to residential properties but this is rarely, if ever, used for commercial sites. Timber simply has too high a maintenance requirement and too short a life cycle to justify its use, with low-quality timber typically used for fencing panels, which are prone to rot and wind damage. Steel wire mesh fencing, on the other hand, offers very little wind resistance and modern wired fencing is also plasticised to give improved resistance to the corrosion typically associated with older forms of this type of fencing. Included in the category of fences should be the provision of access gates and security; this is typically an operational requirement of the building occupier of tenant. As with all construction elements and components, the surveyor should note the presence, type and condition of access gates and even the presence of CCTV cameras. It is unlikely that the surveyor will be able to ascertain the serviceability or suitability of items such as CCTV and they should refrain, unless specifically instructed, from commenting on this. However, there is no harm in commenting on the operation and function of access gates or barriers, where these are seen to be operational during the visit. These are also items whose serviceability the owner, property manager or tenant can be asked to verify.

The perceived weakness to wire mesh fencing is that it can be cut through by intruders and the fixing points with the posts are areas of potential defect. The posts may become loose due to poor initial installation or impact damage, typically from vehicles, which can be an issue with logistics sites. Steel posts may corrode; however, as these are mostly plasticised they are therefore quite resistant to rusting. Pre-cast reinforced concrete posts can be the subject of carbonation and the corroded steel reinforcement invariably leads to spalling of the concrete; if left to persist, this can result in weakness and potential collapse or disintegration of the posts. Corroded steel or damaged concrete posts should be noted in the building survey or TDD report, with appropriate recommendations to repair or replace these accordingly.

Damaged concrete kerb stones and steel mesh-and-post fence caused by heavy goods vehicle impact.

A reinforced concrete fence post with corroded steel, which will continue to decay if left untreated.

Vegetation and landscaping

During the development process, the planting and landscaping of the site is often one of the last processes and this can be an opportunity to 'frame' the building, emphasising the symbiosis between the built and natural environments. However, invariably there is often little funding set aside for the landscaping and planting of a site on completion. As a consequence, planting schemes may seek to achieve maximum green coverage at minimal cost and this is when some inappropriate vegetation may be introduced to the external areas.

The site survey should assess the external vegetation and landscaping with respect to identifying potential short-, medium- and long-term costs, as well as identifying plants or vegetation which may contribute to defects. The link between mature trees and subsidence has long been established, with tree roots drawing moisture from cohesive soils (clays) being one of the main contributors to this. Trees can also prove to be a risk on non-cohesive soils which do not shrink or swell but can be the subject of erosion should tree roots penetrate and crack underground drainage pipes. However, one very important consideration is that tree root damage and subsidence are usually only an issue for 'traditional' low-rise commercial properties, where there are relatively shallow foundations when compared to larger commercial properties. The presence of a basement level is advantageous in reducing the risk of subsidence and more than one level below ground is likely to negate totally the potential for this. Therefore, the presence of mature trees and their proximity to the facades should be noted with interest in the case of low-rise properties. This does not mean that mature trees close to medium- or high-rise properties can be discounted as these can still have an influence on drainage systems. They should also be noted as they may be classified as listed on tree preservation orders (TPOs), meaning that excessive pruning or removal may require prior permission from the authorities. If necessary, information regarding TPOs can be found by consulting the relevant lists held with the authorities.

Significant vegetation issues are the overgrown ivy to the facade, which should be cut back, and a flowering cherry tree which is too close to the facade and has the potential to cause subsidence.

Tree roots below ground can be problematic and above ground their branches or leaves can also have an effect on the built environment. Typically, this occurs when there are loose or rotten branches, which can fall during storm conditions. However, it is not the responsibility of the surveyor to audit and assess the condition of trees on a site and their potential to be affected by storm damage. Logically, it should be supposed that the surveyor will note, where visible, any leaning or obviously defective trees. One thing the surveyor can and should note is the potential for tree leaves to cause blockages of rainwater drainage or gutters. The roof inspection will typically note leaf build-up and debris to gutters, and this may be inevitable to many properties as wind-blown leaves can travel relatively large distances with little way of controlling this. Removal of such build-up then becomes a matter of routine maintenance or property management. However, where trees are in very close proximity to facades, their leaves have the potential to cause serious blockage of gutters and drainage channels, then the surveyor can recommend, subject to verification of a TPO, pollarding or removal of the tree.

The presence of mature trees adjacent to facades which cause leaves to block the rainwater evacuation points (photo courtesy of Widnell Europe).

The majority of trees are deciduous, meaning that they lose their leaves in the autumn and re-growth occurs in the spring. Trees that retain their leaves all year round are known as evergreen trees. Examples of evergreens are pine or spruce; however, these are rarely used in planting schemes for new developments. Leylandii are a very common evergreen tree used to create hedges; however, if left unchecked they can grow very rapidly to significant heights and it is not unheard of for these to reach three to four storeys high. They are relatively shallow-rooted trees but their roots can spread for great distances in search of moisture, meaning that they can be quite destructive on cohesive soils. It would therefore be prudent for the surveyor to note these during a building survey or TDD inspection and advise accordingly on the need to undertake yearly pruning or management of these.

Regarding vegetation on sites, the RICS guidance note pays particular attention to the need for surveyors to note the presence of Japanese knotweed and giant hogweed. These two types of plant have different characteristics but can both grow vigorously to epidemic proportions.

Japanese knotweed

Japanese knotweed is classed as a non-native species and is believed to originate from Japan, Taiwan and northern China. It is a deciduous plant which can be identified by its pink or purple speckled stems with 'ace' shaped green leaves in the summer. It is a vigorous, hardy plant whose stems are hollow, similar to those of bamboo. Its root networks can be destructive to the built environment, as they are capable of penetrating tarmac and other materials used for hardstanding areas.

The image above clearly shows the speckled stem and distinctive leaf shape of Japanese knotweed (photo courtesy of Chris Mitchell, Paragon).

Giant hogweed

Giant hogweed is not known for its destructive root network but can present a hazard to health as contact with the plant and even relatively small amounts of the sap can cause blistering of the skin and extreme discomfort. The is also a plant which is not native to the UK but is now widespread. It has typically been known to grow adjacent to railway lines or water courses but it has the potential to spread into the built environment though seed dispersal. It is not dissimilar in appearance to the native 'cow parsley', which grows in abundance in the spring and summer with similarly shaped flower heads. However, it is significantly taller when fully grown, with heights up to 5 m and thick stems 5–10 cm in diameter; the stems are also covered with sharp spines.

Bamboo

Another non-native plant common in the built environment is bamboo, and this is often selected for landscaping as it has rapid growth with striking vertical stems, which can produce an evergreen screen. It is extremely hardy with vigorous growth and the risk to the built environment is posed by its potentially destructive root networks, which can penetrate roof coverings, asphalt and other components of a building envelope. This plant differs from Japanese knotweed or giant hogweed in that it is actively used as landscaping and not necessarily seen as a problem. Where present, the surveyor should advise on the potential for root damage, seeking to identify whether the plants are isolated or contained within planting boxes, which means there is less potential for the roots to spread.

Vigorous growth of bamboo used as ornamental planting to an office building. The new shoots of the bamboo are spreading in an uncontrolled manner.

Climbers

As discussed in *Chapter 7 – Facades*, climbing plants such as ivy and Virginia creeper can be used effectively to create historic 'charm' to a commercial property, and this is typically evident on older, 'traditional' properties. Climbing plants can also be present to external areas and, as such, are usually found on boundary walls or fencing. As stated regarding their presence on facades, they are extremely hardy and vigorous in their growth. If left unchecked, climbing plants can completely take over external fencing or boundary walls. The default position of the surveyor should be to recommend that these are cut back and managed, unless there is a specific requirement contrary to this.

Summary of external areas

When undertaking an inspection of the external areas, it should be remembered that this does form an important part of the survey. These areas are the interface between the building and its immediate surroundings or the natural environment and some general principles should be noted when reporting upon the external areas:

- The definition of the site boundaries is typically a legal/technical issue and, without the relevant equipment or experience, the surveyor should avoid verifying the site dimensions or boundaries. Should such work be required, then it appears appropriate to instruct a land surveyor.
- Car parking, access roads and areas or hardstanding can equate to very large surface areas and, despite the relatively low-cost materials utilised for these areas, the quantity inevitably means that short-, medium- and long-term costs can be quite high.

- There are important issues regarding the duty of care of a building owner or tenant in terms of safe access to commercial properties. Therefore, it is essential for the surveyor to assess these areas of defects or omissions, which may cause slips, trips or falls.
- Outbuildings and waste areas are of specific concern regarding planning and operation of commercial properties. While the outbuildings themselves should be treated as individual buildings in terms of construction technology, their presence, size and degree of permanence may form a legal issue. Therefore, these should be referred, where necessary, to the legal advisors for discussions on conformity. In particular, waste areas can result in health, fire and environmental risks or issues of non-conformity. It should be acknowledged that most surveyors are neither environmental nor fire experts and, to an extent, common sense is required to assess the situation on site. If there is evidence to suggest that the location, storage or processing of waste presents a health, fire or environmental risk, this should be referred as a red flag issue.
- The construction detail, condition and defects associated with brickwork, blockwork or natural stone boundary walls should be assessed and analysed using the same approach as that adopted for facades executed in these materials.
- An appreciation of the type, size and significance of vegetation associated with the external areas is important with respect to commercial building surveys and TDD. The natural environment is remarkably hardy and resilient; if left unchecked, the growth of foliage and roots can cause significant harm to commercial assets. Therefore, the surveyor is obliged to spend sufficient time assessing the vegetation and landscaping within the external areas. Where there is evidence of damage or suspected defects, the surveyor should, where relevant, commission further investigation or seek specialist advice.

Legal/technical 11

Introduction

Legal/technical is probably one of the most complex areas of the TDD or commercial building survey process and it does not conform to the principle that the surveyor can say what they see. In essence, this relates to aspects of the building or its operation that have a legal obligation with a technical application. This is where there is a professional interface between the legal due diligence and technical due diligence and, invariably, the matters identified or discussed are potential deal-breakers. The answers or solutions to issues raised are not always straightforward and the surveyor must still deliver evidence-based recommendations, resisting any pressure applied by the agent, client and legal advisors to verify conformities when often all of the information is not available.

The RICS guidance note (4th edition, 2010) identifies legal/technical issues as including the following:

- health and safety
- fire precautions
- accessibility
- environmental considerations and sustainability
  - energy efficiency
  - noise and disturbance
  - land contamination and environmental control
- deleterious materials
- cultural heritage
- statutory matters
  - planning consents
  - listing of classification
  - tree preservation orders
  - party wall issues
- rights of way, easements and shared services
- boundaries
- guarantees and warranties
- leasehold and repairing liabilities.

These items are a list of the minimum legal/technical issues that the surveyor should be expected to report upon during the TDD or commercial building survey process. However, as there is often a bespoke nature to the instructions, other items may be added to this list. As these are legal/technical items, there will be a requirement for the surveyor to undertake a review of any available documents relating to these issues and contextualise them in terms of observations made during the site inspection.

The surveyor will invariably be asked to verify legal compliance of aspects or parts of the building as well as its operation. Therefore, the surveyor should always seek to use the same evidence-based guiding principle regarding this. Unless the surveyor has been involved in advising on the building from its conception through to completion and operation, they will not have sufficient knowledge to verify with 100 per cent certainty that the asset is in compliance with all legal prescriptions. This is often something that clients do not appear to comprehend, and part of the initial fee proposal or contract instruction should seek to inform clients of this fact. If the surveyor wants to verify compliance of the property with all relevant regulations, without having all of the evidence to base this opinion on, they are inevitably going to provide incorrect advice in one or more facets of this information. Furthermore, as the surveyor will remain liable for their findings for many years after the completion of the acquisition, such actions do little to mitigate their exposure to risk. Such a 'cavalier' approach may also invalidate any existing insurance cover that the surveyor may have in respect of a claim for professional negligence.

In order to provide the best possible evidence-based opinion, there is a series of actions that the surveyor can follow, which does include reporting observations or possible infringements to the client and the legal advisors. In the case of potential infringement, it is the legal team which will establish if this is 'material'. Therefore, it is important to examine each legal/technical issue in isolation as well as within the context of the building as seen in operation or on completion.

Health and safety

Within most developed countries and societies, there are forms of health and safety regulations covering both the workplace and the public domain. This is an area of legislation which is vast, covering many individual parts of a building, its surroundings and operation.

The surveyor should seek to ascertain the legal grounds for compliance and with health and safety; this is legislation with which it is implied building owners or employers must comply. In some countries, compliance with health and safety at work legislation is made a condition or implied within a building permit. If this is the case, then the signing off of the building by the design team with acceptance reports located in the data room may be a sufficient starting point. However, this legal/technical item does not stop once a building is completed, it continues indefinitely throughout the life cycle and operation of the property. Other useful documents in the data room to help establish compliance are reports of health and safety consultants as well as risk assessments. However, this specific legislation involves is a large amount of common sense, which the surveyor should apply when undertaking the survey, and there are also some health and safety issues which are covered by additional or specific legislation. The RICS guidance note (RICS, 2010, page 18) lists a series of 'obvious' points (*issues that can be verified on site by the surveyor if full access is provided are given in brackets*):

- slips, trips and fall hazards (*can be verified on site by the surveyor*)
- low head heights (*can be verified on site by the surveyor*)
- overloading, including crowd loading requirements (e.g. stadiums) (*typically requires confirmation from suitably qualified persons or authorities, also covered by building regulations and building codes*)
- instability (*typically requires confirmation from suitably qualified persons or authorities, covered by building regulations, although visual evidence to suggest instability should be reported by the surveyor*)
- demolition hazards, presence of potential asbestos-containing materials (ACMs) (*typically requires confirmation from suitably qualified persons or authorities, also covered by asbestos regulations but any suspected visual evidence should be reported by the surveyor*)

- maintenance and other safe access issues (*can be verified on site by the surveyor*)
- confined spaces (*can be verified on site by the surveyor*)
- excavations (*can be verified on site by the surveyor*)
- falls, falling objects and fragile materials (*can be verified on site by the surveyor*)
- edges and barrier protection (*can be verified on site by the surveyor*)
- glazing (*conformity with building regulations and building codes verified by acceptance reports and as-built data sheets*)
- fresh air, temperature and weather protection (*typically requires confirmation from suitably qualified persons or authorities, M&E acceptance reports and evidence of compliance with building regulations or building codes*)
- fire, fire detection and firefighting (*typically requires confirmation from suitably qualified persons or authorities, covered by building regulations or building codes as well as specific fire prevention regulation*)
- emergency routes (*typically requires confirmation from suitably qualified persons or authorities, covered by building regulations or building codes as well as specific fire prevention regulation*)
- welfare facilities (*typically requires confirmation from suitably qualified persons or authorities, covered by building regulations and building codes*)
- vehicle hazards, traffic routes and workplace transport hazards, issues around the perimeter of the building, vehicular access, deliveries and loading and unloading operations (*can be verified on site by the surveyor*)
- hazardous operations or materials (*typically requires confirmation from suitably qualified persons or authorities, covered also by specific regulations, such as the Control of Substances Hazardous to Health Regulations 2002 – COSHH*)
- lighting levels (*typically requires confirmation from suitably qualified persons or authorities, building regulations or building codes and acceptance reports/tests of the M&E engineer*)
- electrical installations (*typically requires confirmation from suitably qualified persons or authorities, building regulations or building codes and acceptance reports/tests of the M&E engineer*).

To an extent, some of the issues associated with health and safety can be assessed by the surveyor in the course of their site inspection or survey; however, there are much more specific issues covered by other regulations. Verification of conformity regarding more complex and specialist issues should be achieved by reviewing documents held in the data room or provided by the building owner or property manager. Where documents confirming compliance are not available, the surveyor should recommend that the necessary reports are commissioned. This should be referred back to the legal due diligence team for them to determine the significance of potential non-compliance. It is the legal due diligence team who should then be pressed to advise the client.

Fire precautions

Probably one of the most difficult issues to resolve during the due diligence process, from a legal and technical perspective, is compliance with fire regulations. This can be addressed by analysing the building with respect to its occupation and operation as well as the physical construction elements.

Most developed societies recognise the requirement for prevention of fire in the workplace and, accordingly, it is necessary to undertake fire safety inspection as well as testing of fire alarms and evacuation processes. In the UK, fire safety in the workplace is covered by the Regulatory Reform (Fire Safety) Order 2005, which effectively places responsibility on the employer to ensure that the building is the subject of annual inspection. This, in essence, involves undertaking a risk assessment into the operations carried out on the site with respect to fire prevention as well as routine inspection of the fire alarms, escape routes and firefighting. Therefore, the TDD or commercial building survey should request copies of the annual reports covering this as a starting point to assessing fire safety precautions at the property.

More important is the actual physical construction elements and the mechanics of the building with respect to fire engineering. Unless the surveyor has been involved in the initial design, execution and acceptance of a building, they will always have to rely on the reports of others, with some limited information of their own which may be gleaned from the survey. In most cases, the construction, renovation and extension of commercial buildings usually requires that these comply with basic building regulations or building codes. Included in these are requirements for fire engineering and, in the UK, these are predominantly covered in Part B of the Approved Documents which make up the Building Regulations (DCLG, 2013).

In other countries there are similar regulations which are also administered by the authorities with intervention from the fire brigade. Works to a property covered by fire regulations will have to be designed, executed and signed off by a suitably qualified individual or organisation. Unless surveyors are 'qualified' to do this, they should rely upon the expertise of others to sign off the works as finished and compliant. Invariably, there is no obligation to apply modern or current fire regulations retrospectively to an existing property, unless it is going to be the subject of a building permit or planning permission. Therefore, when auditing a property as part of a TDD or commercial building survey, it is important to establish a timeline regarding the initial dates of construction or renovation and extension to ascertain if this was the subject of fire regulations. Within the data room it will be necessary to identify the presence of any acceptance reports which have been signed off by the design team and the authorities (fire brigade or fire officer) verifying conformity. This should be a perfectly normal process in which those applying and administering a building permit are integral to the acceptance process. It is impossible for individuals outside the design, execution and acceptance process to verify the 100 per cent compliance of an existing building with the applicable fire regulations.

Surveyors undertaking TDD or commercial building surveys are not normally experts or specialists in fire engineering, therefore they need to establish the existence of and review the acceptance reports of others concerning compliance with regulations. However, with most commercial buildings it is very rare for the surveyor to be auditing buildings which have been the subject of 'recent' acceptance, therefore it is necessary to assess the building physically to note any obvious deficiencies relating to fire safety. This should be done in accordance with the relevant and applicable fire regulations but it should be noted that these concern the minimum requirements. As seen in high-profile building fires, such as the Grenfell Tower fire in 2017, conformity with regulations does guarantee that a building and its occupants are safe or immune from danger. When assessing an existing commercial building with respect to fire safety, the surveyor should note the following:

- fire alarm/alert and means of escape
- internal fire spread
- external fire spread
- fire resistance – structure
- firefighting provision.

Within the context of the TDD or commercial building survey instruction, the surveyor will not undertake an audit confirming or verifying compliance with regulations. This type of auditing is a specific sub-specialism and, also, in the absence of destructive tests, such as accessing suspended ceilings or going into all shafts or service ducting, it will not be possible to fully verify compliance. If such works are to be commissioned, the TDD or commercial building surveying process would be significantly prolonged and costlier, to reflect the fees required to do this. Therefore, the building surveyor should be viewed predominantly as a generalist, reviewing existing acceptance and audit reports of other experts, and noting and omissions or defects where visible.

Fire alarm/alert and means of escape

Concerning the means of alarm and alert, the surveyor should note the presence of a central fire or smoke detection system along with the presence of call points. The testing of fire detection or alarm systems is not normally undertaken during a survey, but the surveyor should seek to obtain copies of test certificates or confirmation from the building owner or property manager that this is fully functional. The design and maintenance or servicing of automatic fire detection systems is a highly specialist M&E discipline and, while covered by the legal/technical part of the survey, it should also be detailed in the M&E survey.

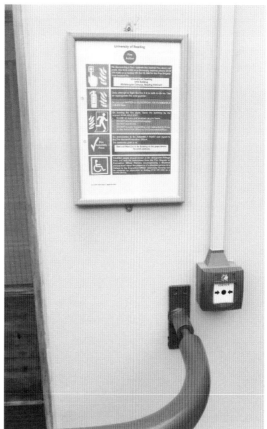

A visual fire alarm (TOP LEFT), a push button 'call point' (RIGHT) and a siren, which is disconnected (BOTTOM LEFT).

When commenting upon the means of escape, the surveyor should note the presence and dimensions of existing emergency staircases. This can be done to assess compliance with current regulations but, in most cases, under building regulations or building codes there is no obligation to enforce retrospective compliance. However, it may be possible under health and safety legislation to recommend remedial action where the situation poses a risk to health and safety.

External ladders and gantries used to form emergency escape routes are not acceptable
(photos courtesy of Widnell Europe).

One principle of fire engineering is that the occupants should be able to turn their backs on a fire and seek an alternative means of escape. However, this principle cannot always be applied when a property or room is engulfed in fire. Therefore, the priority is to raise the alarm using highly efficient and sensitive smoke and fire detectors to ensure rapid evacuation. The governing principle of escape routes is that these should be fire separated from areas of occupation, the floors, walls and ceilings of these should have sufficient fire resistance to allow evacuation and they should be equipped with self-closing fire doors. Fire doors should not be fitted with locks, latches or bolt fastenings and, where there is an access control system in operation, electrically operated doors should return to the unlocked position when the fire alarm is sounded. Doors should open in the direction of travel and doors leading into an escape route should not open into the route itself. Specifically, revolving doors can cause obstruction, therefore they need to either 'flatten' in an emergency or an adjacent alternative should be installed. Escape routes should not have headroom that is limited or restrictive and the floor finishes should have minimal slipperiness when wet. Concerning illumination of escape routes, there should be adequate lighting, which has a sufficient continuous power supply.

Concerning an audit of escape routes, the surveyor can normally verify the presence and serviceability of fire doors or the presence of emergency lighting. However, without destructive testing to ascertain the presence of fire barriers to suspended ceilings or the type of materials concealed by finishes, it will be impossible to verify the fire resistance of these aspects. Where internal emergency staircases are constructed as reinforced concrete core walls, or they have been executed in masonry, it is not unreasonable to suggest that these appear to have sufficient fire resistance. However, certainty can only be assured by analysing as-built plans or data sheets to establish the thickness and types of materials used.

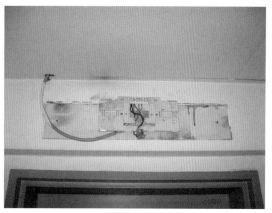

A defective emergency lighting fitting has been removed, which should be reinstated (photo courtesy of Widnell Europe).

Reinforced concrete core walls to an internal emergency staircase create a fire compartment. The stair treads have textured, non-slip nosings and there is a continuous handrail to the open side of the staircase. Wall markings indicate the floor number as well as the direction of travel and lighting is installed, but this is not 'emergency' lighting (photo courtesy of Widnell Europe).

An internal emergency light with direction sign indicates an escape route; this should be a fire compartment but it is not known if there are fire barriers above the suspended ceiling (photo courtesy of Widnell Europe).

Internal fire spread

It is very difficult to verify the compliance of a commercial building concerning the spread of fire internally as this is linked to the fire sensitivity of materials used as well as the placement of fire barriers above suspended ceilings. One important factor concerning the potential spread of fire internally is the construction of fire compartments. This is detailed and governed by building regulations or building codes and, in the UK the maximum size of a fire compartment is 2000 m²; therefore, where the floor space is greater than this, it needs to be separated by fire resistant walls, ceilings and floor with certified fire doors

Generally, there should be vertical and horizontal separation between fire compartments and a good example of where this is done in an office, hotel or residential tower is the provision of separate compartments for the internal staircases, lift shafts, lift lobbies and individual rooms or apartments to hotels and residential buildings, respectively.

The building height is also very important in terms of vertical fire spread. Referred to in *Chapter 6 – Structure* and *Chapter 7 – Facades*, buildings are classed as low, medium and high rise and their technical definition is often included in the relevant building regulations. For example, in the UK, properties with the highest level of occupation above 30 m are classed as high rise, for Belgium the specified height is 25 m and, in both cases, high-rise properties attract the most stringent fire engineering requirements. Without detailed knowledge of the building construction materials or access to all hidden areas, such as shafts and suspended ceilings voids, it is not possible to verify fully that internal fire separation has been achieved. As the survey is in situ and visual, the surveyor should seek to identify the presence of fire stopping materials where pipes, cables or ducts pass through fire compartments. Cement-based mortar is an effective fire seal, but this is rarely used as many holes in compartments are made after completion of the building; these are therefore sealed retrospectively with a range of specialist materials, such as expanding foams or mineral fibre quilting with plaster finish. By accessing shafts or technical rooms and, in particular, the interface between these and other compartments, the surveyor will be able to ascertain if these have been sealed correctly. One common occurrence is that installation works for information technology (IT) often break through fire seals without these being repaired or resealed.

A hole in a fire compartment sealed with 'pink' expanding foam. Pink, red or orange coloured foams are often fire rated materials but data sheets should be checked to verify this. However, damage to the lightweight concrete planks has exposed the reinforcement, which compromises the integrity (photo courtesy of Widnell Europe).

Perfectly sealed fire compartment (technical room) with mineral fibre quilting and plaster (photo courtesy of Widnell Europe).

Remedial placement of IT services has compromised the vertical fire seal in a shaft (photo courtesy of Widnell Europe).

An open shaft, which can be a fire compartment in itself if the door or openings off this are fire rated (photo courtesy of Widnell Europe).

External fire spread

Fire can spread externally and, although the surveyor is not likely to be a qualified fire engineering expert, they should have some appreciation of the risks posed by this. As detailed, the size and limits of fire compartments are determined by building regulations or building codes and, in most cases (dependent upon floor size), each floor may be classed as a compartment. Therefore, these should be separated from the internal spread of fire through the floors, shafts and also via the facade. Vertical separation between floors is stipulated by the relevant fire regulations. These distances can vary in different countries and are typically between 900 mm and 1000 mm (depending on the location of the building). It is therefore critical that the surveyor has sufficient knowledge of the applicable regulations in the countries in which they operate. Vertical fire separation distances have been determined by fire engineering experts; accordingly, it appears that they have concluded that fire passing vertically between compartments (typically via windows) should not be able to bridge a vertical distance of 900 mm or more. It is, of course, assumed to be the case that the facade material itself is inert or has suitable fire resisting properties. When checking for the provision of vertical fire separation, the surveyor should first refer to any specific references or obligations stipulated in the building permit or advice of the fire brigade or fire officer. Furthermore, it will be necessary to verify the presence of acceptance reports signed off by the design team and control organisations. On site measurements may be carried out but, in most cases, it is very difficult to do this in situ on exposed facades, working at height. Therefore, the surveyor should seek to undertake calculations based on measurements taken from the as-built plans. If there is any doubt about the conformity of the facade concerning external vertical separation, the surveyor should raise this as a red flag issue, subject to further confirmation.

A typical 1970s office building with a natural stone cladding; note the 1 m vertical fire separation, which is partially achieved with a fibre cement (asbestos) panel below each window. Even with this relatively 'simple' facade detail, it is difficult to verify the distance during the site inspection (photos courtesy of Widnell Europe).

Fire resistance – structure

Verification of conformity with fire regulations inevitably means that there is a requirement for the surveyor to assess the structure and, as discussed in *Chapter 6*, this is likely to be loadbearing masonry to low-rise properties and either steel, reinforced concrete or timber in single or combined use for medium- and high-rise properties. Masonry walls have relatively good fire resistance but, in the event of a fire, its use in low-rise properties with timber floor structures has much less combined resistance than reinforced concrete. Thicker timber structural members, such as historic oak beams or even modern glulam sections, have good fire resistance and burn at relatively slow rates. Steel is the weakest of the structural materials and can deform or lose strength in relatively short timescales; as a consequence, this requires the application of fire protection.

In almost all situations, the site visit will not be able to verify the complete structure or the fire protection by virtue of the fact that this is almost always covered by the internal finishes or claddings. Therefore, to verify this without suitable evidence would appear irresponsible to both the client and the surveyor. Naturally, where there is evidence of the structure and it is possible to ascertain the type as well as the thickness of fire protection, the surveyor should seek to detail this in the report. This can be reported in phrases such as: 'where visible the structure was noted to comprise … with evidence of … fire protection'. However, without sufficient information relating to the grade, rating or level of the fire protection, it would again appear imprudent for the surveyor to verify conformity. Therefore, the most rational approach is to cross-check the initial requirements relative to the building regulations or building code against evidence of the appropriate acceptance from the design team and control authorities.

The types of protection applied to steel structures are either painted finishes with intumescent paint, sprayed on flock, boxing with fire resistant panels or encasement in concrete. Typically, modern properties have painted finishes, but many early medium- and high-rise buildings constructed from the 1950s to the 1980s have fire protection that incorporates asbestos. While asbestos has relatively good fire protecting properties, it is a deleterious material which has severe implications concerning hazard to health. Therefore, fire protection to steel structures pre-2000 should also be verified via asbestos inventories to determine the composition of the protection.

Sprayed on flock fire protection to a steel-framed 1970s office building, which contains asbestos (photo courtesy of Widnell Europe).

A glulam roof beam has a steel fixing plate at the interface with a reinforced concrete core wall; this has been 'boxed' in using fire resistant panelling (photo courtesy of Widnell Europe).

The concrete cover to steel reinforcement generally provides excellent fire protection, ensuring fire stability at high temperatures and, typically, 25 mm of concrete cover is believed to achieve two hours of fire resistance. However, this does rely upon the quality and consistency of the concrete cover, which is typically better and more stringently controlled with pre-cast elements compared to those cast in situ. Modern concrete typically encompasses reinforcement at a depth of 25 mm or more; however, with older concrete this may not be the case.

Concrete fire resistance – case study

A TDD was undertaken on an office building dating from the 1970s, which had a structure executed in cast-in-situ reinforced concrete. The building was the subject of 'heavy' renovation, including a strip-out to the core in the 1990s. The subsequent planning permission and advice from the authorities was for the building to comply with 'current' fire regulations. This was duly done, and the completed building was signed off by the design team as well as the fire officer. This initial TDD was undertaken in the early 2000s and, when the asset was the subject of a resale some ten years later, the subsequent TDD red flagged the potential for the concrete structure to have insufficient fire protection. Consequently, the TDD recommended the retrospective installing of a full sprinkler system at the cost of the vendor.

Cleary surprised by the absence of such recommendation in the initial TDD, the vendor commissioned testing of the concrete with some sample analysis and widespread x-raying of the structure in situ. Neither of the operations was able to verify with 100 per cent certainty that the structure was either compliant or non-compliant, with varying levels of concrete cover recorded. On reflection, the initial TDD appears to have been correct in its findings; however, this is an example of how difficult it is to carry out TDD. Just one sub-issue of one construction element required weeks of investigation and in-situ testing to form an opinion. TDD and commercial building surveyors are never afforded the luxury of time and, in this case study, the surveyor made an evidence-based assessment in accordance with the information presented during the site visit and the document review.

Firefighting provision

Building regulations and building codes invariably set the requirements for firefighting provision; however, there is also a significant amount of individual input stipulated on a building-by-building basis as there are inevitably localised access issues for many properties. The TDD or commercial building survey should seek to establish the bespoke nature of any advice given regarding the provision of firefighting equipment and any specific access requirements for the emergency services. This can only be done by reviewing evidence of this on documents provided in the data room or verification by the building owner or property manager.

Generally, there is a requirement for the emergency services to have access externally to the property and this may include the provision of a perimeter road capable of withstanding the load of a fire engine. There will also be a requirement for the emergency services to have access to the facades of the property for the positioning of ladders. It is almost impossible to verify conformity with these requirements from the site visit alone, therefore this should be confirmed by reference to historic documents as well as records of present-day fire safety inspections and subsequent correspondence with the relevant authorities.

During this inspection it will be possible to note the presence and serviceability of fire extinguishers, as these should be checked and signed off annually. Furthermore, it may be possible to identify the presence of a sprinkler installation; however, the serviceability of this will normally be confirmed by reference to property management or maintenance documents.

Summary of fire precautions assessment

In conclusion, regarding the role of the surveyor or TDD concerning compliance with fire regulations, this should be a five-stage process:

1. Establish and review existing 'operational' fire safety reports undertaken by the property manager or occupiers.
2. Seek to obtain and review copies of fire engineering advice issued with planning permissions or building permits, including any reference to the relevant fire regulations, building regulations or building code.
3. Seek to identify any evidence of the implementation of regulations though the integration of these into the building design with the acceptance reports signed off by the design team and fire brigade/fire officer.
4. Undertake a visual inspection to ascertain any evidence suggesting deficiencies or anomalies in fire precautions.
5. If in doubt, recommend further investigation to verify conformity, but this may prolong the due diligence process. The further investigation should be undertaken by a suitably qualified organisation or individual.

Accessibility

In the context of a commercial building, accessibility is concerned with the level of inclusivity afforded to persons with disability. Unless specified in the contract instruction, the surveyor does not normally undertake an access audit but will comment upon disabled access provision. This is a relatively straightforward issue and accessibility audits are something that most surveyors undertake, therefore they should have an appreciation or sense of the general accessibility of the property.

Within most developed countries and societies, the provision for disabled access is governed by statute; accordingly, it is possible to research the relevant application of this and any specific dimensions or requirements. This is another example of where surveyors should be comfortable working within their locality as there are inevitably some minor differences in the regulations per individual country.

One important factor to consider is that the majority of legislation, in both the UK and other countries, governing disabled access often stipulates compliance with minimum standards. In the UK, the benchmark for the minimum requirements is Approved Document M (DCLG, 2015) and, as with the provision of fire safety precautions, it is not deemed applicable to force retrospective application of these provisions. However, in the UK, failure to make reasonable adjustments for accessibility may result in a legal process under the Equality Act 2010, therefore it is in the interests of a commercial investor to consider this aspect. Accordingly, surveyors should be expected to have a basic understanding of the legal prescriptions and their application when advising clients.

Another characteristic of disability discrimination legislation, common to both the UK and other countries, is that this is something that has only really been at the forefront of design since the 1990s. While there may be evidence of legislation dating back before this, the principles of inclusivity have only appeared to have gained significance with the rise in awareness of disability in society. Globally, in developed societies, disability has become widely accepted and the achievements of people with disabilities have been showcased with great pride in the Paralympic Games. This incredible 'can do' attitude is inspirational; however, it is regrettable that, despite these advances, the attitude of most commercial property owners or investors is to implement the minimum standards. As a consequence of this relatively recent legislation, it should be noted that many 'older' and certainly historic commercial properties fail to comply fully with disability access legislation. In most cases, the implementation of the legal requirements includes the need for 'reasonable adjustment' and the test of reasonableness can only inevitably be achieved though legal case law. It is therefore critical that building owners, property managers and tenants are aware of the existing provision for disabled access in their buildings.

When undertaking an assessment of accessibility to a commercial property, the surveyor should seek to establish the level of accessibility of the following key requirements:

* access route
* parking
* building entrance
* horizontal circulation
* vertical circulation
* toilets and facilities.

This can be further sub-divided into the following elements and the surveyor should seek to comment on all of these aspects within the TDD or commercial building survey:

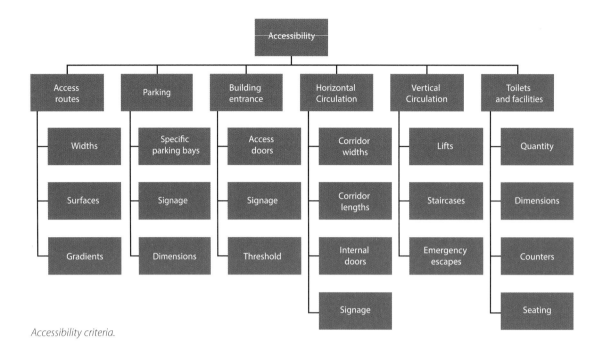

Accessibility criteria.

Environmental considerations and sustainability

A potentially large area of discussion in the due diligence process is the environmental impact of the building. This has become a highly relevant issue since the late 1990s, although the global adoption of worldwide policies to reduce greenhouse gases has been a long and drawn-out process which has taken years to implement. The wider term concerning the effect that the built environment has upon the natural environment and its consequent contribution to global warming, leading to climate change, is sustainability. Buildings, both in construction and in occupation, are one of the largest consumers of energy and creators of waste. This typically relates to the whole cycle, from the 'front end' of the construction process, with the extraction of raw materials used to manufacture building products, through the actual building stage, including the production of waste as well as energy consumption, and finally into occupation. Sustainability seeks to address the negative impacts of construction and real estate on the environment. While there have been some important global landmark agreements, such as those delivered at Kyoto (1997) and Paris (2015), the legislation developed and adopted by individual nations does not appear to have resulted in a significant reduction in greenhouse gases, although it should be noted that these are long-term objectives. The majority of investors and building owners do not appear to be overly concerned about sustainability; their primary focus is the financial return from their capital investment. Tenants may be more concerned about sustainability in relation to the building's energy efficiency, as this impacts directly on their running costs associated with energy consumption.

The majority of acquisitions concern commercial properties within the existing built environment and therefore the initial date of construction or renovation is important in terms of sustainability. Despite the Kyoto agreement occurring in 1997, the majority of significant legislation affecting the construction and renovation of commercial buildings did not come into force until more than ten years later. Therefore, investors who have a keen interest in acquiring buildings which may be perceived as having greater sustainability, are likely to seek to invest in modern properties. Investors concerned about sustainability may also seek to buy into development projects where they can get become actively involved in the application of building design and environmental impact planning.

The most realistic and practical application of sustainability of the existing built environment is the grading or classification of a building according to its environmental impact. This is typically measured in terms of its energy efficiency, which relates directly to the thermal efficiency of the construction elements as well as the means of production of heating, ventilation and air conditioning. Furthermore, there may be specific operations within the building or at the site which may require operating permits or licences.

Increasingly, the due diligence process will include a specific environmental due diligence to cover these issues and this may interface in some capacity with the TDD or commercial building survey, but it does largely remain a specialist audit. However, the surveyor can offer an opinion regarding environmental issues on the asset as a whole with a review of documents made available in the data room and by undertaking some checks on site.

In essence, the surveyor should be able to comment on the following issues:

- energy efficiency
- noise and disturbance
- land contamination and environmental control.

Energy efficiency

Energy performance certificates (EPCs) appear to be the most logical way to benchmark commercial properties in terms of their energy efficiency. Similar to energy ratings for consumer items, such as electrical goods, these offer an almost instant snapshot of the energy consumption (energy efficiency) and CO_2 production (environmental impact) of a commercial property. A visual, colour-coded graph indicating the kWh/m² per annum, which ranges from green (the most efficient) to red (the least), is used to grade the property from A to G, with A being the most efficient. The EPC is also used to calculate the CO_2 production in tonnes, which is also graded A to G. Contained within these short one- or two-page documents are also some recommendations for improving the energy efficiency and reducing environmental impact.

Certain buildings are exempt from having an EPC and these typically include historic or listed buildings, as well as temporary buildings, some agricultural or industrial property and holiday accommodation (subject to usage or occupation). It may be commented that, while this is a good method by which to red flag buildings with poor energy efficiency, the survey itself is not particularly detailed and is not intrusive. Therefore, when assessing items, such as external walls or flat roofs, it may have to make some assumptions regarding thicknesses and types of insulation. Other anomalies, such as having a building connected to a district heating system, can positively skew the EPC rating, and the overall accuracy appears to be the biggest issue with EPCs. This is of particular relevance in England and Wales since the Minimum Energy Efficiency Standards (MEES) came into force in April 2018. The MEES are derived from the Energy Efficiency (Private Rented Property) (England and Wales) Regulations 2015, which stipulate that privately rented property must have a minimum EPC E rating before granting a new tenancy to new or existing tenants (BEIS, 2017).

The Urban and Regional Studies (URS) building at the University of Reading (UK), which was constructed in the early 1970s with concrete facades and aluminium-framed single glazing has an EPC rating of 52 kWh/m² per year. This appears remarkably good and it is believed that the C grade rating is skewed by the fact that this building is connected to a district heating system.

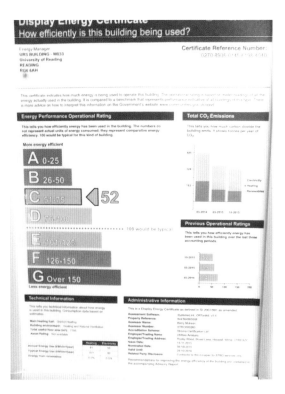

In most cases, the TDD or commercial building survey will not undertake an EPC but will review or comment upon the findings of the existing document. If this document is absent from the data room it will be a red flag issue for the legal and technical due diligence teams. Without sufficient training or knowledge, the surveyor undertaking the TDD will be unable to confirm the U-values of materials used in the construction components for most existing buildings. Therefore, to cross-check or calculate the EPC would appear unwise, based upon so little evidence or information. Sometimes information confirming general U-values of facades or roofs may be stipulated in performance criteria issued by agents or included in the building specification. Furthermore, individual technical data sheets for components such as glazing may be present in the as-built technical files, but in the scope and timescales of a 'normal' TDD, in-depth analysis is not common.

Founded by the Building Research Establishment (BRE) in the 1990s, the Building Research Establishment Environmental Assessment Method (BREEAM) has become a globally recognised system for rating and benchmarking commercial buildings. Buildings are scored out of 100 per cent for a series of criteria including: management, health and wellbeing, energy, transport, water, materials, waste, land use and ecology and pollution. The final assessment is graded as: outstanding, excellent, very good, good and pass. Clearly, the BREEAM rating of a building can be used as a valuable marketing tool and has a positive impact on occupancy as lower running costs are desirable for tenants. As with the EPC review, the surveyor should analyse the BREEAM assessment, where available.

In conclusion, regarding energy efficiency, the surveyor should report the findings of the existing EPC and BREEAM reports. It is not normal to critically appraise these, but any obvious omissions or deficiencies should be reported to the client.

Noise and disturbance

Concerning noise and disturbance, the surveyor should note any observations made during the site inspection regarding noise at the site. In context, certain site operations, such as manufacturing or industrial properties, are noisy as a result of their operation. Likewise, in some technical rooms of commercial properties it can be very noisy due to the generation of heating, ventilation and air conditioning (>95 dB). This type of noise needs to be assessed in terms of the effect that it might have on the occupants of the building or those of neighbouring buildings. In some instances, maximum noise limits, including the permitted times of operation, may be stipulated within operating permits or licences. The surveyor should make note of any requirements or conditions listed on the permits and report these in the TDD findings. Unless the surveyor is suitably qualified, they should not take measurements or readings of noise but should instead ask the building owner, property manager or tenant to disclose any complaints or actions arising from noise at the building or site. In the event that there are recognised issues concerning this, the surveyor may recommend further detailed investigation by a qualified acoustic engineer.

Noise generated adjacent to the property from the surrounding environment or neighbours should also be reported to the client in the TDD or commercial building survey findings. This may come from neighbouring operations or the local environment, such as train or traffic noise if the property is located next to transport infrastructure. Furthermore, if the property is located under the flight path of an airport, this may also cause noise pollution. During the site investigation it may be possible to hear noise in occupied areas and the building owner, property manager or tenant should be consulted as to any reported complaints by staff or occupants. If there are problems with noise inside an occupied building and related to an external infrastructure source, it is very difficult to eradicate this retrospectively.

A further source of nuisance could be surrounding air pollution, such as that generated from manufacturing, city centre traffic or even if a brewery is located close to the property. This may cause unpleasant odours and, regarding buildings situated in polluted cities with poor air quality, may even constitute a hazard to health. The issues, where obvious, need to be highlighted during the TDD or commercial building survey process.

Land contamination and environmental control

A potentially contaminated site or even the ground underneath a building is a major problem for property owners and investors as there is usually a significant cost and disruption associated with removal or treatment of the soil. Establishing historic land use is not normally within the remit of a TDD or commercial building survey as this information is contained in documents that should be reviewed by the legal due diligence team. Such information is usually located on registers or databases held by the authorities and these are primarily requested by the legal advisors as part of the overall due diligence process. However, any evidence of activities or operations on a site which may potentially cause land pollution or the contamination of watercourses should be noted. This sometimes happens with logistics or distribution sites, where there may be pumps for refuelling lorries or other vehicles on the site. Typically, the tanks for these are situated underground in a similar arrangement to most commercial fuel stations. Tanks may also be placed above ground and the obvious risk is that there is a major spillage during the filling of the tanks, which can result in fuel contaminating the adjacent ground or getting into a watercourse from the surface drainage network. Risk prevention for fuel spillage typically amounts to the placement of an 'interceptor' connected to a ring of grill or slot drains around a pump station. Furthermore, an interceptor can be placed at the exit point where the site drains are connected to the public or city drainage. An interceptor comprises a large vessel with a series of compartments which separates petrol, oil and lubricant from water, allowing this to be removed and pumped away. It is impossible for the surveyor to verify the presence of this from the site inspection alone and this will normally be done with reference to the as-built file and any maintenance interventions by the owner, property manager or tenant.

Diesel fuel tanks are also typically used for emergency power generators, as well as large tanks for the storage of oil for heating systems, although this is only normally evident to properties located on remote sites. The survey should seek, where possible, to identify the presence of these but, unless the surveyor is a specialist in environmental engineering, it is unlikely that they will be able to advise on the conformity of the installations to the legal prescriptions.

Other sources of potential land contamination may come from acid used in batteries for fork lift trucks or uninterrupted battery power supplies. Historically, old transformers which were used to provide the electrical supply and distribution to large buildings contained polychlorinated biphenyls (PCBs) which are also classed as a high-risk contaminant. PCBs are not used in modern installations and are likely to be found in technical equipment dating back to the 1970s. While most of this equipment should now be obsolete, it is still necessary to establish historic and current applications.

It is anticipated that the presence of fuel tanks both above and below ground will be the subject of an operating license or permit. Such a document is an example a legal requirement with a technical application and the surveyor cannot be expected to verify compliance with all relevant regulations, as this is usually outside their experience. Often, the equipment or operations listed on permits or licences are also closely linked to the M&E inspection; therefore, the M&E engineer should also be instructed to comment, where possible, on the compliance of these with regulations. Essentially, the surveyor should seek to verify the presence of any listed equipment or site operations through visual inspection but suitably qualified persons are required to verify conformity. In the context of a commercial transaction, the legal advisors and the subsequent sale and purchase agreement will usually place a requirement on the vendor to transfer the building or property in full compliance with all relevant regulations.

Concerning specific equipment or operations associated with commercial properties, the following is a list of those which have a legal/technical connotation and require certificates or test reports verifying this:

- heating – boiler size/capacity: conformity with gas regulations and, where relevant, the storage of fuel oil
- ventilation – the number and capacity of air handling units
- cooling – the capacity and types of coolant
- electricity – high voltage transformer: conformity with electrical regulations
- electricity – low voltage distribution boards: conformity with electrical regulations
- electricity – emergency generators: storage of fuel oil and test reports
- electricity – uninterrupted power supplies: storage of batteries and test reports
- lifts – size and conformity with regulations/periodic test and safety reports
- fire detection/alarm – conformity with relevant regulations and routine testing report
- firefighting – testing and conformity of sprinkler installations and, where relevant, the collection of firefighting water (typical to Luxembourg)
- firefighting – periodic testing and conformity of fire extinguishers
- parking – underground: conformity with fire regulations as well as the placement of an

  interceptor, ventilation and CO detection
- parking – above ground: conformity with planning requirements
- waste storage and disposal – conformity with environmental regulations and regulations governing fire risk of combustible materials and potential for rodents
- hazardous chemical storage – conformity with relevant regulations, such as COSHH.

Above are listed some of the typical legal/technical areas concerning environmental control and, in all cases, the surveyor may note or confirm the presence of these operations/equipment on site. However, without sufficient M&E or environmental engineering training, it appears dangerous and potentially risky for the surveyor to verify conformity.

The placement of a fuel station on a logistics site. Immediate verification is required to ascertain the condition of the underground tanks and what measures are in place to prevent ground water or watercourse contamination, as well as conformity with all relevant legislation.

Toxic chemicals stored in an internal emergency staircase; a situation which requires immediate action (photo courtesy of Widnell Europe).

Deleterious materials

The term 'deleterious' is used to describe not only materials which are toxic or hazardous to health but also those prone to decay or failure. To an extent, all construction materials are prone to becoming fatigued or worn and prone to decay. However, the seriousness of this, in the terms of advising investors, often relates to those defects with health and safety or cost risks. Rushton (2006) comprehensively details the cause and effect of deleterious materials used in construction. Further detailed analysis of specific materials, investigation and remediation are located in numerous Building Research Establishment (BRE) Digests. Many individual defects associated with deleterious materials specific to construction elements have been individually detailed in *Chapters 5* to *10* including the following:
- asbestos (*Chapter 5 – Roofs; Chapter 6 – Structure; Chapter 7 – Facades; Chapter 8 – Finishes*)
- brick slips (*Chapter 7 – Facades*)
- concrete (*Chapter 6 – Structure; Chapter 7 – Facades*)

- CFCs and HCFCs (*Chapter 9 – Services*)
- composite panels (*Chapter 7 – Facades*)
- glulam (*Chapter 6 – Structure*)
- profiled steel sheeting (*Chapter 5 – Roofs; Chapter 7 – Facades*)
- RAAC planks *(Chapter 5 – Roofs)*.

The deleterious material that is of greatest concern to investors and building owners is asbestos and, despite being finally banned in 1999, its widespread use in buildings from the 1950s to 1980s means that it remains a concern for a large quantity of real estate assets. Asbestos comprises three varieties which are defined by colour, with blue and brown posing the greatest risk to health. Consequently, these were banned in 1985. White is the most common application of asbestos and it was widely used in fibre cement construction products.

According to Rushton (2006), asbestos can typically be found in the following construction materials:

- sprayed coatings and laggings
- insulating boards
- ropes, yarns and cloth
- bitumen roofing felts
- damp proof courses
- asbestos-paper backed vinyl flooring
- unbacked (homogenous) vinyl flooring and floor tiles
- textured coatings and paint containing asbestos
- mastics, sealants, putties and adhesives
- asbestos reinforced PVC.

*Typical applications of asbestos: (**TOP LEFT**) pipework insulation (photo courtesy of Widnell Europe); (**TOP RIGHT**) prefabricated school building (photo courtesy of Widnell Europe); (**MIDDLE LEFT**) fibre cement chimney flue (photo courtesy of Widnell Europe); (**MIDDLE RIGHT**) fibre cement rainwater drainage pipe; (**BOTTOM LEFT**) vinyl floor tiles; (**BOTTOM RIGHT**) corrugated fibre cement roofing sheets.*

In most developed countries and societies there is a universal recognition of the dangers posed by asbestos and, accordingly, legislation is in place to deal with this issue. In the UK, there is health and safety regulation as well as COSHH, but the principal legal prescription concerned with asbestos is the Control of Asbestos Regulations 2012. In essence, asbestos should be removed if it is in a state of decay or if there are planned renovation works to the area in which this is located; however, if it is in good condition, secure and in areas of relative inoccupation, it can remain in situ. There is a requirement for all known asbestos within a property to be placed on an asbestos register or inventory. The inventory should be held at the building and a copy with the owner, property manager and tenant in order to provide a reference document which should be consulted prior to any works. Asbestos in situ should be the subject of routine inspection.

A toilet and cistern dating from the 1970s. The presence of asbestos has been clearly noted to the cistern and this is also recorded on a corresponding asbestos inventory.

Most surveyors are not qualified asbestos experts and, although it is 'normal' to note any obvious or visible evidence of materials potentially containing asbestos, unless proven through testing, this remains a suspicion. The TDD or commercial building survey process will review or summarise the remarks contained within the asbestos inventory but it will not normally carry out any tests or sample analysis. If there is no asbestos report or inventory contained within the data room documents, the surveyor should recommend immediate commissioning of this as a red flag issue. For buildings constructed post 1999, it should be possible to receive a certificate or attestation verifying that no asbestos-containing products have been used and that the property is free from asbestos.

The cost of asbestos removal and disposal is potentially very high, therefore it is prudent for the surveyor to specifically exclude this from any budget cost estimates in preference of seeking quotes from specialist asbestos contractors to address this issue.

Cultural heritage

When undertaking a TDD or commercial building survey, the surveyor may need to consider the cultural heritage of the building or the location in which this is situated. Legal prescription exists to protect buildings of historical, architectural or cultural significance and this should be noted within the legal searches associated with the legal due diligence. However, this information should normally be within the public domain and it should therefore be possible to find this by accessing records held by the authorities via an online search.

Listed or classified commercial buildings are not always those deemed to be classic historic properties as there is an increasing recognition of the need to protect modern architecture. Famous concrete framed and clad buildings attributed to brutalism and constructed from the 1960s have achieved worldwide listed status. As a consequence, repair, renovation or refurbishment of these properties may be limited or restricted by the listing, which may require the use of specific materials, techniques or style. Such works may prove costlier than 'conventional' repairs, particularly regarding the upgrade of thermal efficiency, which may be restricted to the internal surfaces of external walls.

The consequence of listed status to historic commercial properties may be even more profound and the surveyor should carefully consider any cost implications of this for the client when recommending repairs to defects. If necessary, the surveyor may seek the advice of specialist contractors accordingly.

Cultural heritage is also relevant to the location of the property as this may be within a conservation area, national park or world heritage site. This will place an additional burden on the building owner to comply with the requirements of the authorities when repairing, renovating or extending a building, irrespective of whether or not it is listed.

Statutory matters

As with verifying a building's compliance with fire regulations, confirming conformity with certain statutory matters can be difficult to achieve. Within the scope of works and contract instruction it is important to clarify or inform the client of the limitation of the survey mission. When assessing statutory matters, the following items may be considered:
- planning consents
- listing or classification
- tree preservation orders
- party wall issues.

Planning consents

The legal due diligence team will inevitably identify the presence of planning permissions or building permits for a commercial asset. These should be sorted or arranged in chronological order to enable the surveyor to understand and report the history of events leading to the current status of the property. As part of the document review, it is important to establish what was permitted, to whom and when. With most planning permissions or building permits there should also be a set of approved plans. It is important to review these plans and, in particular, these should be copies certified by the authorities.

As discussed in *Chapter 3 – Contract instruction*, it is impossible to verify with 100 per cent certainty that the as-built situation matches that permitted on the authorised plans by virtue of the fact that, within the time constraints and qualifications of the surveyor, it will not be possible to measure the building height, full internal and external dimensions as well as the positioning of the building on the plot. Such works are usually carried out by a land surveyor and, depending on the size or complexity of the building, can take weeks to complete.

Therefore, the surveyor should request copies of the certified as-built plans, sections and elevations, allowing for a comparative analysis to be made between the permitted and as-built situations. They can use the site visit and photographic records to illustrate any differences between the two situations. It is important to refer any discrepancies to the legal due diligence team for them to deliver an opinion on the 'significance' of the findings and whether these are 'material'.

It should be the responsibility of the permit holder or the design team to verify that the building has been executed in accordance with all permit conditions. It is also noted that, in many countries (but not all), there are mechanisms for the authorities to sign off buildings that have been completed in accordance with their planning permissions or permits. The surveyor should note the presence of such approvals in the report but should also confirm items or issues where visual infractions have been observed.

Listing or classification

As previously suggested, the listing of a commercial building or its location within a conservation area should normally be discovered as part of the legal searches associated with the legal due diligence. The significance of this will be realised and advised by the surveyor in respect of the condition of the property and any subsequent recommendations for repair or refurbishment which might be in contradiction to the listed status.

There is normally a hierarchy to the listing of buildings of historical, architectural or cultural importance and, in England, the categories are as follows:

- Grade I
- Grade II*
- Grade II.

According to Historic England (2018a), 2.5 per cent of all listed buildings are of Grade I status, with 5.8 per cent being Grade II* and 91.7 per cent Grade II. Naturally, the higher the category of listing, the more stringent the requirements will be on conserving this, with Grade I listed properties being the most 'protected'. There are commercial buildings which have Grade I and Grade II* listed status, but the majority of listed commercial properties are likely to be Grade II. While listing may be associated with 'historic' buildings, increasingly more 'modern' buildings are being given this status.

The URS building at the University of Reading, which was listed in 2016 with Historic England citing, among other things, the 'expressive use of structure to enclose space with reference to traditional Japanese construction' (2018b).

In real terms, the URS building (above) may be considered a fine example of modern architecture but, pragmatically, it is afflicted with many of the typical problems associated with buildings embodying exposed reinforced concrete structure, panels and aluminium-framed, single glazed windows. The challenge for owners, occupiers and investors in this type of property is maintaining the architectural characteristics or features of the property, while ensuring that it is fit for purpose in the context of its workable space and internal comfort.

Tree preservation orders

Similar to the listing process, the presence of mature or rare species of trees is an important consideration in the TDD or commercial building survey process. This is of particular relevance if the root networks associated with these are causing damage to the building, as removal of the tree or any remedial pruning may be in direct contravention of the protection afforded to it. When undertaking the site inspection, it is necessary for the surveyor to note the presence of large or unusual trees, therefore the surveyor will need to have some basic knowledge of native tree species. This can be cross-referenced to records held by the authorities but it is an issue that the legal due diligence team should pursue.

Party wall issues

The presence of adjoining buildings, and particularly those which share a boundary or connecting wall, should be noted during the survey as the owner of the property is obliged to comply with party wall legislation if they intend to do works to this in the future. The survey should also take into account any visual evidence of ongoing or historic works to a party wall, and the vendor should be asked to disclose any knowledge of completed or planned works to the party wall which have been or are likely to be the subject of a party wall award.

It should be noted that, in the course of a normal TDD or commercial building survey, the surveyor is not expected to carry out a survey or schedule of condition which could be used for a party wall award. This is a specific and separate type of instruction, which does utilise a similar skill set to that required by a TDD or commercial building survey, but which is governed by completely different rules of engagement. Furthermore, there are specific RICS guidance notes concerning engagement as a party wall surveyor.

Rights of way, easements and shared services

One complicated legal/technical issue is that concerned with easements or rights of way. This is often identified during the legal due diligence and sometimes the TDD or commercial building survey is asked to comment on this issue. In order to understand the context of an easement or right of way, it is necessary to receive a copy of the relevant documents concerned with this and also advantageous to see this on an approved or certified plan. The consequence of easements and rights of way are typically for these to be maintained, therefore the costs of this should be incorporated in the reporting process, to include any subdivision of this requirement according to that apportioned by the co-ownership or set out in the title deeds. The presence of an easement may also restrict the extension or development of the site or property if the proposed works alter or remove this, therefore the presence of an easement may be seen as a negative.

Other easements or rights of way may be afforded to utility companies for access and maintenance of water, gas or electrical supplies and, often, such installations are located within the site boundaries as either above or underground cables, pipework or ducting. Some city or town centre properties may also have equipment, services or facilities owned by a separate entity in the basement or on the roof and, accordingly, there may be an easement for access to these.

The surveyor may experience sites or buildings where there are shared services, and these can range from something relatively simple, such as rainwater evacuation or drainage, through to much more complex issues, such as the production of heating or cooling. In some situations, there may be buildings or sites supplied with district heating; accordingly, there will be a legal obligation or performance criteria to ensure security and continuity of the shared services. Typically, this interface between the legal and technical due diligence teams will be the subject of discussion between the two sets of advisors. The surveyor will also have to largely rely on the opinion of the M&E consultant to advise the client about the issue, offering sufficient options to allow them to make an informed decision.

Boundaries

Some of the potential issues concerned with boundaries have been discussed in *Chapter 10 – External areas* and, in order to deliver an evidence-based opinion, the surveyor should have access to the site or building plans. The boundaries should be defined on the plans and this information is usually obtained by the legal due diligence team through their searches. Checking and verifying boundaries is not something that is normally done as part of a TDD or commercial building survey; the work is typically performed by a land surveyor. Therefore, the site inspection should seek to establish the presence, type and condition of the site boundaries but, without as-built site plans or copies of the title deeds, it will be difficult to deliver an opinion on this matter.

Guarantees and warranties

Disclosure of existing guarantees and warranties is something that is usually requested as part of the legal due diligence. There may be a need to analyse these from a technical perspective, and this will typically be the case with technical equipment. The M&E consultant is probably the most appropriately qualified individual or organisation to critically appraise guarantees and warranties associated with technical installations. The surveyor may typically receive a request to advise on any warranties associated with the building structure, envelope or finishes, and this will be done with respect to the condition of these items observed during the inspection and any documentary evidence of existing guarantees or warranties.

Leasehold and repairing liabilities

Similar to undertaking a dilapidations instruction, the surveyor may be requested to assist the legal due diligence in the undertaking of a review of any repair liabilities contained within lease agreements. This is more of a desk survey and it is necessary to have access to the lease agreements as well as any attached plans identifying the demise or leased area. Also of use will be copies of any schedules of condition which are appended to the lease agreements. These may reveal important information about the existing building features, finishes and fit out at the commencement of the lease and will help to contextualise the current situation at the building.

Repairing liabilities are important as these may affect the overall long-term cost planning for the property. While the survey should seek to identify the location, construction and condition of all building elements, the cost of maintaining, refurbishing or replacing these may need to be apportioned partially to the tenants, depending on conditions set out in the lease.

When reporting on a TDD or commercial building survey, the use of a schedule of condition/ defects and costs is the preferred method to illustrate the investor's ten-year cost risk. Much like a Scott schedule used for dilapidations, a 'remarks' column can be added to the schedule in which the current building owner, in discussion with the due diligence team, can notify the potential buyer of items covered by existing contract or lease liabilities. The cost schedule can be altered accordingly to illustrate all anticipated costs, with further information included to identify the overall cost reduction and the 'net' investment required by the investor.

References

BEIS. 2017. *The non-domestic private rented property minimum standard: Guidance for landlords and enforcement authorities on the minimum level of energy efficiency required to let non-domestic property under the Energy Efficiency (Private Rented Property) (England and Wales) Regulations 2015*, Department for Business, Energy & Industrial Strategy, London, Crown copyright.

DCLG. 2013. *The Building Regulations Approved Document B: Fire Safety. Volume 2: Buildings other than dwelling houses (2006 edition incorporating the 2010 and 2013 amendments)*, Newcastle upon Tyne, NBS.

DCLG. 2015. *The Building Regulations Approved Document M: Access to and use of buildings. Volume 2: Buildings other than dwelling houses*, Newcastle upon Tyne, NBS.

Historic England. 2018a. *What is listing? Listed buildings*. Available online at: https://historicengland.org.uk/listing/what-is-designation/listed-buildings.

Historic England. 2018b. *URS building: List entry summary*. Available online at: https://historicengland.org.uk/listing/the-list/list-entry/1435127.

RICS. 2010. *Building surveys and technical due diligence of commercial property* (4th edition, guidance note), London.

Rushton, T. 2006. *Investigating hazardous and deleterious building materials*, RICS Books, London.

Index

Note, page numbers in *italic* refer to illustrations.